Studies in Systems, Decision and Control

Volume 433

Series Editor

Janusz Kacprzyk, Systems Research Institute, Polish Academy of Sciences, Warsaw, Poland

The series "Studies in Systems, Decision and Control" (SSDC) covers both new developments and advances, as well as the state of the art, in the various areas of broadly perceived systems, decision making and control–quickly, up to date and with a high quality. The intent is to cover the theory, applications, and perspectives on the state of the art and future developments relevant to systems, decision making, control, complex processes and related areas, as embedded in the fields of engineering, computer science, physics, economics, social and life sciences, as well as the paradigms and methodologies behind them. The series contains monographs, textbooks, lecture notes and edited volumes in systems, decision making and control spanning the areas of Cyber-Physical Systems, Autonomous Systems, Sensor Networks, Control Systems, Energy Systems, Automotive Systems, Biological Systems, Vehicular Networking and Connected Vehicles, Aerospace Systems, Automation, Manufacturing, Smart Grids, Nonlinear Systems, Power Systems, Robotics, Social Systems, Economic Systems and other. Of particular value to both the contributors and the readership are the short publication timeframe and the world-wide distribution and exposure which enable both a wide and rapid dissemination of research output.

Indexed by SCOPUS, DBLP, WTI Frankfurt eG, zbMATH, SCImago.

All books published in the series are submitted for consideration in Web of Science.

More information about this series at https://link.springer.com/bookseries/13304

Elena Lysenko · Alexander Rogachev ·
Oldřich Starý
Editors

Recent Developments in the Field of Non-Destructive Testing, Safety and Materials Science

Springer

Editors
Elena Lysenko
Tomsk Polytechnic University
Tomsk, Russia

Oldřich Starý
Czech Technical University
Prague, Czech Republic

Alexander Rogachev
Research Institute of Physics and Chemistry
Francisk Skorina Gomel State University
Gomel, Belarus

ISSN 2198-4182 ISSN 2198-4190 (electronic)
Studies in Systems, Decision and Control
ISBN 978-3-030-99059-6 ISBN 978-3-030-99060-2 (eBook)
https://doi.org/10.1007/978-3-030-99060-2

This Springer imprint is published by the registered company Springer Nature Switzerland AG
The registered company address is: Gewerbestrasse 11, 6330 Cham, Switzerland

Preface

With technological advances, the modern world is on the verge of a new industrial revolution, being in the stage of digital transformation, when innovations from different industries interpenetrate and complement each other. The School of Non-Destructive Testing, Tomsk Polytechnic University, Russia, promotes scientific research and industrial application of non-destructive testing and materials science technologies and related tests, as well as methods, to ensure safe manufacturing processes.

Today, research and technology advancement is driven by innovations, and there is a need for publications to stimulate the formation and continuous training of specialists in non-destructive testing, materials science and safety. This book can be used as a complementary technical document to upgrade the skills of specialists in non-destructive testing, materials science and safety, and as a primary resource for training managers and decision makers in various industries.

The current timely book presents the latest advances and emerging trends in research and industrial applications in non-destructive testing, manufacturing and process safety, and diagnostics and materials science.

Innovations in the fields of non-destructive testing, production and process safety, diagnostics and materials science and books that highlight the best and instructive are central to our technological world.

I am pleased to see this comprehensive book taking shape and advancing this field to the next generation of scientists seeking for new research opportunities.

Tomsk, Russia — Elena Lysenko
Gomel, Belarus — Alexander Rogachev
Prague, Czech Republic — Oldřich Starý

Contents

Innovation in Non-destructive Testing

Investigation of the Impact of the Size Effect of the Viscosity of Polar Liquids on Their Penetrating Ability

Irina Lobanova, Aleksey Kalinichenko, and Ekaterina Maryasova

Abstract Communication of the defect cavity with the medium surrounding the test object is critical for implementation of liquid penetrant testing techniques. Therefore, the test surface is pre-cleaned prior to the penetrant application to provide liquid access deep into defects with contaminated mouths. The faster and more completely the defects are filled with liquid, the more effective the control process. It is known that the better the inner surface of the defect is wetted, the deeper the liquid penetrates into the defect, and the lower the liquid viscosity, the faster the process proceeds. It was experimentally proven that when polar liquids move through ultrathin quartz capillaries with sizes less than 1 μm, the effective viscosity increases significantly. However, a number of other factors affect the penetration kinetics, such as the sample material, surface purity and roughness, and liquid density and wetting angle. Thus, the study attempted to determine the impact of the size effect of the liquid viscosity on absorption kinetics (the depth of liquid penetration into defect cavities and the time), with regard to the liquid parameters, capillary size, and surface roughness.

1 Introduction

To improve the efficiency of penetrant testing [1, 2], it is necessary to study the properties of liquids that affect their penetrating ability. These factors include liquid viscosity and density, wetting angle, surface roughness, and size effect.

Viscosity is the property of liquids to resist the flow caused by external forces, and it is commonly distinguished as kinematic and dynamic viscosity. Dynamic viscosity is a measure of resistance to flow or deformability of a liquid, and kinematic viscosity is a measure of resistance to the liquid flow under the influence of gravity and depends

I. Lobanova (✉) · A. Kalinichenko · E. Maryasova
Division for Testing and Diagnostics, National Research Tomsk Polytechnic University, 30, Lenin Avenue, Tomsk, Russia 634050
e-mail: konarevai007@tpu.ru

A. Kalinichenko
e-mail: lex-k@tpu.ru

E. Lysenko et al. (eds.), *Recent Developments in the Field of Non-Destructive Testing, Safety and Materials Science*, Studies in Systems, Decision and Control 433,
https://doi.org/10.1007/978-3-030-99060-2_1

on the liquid density [3]. An increased viscosity parameter of the liquid increases the time of liquid penetration into capillaries of test objects, and, hence, the liquid penetration depth per conditional unit of time decreases.

The liquid density is directly related to the dynamic coefficient of viscosity. This is confirmed by the fluidity of the substance—the value of the inverse dynamic viscosity. The lower the liquid density, the greater its fluidity, and the shorter the time of its penetration into the capillary at a certain length [3].

Wetting is manifested through interaction of molecules in a solid, liquid gaseous (or second liquid) phase, which results in liquid spreading over the surface of a solid. For quantitative interpretation, the concept of a limiting wetting angle is used, which is determined by the cosine of the angle between the tangent to the liquid surface at the liquid–solid–gas media interface, and surface tension. The dependence of the wetting ability on liquid penetration is directly proportional; as the wetting parameter increases, the penetration ability of liquids also increases [4–9].

Cleanliness and microrelief of the surface have a strong effect on wetting, and, hence, on the penetrating ability of the liquid. The limiting wetting angle on a rough surface is calculated using the Wenzel–Deryagin equation, which shows the interrelation between the wetting angles of the test liquid on rough and smooth surfaces and the ratio of the area of the true surface of a rough body to its projection onto the plane [10–12].

The size effect means that the viscosity parameter of polar liquids increases during their flow through microcapillaries, when the radius of the latter reduces [13, 14]. Possible causes of the size effect include liquid polarity (it should be noted that most penetrants are polar liquids), a decrease in the radius of the capillary through which the liquid flows, the temperature of the capillary, and a decrease in the average radius of the pores of the medium. In [13], it is shown that in microcapillaries with a radius of about 10–6 m, the parameter of liquid viscosity can be higher than the table values by 60–70%. This is due to translational and rotational degrees of freedom of microelements of polymolecular associates [14]. In [13], P.P. Prokhorenko and N.P. Migun show the relationship between the relative viscosity of penetrants with regard to their polarity and the size of the capillary through which they flow. It is shown that an increase in the effective viscosity of liquid caused by the decrease in the capillary radius is more pronounced in water-based penetrants, and is less evident in those based on alcohols.

The study employs simulation of the liquid flow in a capillary with flat parallel walls to investigate the size effect of liquid viscosity on absorption kinetics with regard to the parameters of the liquid, the size of the capillaries through which this liquid flows, the surface roughness, and the sample material.

2 Simulation of the Liquid Flow in Slit Capillaries with Sizes Less Than 1 μm

Simulation of the liquid flow in capillaries with sizes less than 1 μm was carried out using the Project 1 software.

The program is designed to simulate the liquid flow in dead-end and in through slit capillaries. The interface of the complex for computer simulation comprises components for entering the task parameters: capillary size, liquid density, dynamic viscosity coefficient, surface tension coefficient, wetting angle, capillary tilt angle relative to the horizontal, surface roughness coefficient, time interval for observing the liquid flow, atmospheric pressure, dead-end capillary depth, integrator step, and the number of points on the graph. In addition, the user interface includes a field for graphical representation of simulation results and a field for entering current values of time and depth of capillary filling. These data can be used if necessary [15].

A simulation experiment was carried out to determine the time of filling of steel and glass capillaries less than 1 μm in size with different liquids to a certain depth. Water, kerosene and isopropyl alcohol were chosen since they are part of the majority of indicator penetrants used in liquid penetrant testing. Water was of particular interest for the experiment since the hydrodynamics of water-based liquids in capillaries with a diameter of more than 1 μm is well studied [13].

For the study, a fixed observation time for the liquid flow was equal to 10 s, and the following parameters were controlled:

1. Capillary size: 0.5 μm and 0.3 μm;
2. Capillary material: steel and glass;
3. Capillary type: through and dead-end;
4. Liquid: kerosene, isopropyl alcohol, water.

Table 1 summarizes the main parameters of the problem.

Table 1 Values of the main parameters of the problem

Liquid	Dynamic viscosity coefficient, [Pa × s, 10^{-3}]	Surface tension coefficient, [N/m, 10^{-3}]	Wetting angle, [rad]		Liquid density, [kg/m^3]	Fatigue strength surface condition factors, [mm]	
			Steel	Glass		Steel	Glass
Kerosine	1.85	24.0	0.64	0.43	800	0.95	1.00
Isopropyl alcohol	2.43	21.7	0.55	0.29	786		
Water	0.89	73.0	1.13	0.82	1000		

Fig. 1 Time dependences of the depth of liquid penetration into through glass capillaries

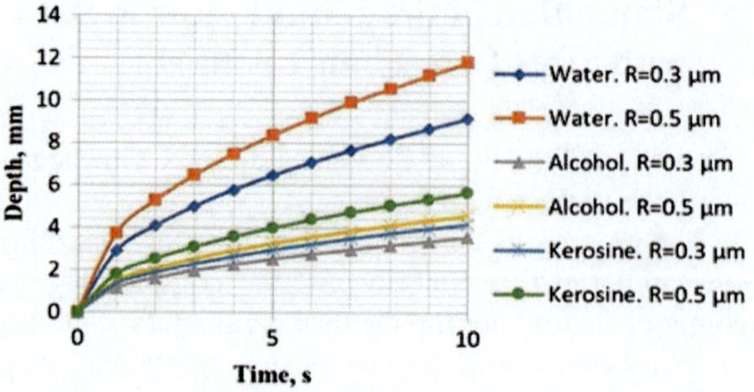

Fig. 2 Time dependences of the depth of liquid penetration into steel capillaries

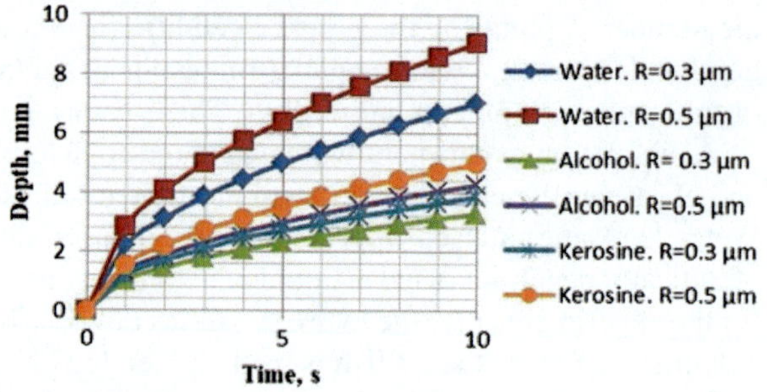

2.1 *Result*

2.1.1 Opening Capillaries

Figures 1 and 2 present the results obtained in simulation of penetration of water, isopropyl alcohol and kerosene into steel and glass capillaries 0.5 and 0.3 μm in size.

Analysis of the time dependences of the liquid penetration depth revealed the effect of the capillary size on the penetrating ability, namely, a decrease in the penetration depth as the capillary size decreases. This effect is more obvious for water than for kerosene and isopropyl alcohol. The curves show the effect of the sample material on the penetration depth. That is, liquid flows faster in glass capillaries than in steel capillaries.

2.1.2 Dead-End Capillary

Figures 3, 4, 5, 6 illustrate the results obtained in the simulation experiment with dead-end steel and glass capillaries 10 mm long.

The obtained simulation dependences indicate an insignificant discrepancy in the results of filling a dead-end capillary with liquids with regard to the capillary size effect. The effect of the sample material on the penetrating ability of liquids is consistent with the known data.

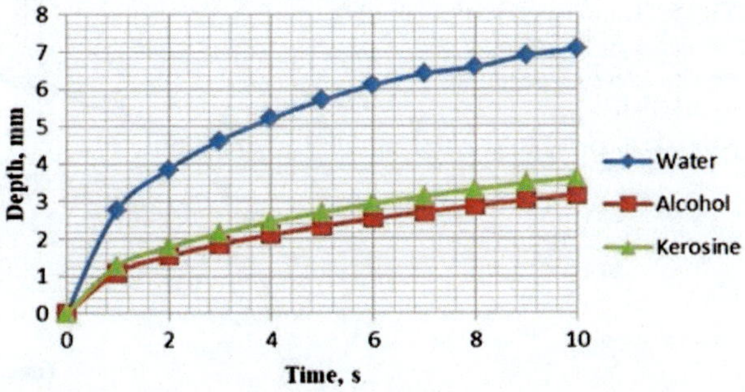

Fig. 3 Time dependences of the depth of liquid penetration into dead-end glass capillaries. Size—0.3 µm

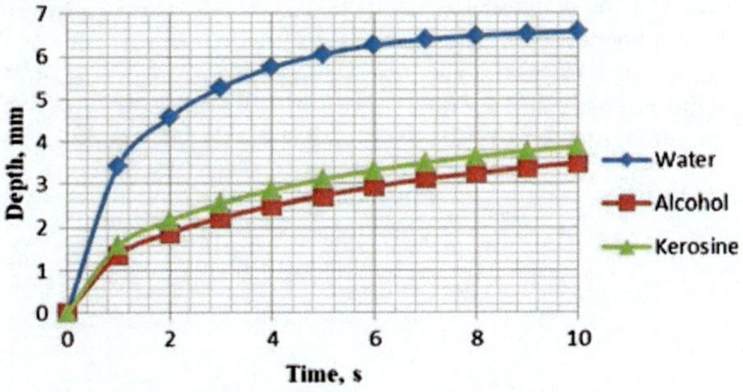

Fig. 4 Time dependences of the depth of liquid penetration into dead-end glass capillaries. Size—0.5 µm

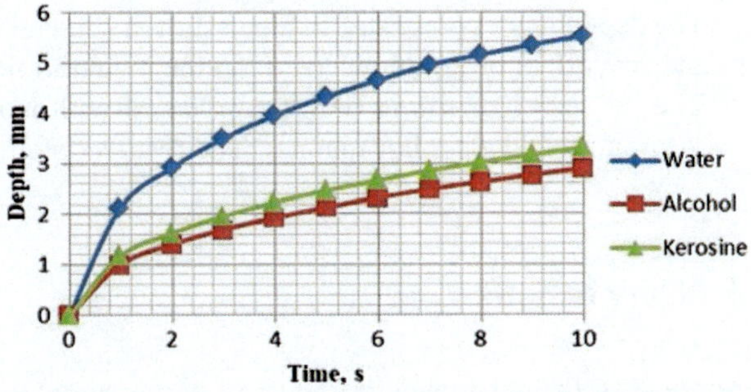

Fig. 5 Time dependences of the depth of liquid penetration into dead-end steel capillaries. Size—0.3 µm

2.1.3 Study of the Size Effect of Viscosity on the Penetrating Ability

Assume that the size effect of liquid viscosity affects the kinetics of liquid absorption into a capillary.

Compare two cases to assess the assumption. All other things being equal, simulate the process of liquid flow in the Project 1 program, namely, water flow in a dead-end (10 mm) glass capillary 0.5 µm in size, with different viscosity parameters (a reference value, and assuming an increase in liquid viscosity by 70% at the decreased capillary size as described in [13]). The obtained dependencies are shown in Fig. 7.

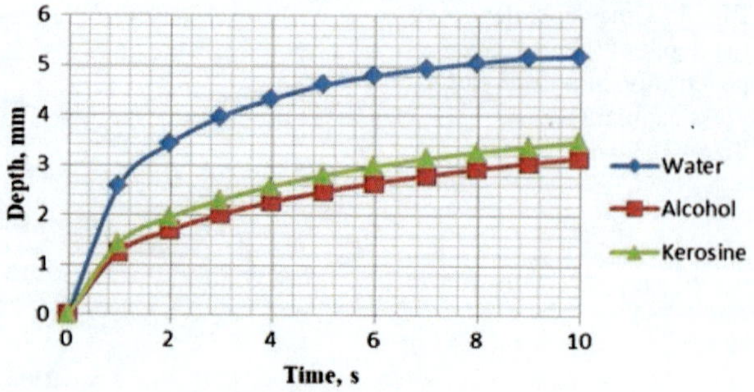

Fig. 6 Time dependences of the depth of liquid penetration into dead-end steel capillaries. Size—0.5 μm

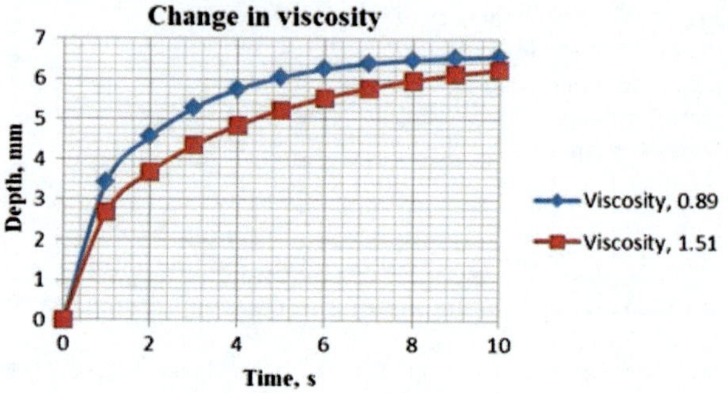

Fig. 7 Time dependence of the fluid penetration depth for the standard value of liquid viscosity and for the size effect of viscosity

The dependences presented in Fig. 5 show a slight discrepancy (about 16%) caused by the size effect of liquid viscosity on the penetration depth.

The processes of filling of slit capillaries with isopropyl alcohol and kerosene were simulated in a similar manner. The results obtained are similar.

3 Conclusions

Analysis of the dependencies obtained during simulation revealed the following:

- With the sizes of through and dead-end capillaries being similar, water spreads almost two-fold faster than kerosene and isopropyl alcohol in both steel and glass capillaries due to higher polarity.
- The flow rate of liquids in glass capillaries is higher than that in steel capillaries.
- In 0.5 and 0.3-μm capillaries, all other things being equal, the liquid penetration depth decreases as the capillary size decreases, and this is consistent with the experimental data reported by P.P. Prokhorenko and N.P. Miguna [13].
- The depth of penetration of polar liquids into the capillary decreases by 16%, which indicates the size effect of viscosity at the given parameters. Thus, the size effect reveals itself, but it does not significantly affect the kinetics of liquid absorption into the capillary.

References

1. Maslov, B.G.: Defectoscopy by penetrating substances. Higher school, Moscow (1991)
2. Lobanova, I.S.: Development and research of samples from non-metallic material for testing means of capillary non-destructive testing: Dissertation for the degree of candidate of technical sciences: spec. 05.11.13, Tomsk (2018)
3. GOST 33768–2015 Method for determination of kinematic viscosity and calculation of dynamic viscosity of transparent and opaque liquids. Standartinform, Moscow (2019)
4. Huhtamäki, T., Tian, X., Korhonen, J.T., Robin, H.A.R.: Surface-wetting characterization using contact-angle measurements. Nat. Protoc. **13**, 1521–1538 (2018). https://doi.org/10.1038/s41596-018-0003-z
5. Hebbar, R.S., Isloor, A.M., Ismail, A.F.: Contact angle measurements. Membrane characterization **12** (2017). doi: https://doi.org/10.1016/B978-0-444-63776-5.00012-7
6. Glazkov, Yu.A.: Question of evaluation of the wettability of liquids for penetrant inspection. The Soviet journal of nondestructive testing **26**(11), 801–806 (1991)
7. Sum, B.D., Goryunov, Yu.V.: Physico-chemical fundamentals of wetting and spreading, chemistry, Moscow (1976)
8. Zhdanov, K.V., Erokhin, M.S., Stepkina, M., Zhirnov, A.A., Kudryashova, O.B.: Investigation of the effect of electrostatic surface charge on the wetting angle. Polzunovsky Bull **2** (2016)
9. Samsonov, V.M., Kaplunov, I.A., Ivanov, A.M., Talyzin, I.V., Tretyakov, S.A., Pushkar', M.Y.: Effect of treating a quartz surface on its wetting with molten germanium. Bull. Russ. Acad. Sci. Phys. **82**(7), 905–908 (2018). https://doi.org/10.3103/S1062873818070341
10. Hideo, N., Ryuichi, I., Yosuke, H., Hiroyuki, S.: Effects of surface roughness on wettability. Acta Materialian **46**(7), 2313–2318 (1998). https://doi.org/10.1016/S1359-6454(98)80012-8
11. Kubiak, K.J., Wilson, M.C.T., Mathia, T.G., Ph., Carval: Wettability versus roughness of engineering surfaces. Wear **271**, 523–528 (2011). https://doi.org/10.1016/j.wear.2010.03.029
12. Lobanova, I.S., Vodopyanov, A.S., Kalinichenko, A.N.: Investigation of the liquid flow on rough surfaces to solve the problems of liquid penetrant testing. Studies in Systems, Decision and Control **351**, 89–99 (2021). https://doi.org/10.1007/978-3-030-68103-6_9
13. Prokhorenko, P.P., Migun, N.P.: Introduction to the theory of capillary control. Science and technology, Minsk (1988)
14. Migun, N.P., Volovich, I.V.: Soaking of polar indicator liquids into dead-end microcracks. Proceedings of the National Academy of Sciences of Belarus, Series of Physical and Technical Sciences **2** (2011)
15. Lobanova, I.S., Meshcheryakov,V. A., Kalinichenko, A.N.: Modeling of liquid flow in surface discontinuities. IOP Conf. Series: Materials Science and Engineering **289**, 012023 (2018)

X-ray Inspection Systems with Sandwich Radiation Detectors: A Survey

Victor Udod and Svetlana Nazarenko

Abstract At the present time, scanning systems for digital radiography (SSDR) are often used for inspection, as well as for defectoscopy and diagnostics of materials and products. The range of tasks solved on the basis of SSDR is very wide and includes, in particular, the tasks of detecting explosive, toxic and radioactive substances, as well as explosive devices, all types of weapons and damaging elements For the recognition of materials in the SSDR, intended for inspection control in order to ensure the safety of transportation and suppress attempts to smuggle illegal items, the method of dual energies (MDE) is widely used today. This method makes it possible to simultaneously evaluate two parameters of the object of control (OC) – the effective atomic number of the OC material and its mass thickness. This paper reports the state of X-ray inspection systems with sandwich radiation detectors. Besides, this paper presents description of dual energy method with using x-ray sandwich detectors in inspection control system for material recognition.

1 Introduction

The aim of non-destructive quality control of substances, materials and products is not only to detect defects, but also to recognize them due to the different level of potential danger of defects of different types. Recognition of detected defects significantly increases the information content of control, which is objectively necessary to optimize the management of technological processes and product quality.

For inspection, as well as for defectoscopy and diagnostics of materials and products, scanning systems of digital radiography (SSDR) (Fig. 1) are widely used today.

V. Udod
Department of Information Technologies, National Research Tomsk State University, 36, Lenin Ave, Tomsk, Russia 634050

S. Nazarenko (✉)
Division for Testing and Diagnostics, National Research Tomsk Polytechnic University, 30 Lenin Ave, Tomsk, Russia 634050
e-mail: svetanaz@mail.ru

E. Lysenko et al. (eds.), *Recent Developments in the Field of Non-Destructive Testing, Safety and Materials Science*, Studies in Systems, Decision and Control 433,
https://doi.org/10.1007/978-3-030-99060-2_2

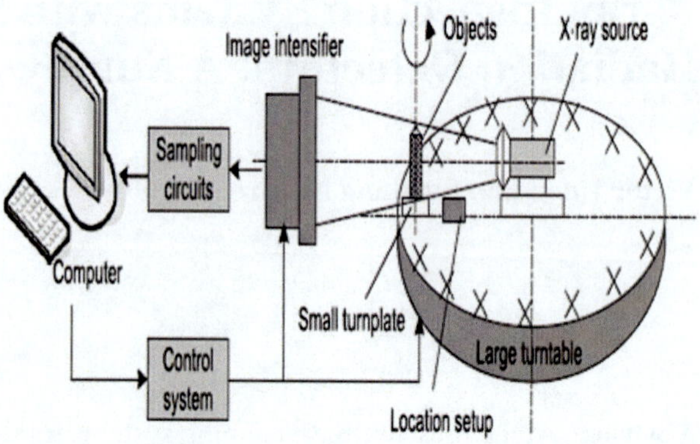

Fig. 1 Scheme of scanning systems of digital radiography

The principle of operation of these systems is that the control object (OC) in the process of its movement is illuminated by a fan beam of radiation, and the radiation transmitted through the OC is recorded by a line of detectors.

At the present time, the Dual Energy Method (MDE) is used frequently to identify inspection materials in SSDRs to ensure the safety of transport and to prevent attempts to smuggle illegal items. This method makes it possible to simultaneously evaluate two parameters of the object of control (OC)—the effective atomic number of the OC material and its mass thickness. This is achieved by solving (with respect to these parameters) a system of two equations, which represent the equalities of the theoretical and experimental radiation transparency of the OC for two different maximum (or effective) energies of X-ray radiation.

Considering that SSDRs with sandwich radiation detectors are widely used for screening baggage and carry-on baggage at airports, a challenge of a theoretical assessment of the potential accuracy of the MDEs used in these systems for recognizing the materials of inspected objects is arised.

2 Material Recognition by the Dual Energy Method in a Security Inspection System with Sandwich Detectors

Currently, the method of dual energies (MDE) is widely used for material recognition in scanning digital radiography systems (SSDR), designed for inspection control, in order to ensure the safety of transportation and suppress attempts to smuggle prohibited items (Fig. 2) [1–7]. This method allows one to simultaneously evaluate two parameters of the object of control (OC) – the effective atomic number of the OC material and its mass thickness [4, 8, 9]. Formally, this is achieved by solving (with respect to these parameters) a system of two equations, which are equalities of theoretical and experimental radiation transparencies of the OC for two different maximum (or effective) energies of X-ray radiation [4, 8].

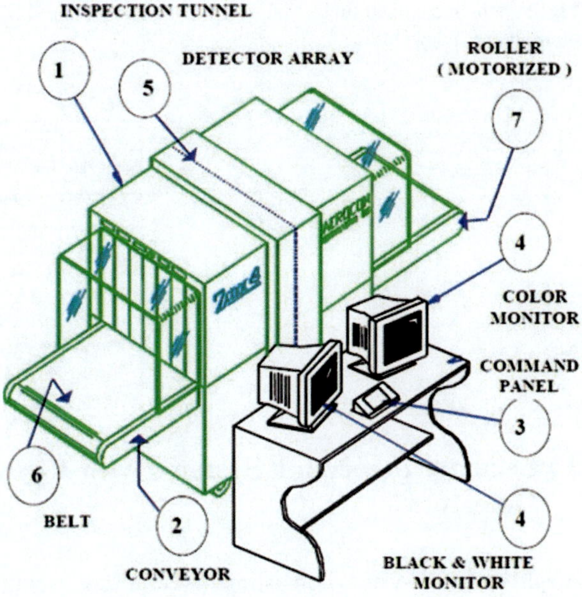

Fig. 2 Scheme of installation of baggage control at the airport

In addition to inspection control based on SSDR, MDE is actively used for medical purposes [10], as well as for diagnostics of materials and products using X-ray computed tomographs [11–15].

The MDE implementation schemes can be different. The traditional (classical) scheme is expressed in the fact that the OC is transmitted through twice, and each time at a different maximum (effective) energy in the X-ray spectrum [4].

Another scheme for implementing the MDE consists in a single scanning of the OC with a radiation beam created by a source that alternately generates high and low energy pulses [16]. Another scheme for the implementation of the MDE consists in a single scanning of the OC and registration of the radiation transmitted through it by sandwich detectors, which are combinations of two detectors located one after another in the direction of propagation of radiation quanta and separated by a filter (usually in the form of a copper plate) [1, 4, 10, 13]. Radiation is reorded by detectors, radiation-sensitive volumes (RFO) of which are located one after another in the direction of propagation of primary photons. With the proper selection of RFO materials and their thicknesses, the first detector registers predominantly the soft component of the radiation, and the second detector predominantly detects the hard component of the radiation. To enhance the effect of separation of the soft and hard components, the X-ray radiation transmitted by the first RFR is filtered by an intermediate filter (Fig. 3).

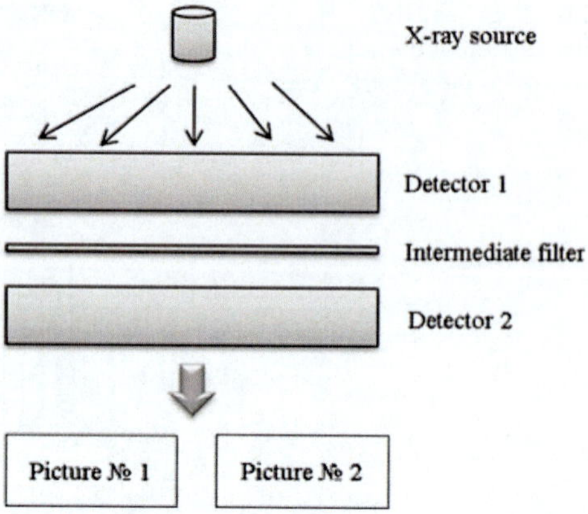

Fig. 3 Schematic diagram of a sandwich detector

3 Security Inspection Systems with X-ray Sandwich Detectors

A split detector system consisting of a thin CaF2 scintillator followed by a thick NaI crystal was installed in the EMI Mark I. Energy separation is possible because the CaF2 crystal reacts mainly to low-energy photons; images with two energies can be obtained from one scan. Using a split detector, it was possible to distinguish between two weak solutions CaCl2 and KI, which had the same computed tomography number (44 H), under normal scanning conditions [17].

A double layer detector for simultaneous scanning at two energy levels in was first proposed by Brooks and Chiro in 1978 year. The two-layer scintillator structure, called the split detector, consisted of a thin CaF2 scintillator, followed by a thick NaI crystal, and made it possible to read information separately from each of the structures. Energy separation is possible because the CaF2 crystal reacts primarily to low-energy photons; dual energy images (dual energy images) could only be obtained from one scan. Calculations and preliminary experiments have shown that the split detector separates energies better than the method based on the use of peak kilovoltages of 100–140 kVp. Using the split detector, it was possible to distinguish between two weak solutions CaCl2 and KI, which had the same computed tomography number (44 H), under normal scanning conditions.

The technique, which permits dual-energy scanning digital radiographic projection without switching the voltage of the X-ray tube, uses a high / low atomic number multilayer detector that simultaneously separates the X-ray beam into low and high energy components. To test the method, the response of a scanning line array of energy-sensitive detectors was modeled and images of the bones and soft tissues of an anthropomorphic chest phantom were obtained at 140 kVp. They were compared with similar images obtained by switching the X-ray tube voltage from 80 kVp to a heavily filtered 140 kVp. For comparable input to the skin, dual energy detector images required less tube loading and resulted in higher noise levels. The latter is due to the fact that the energy separation between the high and low energy components is less with a dual energy detector than with the voltage switching method, and problems with misregistration associated with the simulation methodology. The design of a detector that will lead to improved energy separation and reduced noise levels is also discussed. In view of with this capability and the advantages of tube loading, the method looks promising for digital scanned projection radiography [18].

The breakthrough came relatively soon with the advent of the so-called sandwich detectors. Sandwich detectors consist of two (or more) detector layers, where the upper layer predominantly detects low-energy photons, and the lower layer detects a filtered and therefore harder spectrum. In some implementations, filtering includes only the top layer of the detector, while in some cases a thin metal filter, most often a copper filter, is inserted between the layers of the detector, which increases spectral separation but decreases dose efficiency.

The Fuji system is a computed radiography (CR) system with two phosphor plates and a 0.8 mm copper filter. This sandwich detector system was commercially implemented by Philips in 2015 [10]. The detector consists of an yttrium garnet scintillation layer over a layer of gadolinium oxysulfide. A thin front-illuminated photodiode is mounted on the side of the scintillator and hidden under an anti-scattering grating so as not to reduce the active area of the detector. There is no additional filtering between the scintillator layers, which maximizes dose efficiency, but spectral separation can be lower than with other methods.

One way to reduce the count rate per channel and at the same time provide additional energy resolution for photon counting detectors is to segment the multi-electrode sensor in the depth direction so that the electrodes at greater depths detect harder spectra, thus, in a way, combining the technology of photon counting and sandwich detectors. Another way to improve energy resolution is to remove photons detected in every other cell so that spectral separation between cells that used for spectral analysis, increases. This approach, somewhat similar to the introduction of filters in sandwich detectors, increases quantum noise, but has the potential to improve spectral characteristics [10].

This article describes a multilayer ("sandwich") configuration detector, consisting of two flat panel X-ray imaging detectors for one-time dual energy imaging. An intermediate copper filter is used to increase the spectral separation between the two detectors to improve contrast at the expense of image noise. Monte Carlo and cascade systems analyze signal and noise characteristics that quantify. The quality of the dual energy images obtained with the prototype sandwich detector system is assessed using a quality score defined as the squared contrast-to-noise ratio normalized by X-ray irradiation for a mouse phantom for preclinical imaging. Demonstration images of dual-energy bones and soft tissues in a dying mouse were obtained using a prototype system. Although the fidelity score with a single shot detector is lower than with the conventional double shot (double kV) method, a single shot approach may be preferred when imaging speed or insensitivity to motion artifacts is a major concern.

An alternative method is to use a one-shot approach to obtaining dual energy imaging by simultaneously acquiring two images, for example, by combining photo-stimulated phosphors in a multilayer configuration. The front layer absorbs (mainly) low-energy X-ray photons, while the back layer absorbs (mainly) high-energy photons. The use of various detector systems has previously been investigated. Used pairs film screens, but found that their dynamic range and speed were insufficient for practical use. The design of a split xenon detector was also investigated, which demonstrated single images using a pair of linear scintillator coupled photodiode arrays, which required a lengthy scanning procedure for images of the area [19].

The new sandwich type detector was developed by combining two flat panel detectors [19]. Each flat panel detector consists of a commercially available terbium doped gadolinium oxysulfide (Gd2O2S: Tb) scintillator optically coupled to a complementary CMOS photodiode array (RadEye1TM, Teledyne Rad-icon Imaging Corp., Sunnyvale, California). The anterior and posterior scintillators are Min-RTM 2000 (34 mg / cm^2) and Lanex TM Fast (48 mg / cm^2), respectively. The rear scintillator is thicker to achieve high quantum efficiency with a higher energy spectrum. The CMOS sensor has 0.048 mm pixels arranged in a 1024 × 512 format to provide an image area of approximately 50 × 25 mm. A thin sheet of copper (Cu) was placed between two flat panel detectors to improve spectral separation as described below. The sandwich detector was installed in an opaque aluminum (Al) box with a 1 mm thick polycarbonate entrance window [19].

An energy separation detector for dual energy radiography can be configured as a two-layer sandwich, where the average energy of the photons detected by the two layers is different. To characterize the quantum noise of such a detector, it is necessary to know the noise covariance between the two layers in addition to the noise variance in each layer. A theory is presented that allows one to calculate the noise covariance, and it turns out to be negligible. The experimental results based on measurements with a sandwich detector Na1 and a radiation source are presented and shown to confirm the theory. Quantum noise in the layers is independent and Poisson [20].

X-ray imaging with two energies can be achieved by changing the potential of the X-ray tube, by arranging the recording of the beam in sandwich detectors (Knoll, 2000), which have different sensitivity to photons of high and low energy rays, or by using so-called K-edge filters [21]. In X-ray beams filtered by elemental media, the attenuation coefficients show discontinuities, the energy photons of which correspond to the minimum binding energy of electrons and are sufficient for the release of electrons. Thus, for example, slightly higher than the binding energy of electrons with the K-shell, electrons have an increased probability of absorbing photons in comparison with the subthreshold photon energy. The photon energy at which there is a sudden change in the absorption probability is called the K-edge, and the coupled differential absorption is sufficient to separate the original spectrum into high and low energy components. In practical terms, this approach suffers from the limited energy resolution offered by the K-edge filters, as well as from photon starvation, which leads to increased noise levels in the raw images, especially if multi-element filters are used to further decompose the raw spectrum into several quasi-monoenergy regions. The consequence of this filtering is the need for more powerful X-ray sources and, therefore, a greater degree of shielding.

Features that can be extracted from dual-energy CT scans include the spatial distributions of both the (electron) density and variations in the mean atomic number.

Therefore, all other things being equal, a dual-energy CT scanner would be expected to have ROC detection efficiency, which is superior to the single-energy scanner type simply because it provides more physical performance [21].

A method for obtaining molecular interference function (MIF) of unknown fluid for screening safety is presented in article [22]. Based on the effective atomic number retrieved from dual energy computations, tomography (CT) scans are used to estimate the equivalent molecular formula of the liquid. After a series of optimizations, the MIF and the new effective atomic number are finally obtained from the X-ray diffraction profile. The proposed method gives more accurate results with less sensitivity to noise.

4 Conclusion

The paper provides an analytical review of the current state of inspection control systems with sandwich X-ray detectors, intended for inspection control, in order to ensure the safety of transportation and suppress attempts to smuggle illegal items, as well as actively used for medical purposes. The development of SSDR with sandwich radiation detectors of objects is mainly related with the improvement of sources and detectors of X-ray radiation, and also with the development of highly efficient algorithms and software for digital processing of radiometric signals and radiographic images.

References

1. Rebuffel, V., Dinten, M.: Dual-energy X-ray imaging: Benefits and limit. Insight-non-destructive testing and condition monitoring **49**(10), 589–594 (2007)
2. Park, J.S., Kim, J.K.: Calculation of effective atomic number and normal density using a source weighting method in a dual energy X-ray inspection system. J. Korean Phys. Soc. **59**(4), 2709–2713 (2011)
3. Duvillie, J., Dierick, M., Dhaene, J., Van, L.D., Masschaele, B., Geurts, R., Hoorebeke, L.V., Boone, M.N.: Inline multi-material identification via dual energy radiographic measurements. NDT and E Int. **94**, 120–125 (2018)
4. Ogorodnikov, S.A.: Recognition of materials during radiation customs control based on a linear electron accelerator. (Dissertation of Candidate of Technical Sciences). St. Petersburg, Russia. D.V. Efremov Institute of Electrophysical Apparatus. (In Russian) (2002)
5. Gavrish, Yu.N., Berdnikov, Ya.A., Spirin, D.O., Perederiy, A.N., Safonov, M.V., I.V. Romanov, I.V.: Software package for restoration of introscopic images using the dual energy method. Problems of atomic science and technology. Series: Nuclear Physics Investigations **3**(54), 123–125. (In Russian) (2010)
6. Gil, Y., Cho, M., Namkung, W.: Radiography simulation on single-shot dual-spectrum X-ray for cargo inspection system. Appl. Radiat. Isot. **69**(2), 389–393 (2011)
7. Macdonald, R.D.R.: Design and implementation of a dual-energy X-ray imaging system for organic material detection in an airport security application. Proc. SPIE **4301**, 31–41 (2001)
8. Osipov, S.P., Udod, V.A., Van, Y.: Recognition of materials by the dual energies method during radiation monitoring of objects. J. Nondestr. Test. **8**, 33–56. (In Russian) (2017)

9. Osipov, S.P., Chakhlov, S.V., Kamysheva, E.N., Usachev, E.Y., Shchetinkin, S.A.: Selecting parameters of detectors when recognizing materials based on the separation of soft and hard X-ray components. Russ. J. Nondestr. Test. **54**(11), 797–810 (2018)
10. Fredenberg, E.: Spectral and dual-energy X-ray imaging for medical applications. Nucl. Instrum. Methods Phys. Res., Sect. A **878**, 74–87 (2018)
11. Zhengrong, Y., Ram, N., Crawford, C.R.: Dual energy computed tomography for explosive detection. J. Xray Sci. Technol. **14**, 235–256 (2006)
12. Chang, S., Lee, H.K., Cho, G.: Application of a dual-energy monochromatic x-ray CT algorithm to polychromatic x-ray CT: A feasibility study. Nucl. Eng. Technol. **44**(1), 61–70 (2012)
13. Iovea, M., Neagu, M., Duliu, O.G., Oaie, G., Szobotka, S., Mateiasi, G.A.: Dedicated On-Board Dual-Energy Computer Tomograph. J Nondestruct Eval. **30**, 164–171 (2011)
14. Alves, H., Lima, I., Lopes, R.T.: Methodology for attainment of density and effective atomic number through dual energy technique using microtomographic images. Appl. Radiat. Isot. **89**, 6–12 (2014)
15. Jia, H., Kejun, K., Li, Z., Zhiqiang, C.: A novel image optimization method for dual-energy computed tomography. Nucl. Instrum. Methods Phys. Res., Sect. A **722**(11), 34–42 (2013)
16. Svistunov, Yu.A., Vorogushin, M.F., Petrunin, V.I. , Sidorov, A.V., Gavrishin, Yu.N., Fialkovsky, A.M. Development of work on the creation of X-ray and nuclear-physical inspection complexes at D.V. Efremov Institute of Electrophysical Apparatus. Problems of atomic science and technology, 3, 171–173. (In Russian) (2006)
17. Brooks, R.A., Di, C.G.: Split-detector computed tomography: A preliminary report. Radiology **126**, 255–257 (1978)
18. Barnes, G.T., Sones, R.A., Tesic, M.M., Morgan, D.R., Sanders, J.N.: Detector for dual-energy digital radiography. Radiology **156**, 537–540 (1985)
19. Han, J.C., Kim, H.K., Kim, D.W., Yun, S., Youn, H., Kam, S., Tanguay, J., Cunningham, I.A.: Single-shot dual-energy x-ray imaging with a flat-panel sandwich detector for preclinical imaging. Curr. Appl. Phys. **14**(12), 1734–1742 (2014)
20. Sones, R.A., Barnes, G.T.: Noise correlations in images acquired simultaneously with a dual-energy sandwich detector. Med. Phys **16**, 858–861 (1989)
21. Wells, K., Bradley, D.A.: A review of X-ray explosives detection techniques for checked baggage. Appl. Radiat. Isot. **70**(8), 1729–1746 (2012)
22. Zh., Li, Tianyi, Y.D.: Determination of liquid's molecular interference function based on X-ray diffraction and dual-energy CT in security screening. Appl. Radiat. Isot. **114**, 179–187 (2016)

Control Method for Calibration Interval Adjustment of Speed Meters Based on the Statistical Analysis of Previous Calibrations Data

Natalya Natalinova, Sergei Nizkii, and Aleksanov Vladislav

Abstract Currently, one of the ways to the visual control of traffic rules on the roads is the installation of measuring devices that automatically record violations of the car speed requirements on the road. The reduction of accidents due to speeding is an important and urgent problem. The main means of control are speeding radar devices. Measuring devices with photo and video recording are complex technical devices. They include major components such as photoradar sensors, transmission equipment data (converters, routers, modems etc.), data storage servers, communication server. Radar is a measuring device for determining the speed of the vehicle. Therefore, ensuring the metrological reliability of this class of devices is an important task. One of the components of this problem is the precise definition of the calibration interval. The universal recommendations for the selection of methods of determining or adjusting the intervals between inspections (calibrations) and a comprehensive critical analysis of these recommendations are presented in the document, OIML D10 - 2007, Guidelines for the Determination of the Calibration Intervals of Measuring Instruments. One of the most significant decisions regarding the calibration is "When to do it" and "How often to do it". A large number of factors influence the time interval that should be allowed between calibrations and should be taken into account by the laboratory. The paper studied the calibration interval adjustment of speed meters "Vizir" based on statistical observations of four years. The method of adjustment of calibration interval group of measuring instruments, in conducting verifications recorded only an alternative sign measuring instruments.

N. Natalinova (✉)
Department of Electronic Engineering, Tomsk Polytechnic University, 30, Lenin ave, Tomsk, Russia 634050
e-mail: natalinova@tpu.ru

S. Nizkii
Federal Scientific Center (All-Russian Research Institute of Soy), 19, Ipatievskoe highway, Blagoveshchensk, Russia 675000
e-mail: agrofak06@mail.ru

A. Vladislav
Joint Stock Company (Scientific and Production Center-Polyus), 56v, Kirov ave, Tomsk, Russia 634050

E. Lysenko et al. (eds.), *Recent Developments in the Field of Non-Destructive Testing, Safety and Materials Science*, Studies in Systems, Decision and Control 433,
https://doi.org/10.1007/978-3-030-99060-2_3

It is shown that the calibration interval must be significantly reduced in comparison with the passport data of the manufacturer.

1 Introduction

At present, the number of road accidents in Russia that caused by drivers is unacceptably high [1]. According to the website of the State Traffic Inspectorate of the Ministry of Internal Affairs of Russia, for this reason, 19,011 people died in 2020. Most of the violations of traffic rules, which entailed severe consequences, is an excess of the speed of the car. Generally, speed mode monitoring is performed using radar speed meters. A significant proportion of the park speed meters are radar video recording speed meters "Vizir". Since the fines for speeding are imposed based on recorded readings, it is necessary to be sure in their metrological perfect (flawless) state. A measuring instrument in the metrological perfect (flawless) state is a measuring instrument which has everyone standardized metrological characteristics meet the established requirements, i.e. do not exceed the allowable values.

There are three possible approaches to reducing the risk associated with metrological failure of a measuring instrument:

- increasing the limiting indicator of metrological accuracy of a measuring instrument - errors and uncertainties;
- improvement of the selected interval of application of the measuring instrument before the subsequent verification of its accuracy;
- creation of a system of periodic checks or continuous monitoring of the deviation of error and (or) uncertainty in the course of using the measuring instrument.

To control to the metrological perfect state of measuring instruments periodic checks and calibrations are performed. At these periodic inspections, metrological characteristics are monitored. The calibration interval is the most important parameter that determines the level of metrological reliability of measuring instruments. On the one hand, the larger the calibration interval, the lower the metrological reliability. On the other hand, the shorter the calibration interval, the greater the financial cost of performing calibrations. You also need to take into account the logistics costs that are associated with the transfer of measuring instruments to the calibration locations. Therefore, it is necessary to determine the optimal value for the calibration interval of the measuring instruments [2–8].

2 Research Method

OILM D10-2007 corresponds to the requirements of universality of recommendations for the selection of methods and level of detail of critical analyses of these methods in different fields of highly diverse applications of measuring instruments in our opinion [1]. The document presents the most general approach to the description of recommended methods of establishing and adjusting the inter-calibration intervals of measuring instruments. Methods based on statistical analysis of an individual instrument or instrument type can also be a possible approach. These methods are gaining more and more interest, especially when used in combination with adequate software tools. When large numbers of identical instruments (i.e. groups of instruments) are to be calibrated, the calibration intervals can be reviewed with the help of statistical methods.

When determining the calibration interval for sets of the same type of measuring instruments, as a rule, a single interval is assigned for all instruments, regardless of their age and the number of the calibration. The primary value of the calibration interval, determined by the developer of the measuring instruments, is entered into the operational documentation (technical documentation for the verification procedure). Then approve this value during type approval tests. During operation, this value should be corrected by organizations performing verification (calibration) of measuring instruments, taking into account its results. The calibration interval is set in calendar time for measuring instruments:

- the change in the metrological characteristics of which is due to: aging (i.e., does not depend on the intensity of operation of the instruments) and in the values of the operating time for measuring instruments,
- the change in the metrological characteristics of which is a consequence of the wear of the elements of the measuring instruments (depending on the intensity operation).

The determination of the calibration interval is carried out on the basis of modeling the dependence of the accuracy or metrological reliability of measuring instruments on the time (operating time) that has passed since the last verification (calibration).

The proposed method for investigating the control of the calibration interval refers to the method of correcting the calibration interval at the stage of operation of measuring instruments.

Step 1. Group the results of the calibrations according to the serial numbers of the calibrations that passed after the release of the measuring device from production or repair:

1st group—measuring instruments received for the 1st verification after manufacture or repair;

2nd group—measuring instruments submitted for the 2nd verification, etc.

Each group will contain several results of measurements of the metrological characteristics of the measuring device (or its instability for the previous calibration procedure).

Step 2. Statistical processing of the grouped verification results is carried out sequentially for each group.

Based on the results of the statistical analysis of the data, the calibration interval of the group of measuring instruments is adjusted. The calibration interval is considered the most important parameter of the metrological service of measuring instruments directly affect the level of uniformity of measurements.

3 Description of the Installation

The most suitable criterion for the optimality of the calibration interval is the economic criterion, which is determined by the conditional minimum of the economic costs of operating the measuring instrument. It consists of losses due to insufficient measurement accuracy (type 1) and the costs associated with carrying out checks, calibrations and repairs of measuring instruments rejected during verification (type 2). The easiest way to estimate the costs of the 2nd type. For example, for the Tomsk region for the "Vizir" measuring Instrument Park, the costs, taking into account the prices for the calibration of an organization accredited to provide calibration services, depending on the duration of the interval between calibrations, are shown in Fig. 1—curve 1. It should be noted that the costs of transportation and repair of devices were not taken into account. It is not possible to accurately determine the costs of the 1st type, but it is possible to make an approximate estimate. For example, according to the traffic police of the Tomsk region, the total number of traffic violations in 2020 was 152,534 cases. There is no data on the percentage of traffic violations, so we estimate the speeding at 30% of the total number of violations (50,000). The "Vizir" devices record 25% of violations (12,500). The fine for such a violation is 500 rubles. Thus, the total amount of fines is 6,250 thousand rubles. After the expiration of the calibration interval, up to 5% of metrological faulty instruments are allowed (specified risk level). Let us assume that all metrological malfunctions consist in the error in determining the vehicle speed. Then the amount of illegally collected fines will be 5% of the total amount (according to 2020) - 312 thousand rubles. Figure 1 shows curve 2.

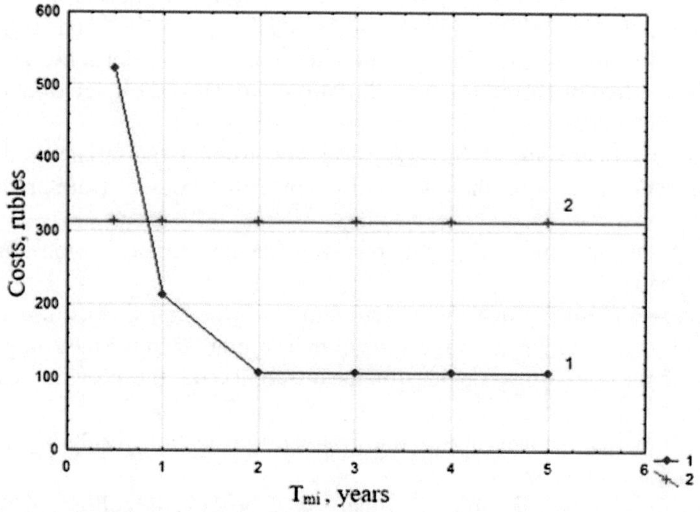

Fig. 1 Graphical presentation of the maintenance costs of measuring instruments, which are determined by economic costs

Table 1 Verification data

Number of verification	The number of measuring devices that have successfully verificated
0	56
1	51
2	39

In the case of an overestimated value of the calibration interval, the percentage of metrologically faulty devices at the end of the calibration interval will increase, therefore, costs of the second type will increase by the same amount. In this regard, it is necessary to adjust the calibration interval on the basis of the verification data of organizations accredited for the right to carry out verification. In this work, we used the verification data of "Vizir" radar devices for 4 years (Table 1).

Since the verification results do not contain valid metrological characteristics, the calibration interval of a group of measuring instruments will be corrected according to an alternative criterion. Let us assume a normal distribution of the error. Then the calculations are carried out according to the following algorithm.

Firstly, the devices are grouped according to the serial numbers of the calibrations that passed after the release of the radar from production or repair: 1st group - radars that entered the 1st verification after manufacture or repair, 2nd group - radars that entered the 2nd verification, etc. Thus, each i-th group will contain Ni of the measurement results of the measuring device error or its instability for the previous calibration interval *xj, (j = 1,..., Ni).*

Then the statistical probabilities are calculated that the measuring devices will be recognized as metrologically suitable according to the results of the *i*-th verification. Let us consider two variants of changing the law of error distribution of the measuring instrument - "fan-shaped" and "linear".

For a "fan-shaped" random process, the change in the parameters of the distribution of the error of the measuring device in time has the form:

$$m(t) = 0; \sigma^2(t) = \beta_1 + \beta_2 t \tag{1}$$

Next, an estimate of the calibration interval for the "fan-shaped" random process is determined – T_1. Before that, we will find statistical estimates of the parameters of the drift of the error under the assumption:

$$m(iT) = 0, \sigma(iT) = \Delta/\lambda_{0.5(1+p_i)}$$

where i – verification number; $\lambda_{0.5(1+p_i)}$ – normal distribution quantile, Δ – half-width of the interval of the level of permissible values of the monitored metrological characteristic.

Further, according to the calculated values of $\sigma^2(iT)$ we estimate the regression coefficients using the least squares method. And, finally, we determine the value of the duration of the calibration interval T_1.

For a "linear" random process, the change in the parameters of the distribution of the error of the measuring device in time has the form:

$$m(t) = m_1 + m_2 t; \sigma^2(t) = \sigma^2(0) \quad (2)$$

Next, we determine the estimate of the calibration interval for the "linear" random process – T_2. Then we find the statistical estimates of the drift parameters:

$$m(iT) = \Delta - \lambda_{p_i}\sigma(0), \sigma(iT) = \sigma(0)$$

where λ_{p_i} – normal distribution quantile, $\sigma(0)$ – root-mean-square deviation of the distribution of the error of the measuring device at the initial moment of time (if unknown, accept $\sigma(0) = \Delta/3$). Further, using the values $m(iT)$ we estimate the regression coefficients using the least squares method. Finally, we determine the value of the duration of the calibration interval T_2.

As the recommended value of the calibration interval, we take the minimum value of the obtained estimates: $T = min\ [T_1, T_2]$.

4 Results and Considerations

For "Vizir" radars, the control is carried out for the compliance of their relative error with the permissible error limit $\Delta = \pm 0.4\%$. The service life of these radars is set at 6 years. 56 items of measuring instruments were in operation. The calibration interval (defined by the manufacturer in the technical documentation for the device) is 2 years. At the same time, measuring instruments put into operation instead of rejected ones are not taken into account. The calculation data are presented in Table 2.

For the hypothesis of a "fan-shaped" random process, it is assumed $m\ (iT) = 0$. For the hypothesis of a "linear" random process, we assume $\sigma\ (iT) = 0.133$. Using the least squares method, we find the mathematical expectation for the "linear" drift model: $m_1 = 0.0184$; $m_2 = 0.0829$. In a similar way, we find the variance for the "fan-shaped" drift model: $\beta_1 = 0.0082$; $\beta_2 = 0.0334$.

Table 2 Selective estimates of the parameters of the drift models of the "Vizir" velocity meters

Verification number	Nuber of erified measuring instruments	p_i	$0,5(1+p_i)$	"Fan-shaped" random process			"Linear" random process	
				$\lambda_{0.5(1+p_i)}$	$\sigma(iT)$	$\sigma^2(t)$	λ_{p_i}	$m(iT)$
0	56	1	1	3	0,133	0.018	3	0
1	51	0.9107	0.9553	1.6954	0.236	0.056	1.345	0.221
2	39	0.6964	0.8482	1.0287	0.389	0.151	0.514	0.332

Let us find the conditional probability of metrological serviceability of the measuring device, provided that the operating time of the measuring device is limited by its service life (in our case, 6 years), according to the following formula:

$$P_{mi}(T) = \frac{n+1}{n} \cdot \frac{\sum_{s=1}^{n} Q_s(T)}{1 + \sum_{s=1}^{n} Q_s(T)} \tag{3}$$

where $n = T_{serv}/T$ – ratio of service life to calibration interval, T_{serv} – service life of a measuring device specified by the manufacturer, T – calibration interval, $Q_s(T)$ – probability of metrological serviceability of devices of the s-th group ($s = 1, 2, \ldots$).

By varying T we will get different values of $P_{mi}(T)$. For example, if you choose the value of the calibration interval according to the technical documentation equal to 2 years, then the following algorithm is used to calculate $P_{mi}(T)$.

Let $T = 2$ years, $n = 6/2 = 3$. We find the mathematical expectation and variance for the "fan-shaped" model:

$m_1(iT) = 0, \sigma(iT) = \sqrt{0.0082 + 0.0334i \cdot T}, i = 0, 1, 2, 3.$

Then we find the probabilities of metrological serviceability of measuring instruments with a "fan" model for instruments of the s-th group by the formula:

$$Q_s(T) = F\left[\frac{0.4}{\sqrt{0.0082 + 0.0334 \cdot s \cdot T}}\right] - F\left[-\frac{0.4}{\sqrt{0.0082 + 0.0334 \cdot s \cdot T}}\right]$$

where F(x) – cumulative standard normal distribution function, $T = 2$ years s – 1,2,3. The calculation results are shown in Table 3.

The value of the probability of metrological serviceability is calculated with a "fan-shaped" model of drift $P_{mi}(T)$ according to the formula (3). The $P_{mi}(T)$ value of 0.915 was obtained. Then we compare $P_{mi}(T)$ with the normalized value of the criterion $P_{mi} = 0.95$. If the metrological reliability of the measuring device turns out to be what is required, and then the calibration interval is increased, otherwise the calibration interval is decreased. The calibration interval values are selected from the members of a series of preferred numbers – 0,25; 0,5; 0,75; 1; 2; 3; 4; 5; 6 years. In our case, the metrological reliability of the measuring device is lower than the required one (0.915 < 0.95), therefore, we choose the value $T_{new} < T$, which is closest to T

Table 3 Calculation data $P_{mi}(T)$ for $T = 2$ years

Verification number	$m_1(iT)$	$\sigma(iT)$
1	0	0.273
2	0	0.377
3	0	0.457
$Q_1(T)$	$Q_2(T)$	$Q_3(T)$
0.856	0.712	0.619

from below – 1.75 years. Then the process is repeated until, P_{mi} $(T) \approx P_{mi}$. Then we find the mathematical expectation and variance for the "linear" model:

$m_2(iT) = 0.0184 + 0.0829 \cdot i \cdot T, \sigma(iT) = 0, 133, i = 0, 1, 2, 3.$

The probabilities of metrological serviceability of measuring instruments with a "linear" model for devices of the s-th group are determined by the formula:

$$Q_{s_2}(T) = F\left[\frac{0.4 - (0.0184 + 0.0829 \cdot \mathrm{i} \cdot \mathrm{T})}{0.133}\right] - F[-3]$$

where $F(x)$ – cumulative standard normal distribution function, $T = 2$ years, s–1,2,3. The results are shown in Table 4.

The value of the probability of metrological serviceability is calculated for the "linear" drift model $P_{mi}(T)$ according to the formula (3). As a result, we get the value $P_{mi}(T) = 0.855$.

Compare $P_{mi}(T)$ with the normalized value of the criterion $P_{mi} = 0.95$. If the metrological reliability of the measuring device turns out to be higher than the required one, then the calibration interval is increased, otherwise the calibration interval is decreased. The values of the calibration interval are selected from the members of a number of preferred numbers - 0.25; 0.5; 0.75; 1; 2; 3; 4; 5; 6 years. In our case, the metrological reliability of the measuring device is below the required value (0.855 < 0.95), then we choose the value closest to T from below - 1.75 years. Then the process is repeated until, $P_{mi}(T) \approx P_{mi}$*. The calculation results are shown in Table 5.

Table 4 Calculation data P_{mi} (T) for $T = 2$ years

Verification number	$m_2(iT)$	$\sigma(iT)$
1	0.184	0.133
2	0.350	0.133
3	0.516	0.133
$Q_1(T)$	$Q_2(T)$	$Q_3(T)$
0.947	0.646	0.193

Table 5 Calculation of the calibration interval for speed meters "Vizir"

Calibration interval, years	$P_{mi}(T)$ "fan-shaped" model	$P_{mi}(T)$ "linear" model	*min{*$P_{mi}(T)$ "fan-shaped", $P_{mi}(T)$ "linear"}
2	0.915	0.855	0.855
1.75	0.926	0.850	0.850
1.5	0.936	0.896	0.896
1.25	0.947	0.910	0.910
1	0.957	0.934	0.934
0.75	0.968	0.952	0.952

5 Conclusion

The results of statistical data processing show that the optimal value according to the $P_{mi} = 0.95$ criterion is the value of the calibration interval $T = 0.75$ years (9 months). A decrease in the calibration interval will lead to high financial costs for carrying out checks, as well as to logistics costs associated with the removal of measuring instruments from their places of direct operation. More frequent checks will reduce the number of devices in which the error goes beyond the limits established by the documentation. Car owners will be confident in the accuracy of fixing the speed, which will reduce the number of disputes and conflicts associated with fixing the speed of vehicles. Adjustment of the calibration interval will improve the quality of traffic management, reduce accidents and injuries on the roads.

During the operation of measuring instruments, the calibration interval established in the technical documentation for the verification procedure can be adjusted taking into account the modes and operating conditions in the direction of both decreasing and increasing. The basis for establishing new values of the calibration interval is the results of previous calibrations of measuring instruments of this group [9–15].

Usually, the adjustment of the calibration interval is carried out by the organization of the state metrological service or another organization that calibrates the measuring instruments of this group. The new value of the calibration interval is agreed with the enterprises (organizations) that use the devices of this group. In cases where the enterprise using measuring devices does not agree with the proposal to change the calibration interval, the results of studies aimed at correcting it are submitted to the metrological centers for consideration. In Russia, such a center is the State Scientific Metrological Center. It can be another organization authorized by the governing body of the state metrological service. An organization should provide an inspection of the submitted materials and, based on its results, make a final decision. Scientific and methodological guidance of work related to the establishment of primary calibration intervals and their adjustment in the countries-participants of the Agreement is carried out by state scientific metrological centers or other organizations authorized by the governing body of the state metrological service.

References

1. International Laboratory Accreditation Cooperation: OIML D10– 2007. Guidelines for the Determination of the Calibration Intervals of Measuring Instruments (2007).
2. Carvajal, S.A., Medina, A.F., Bohórquez, A.J., Sánchez, C.A.: Estimation of calibration intervals using Bayesian inference. Measurement: J. Int. Measur. Confederation **187**,110316 (2022). doi: https://doi.org/10.1016/j.measurement.2021.110316
3. Velásquez, C., Castro, M.A., Rodríguez, F., Espín, F., Falconi, N.: Optimization of the calibration interval of a luminous flux measurement system in hid and ssl lamps using a gray model approximation. ETCM 2021—5th Ecuador Technical Chapters Meeting (2021). doi: https://doi.org/10.1109/ETCM53643.2021.9590764

4. Chunovkina, A.G., Pokhodun, A.I., Sulaberidze, V.S.: The problem of determining and adjusting the inter-calibration intervals of measuring instruments. Meas. Tech. **62**(10), 863–868 (2020). https://doi.org/10.1007/s11018-020-01706-2
5. Nunzi, E., Panfilo, G., Tavella, P., Carbone, P., Petri, D.: Stochastic and reactive methods for the determination of optimal calibration intervals Conference Record—IEEE Instrumentation and Measurement Technology Conference, 106–109. (2004). doi: https://doi.org/10.1109/TIM.2005.851501
6. Natalinova, N., Ilina, N., Frantcuzskaia, E.: Calibration Interval Adjustment of a Measuring Instrument in Industries during Long-term Use. IOP Conference Series: Materials Science and Engineering **132**(1) (2016). https://doi.org/10.1088/1757-899X/132/1/012029
7. Pashnina, N.: Determination of optimal calibration intervals by balancing financial exposure against measurement costs. Flow Meas. Instrum. **60**, 115–123 (2018). https://doi.org/10.1016/j.flowmeasinst.2018.02.001
8. Wang, J., Zhang, Q., Jiang, W.: Optimization of calibration intervals for automatic test equipment. Measurement: J. Int Meas. Confederation **103**, 87–92. (2017). doi: https://doi.org/10.1016/j.measurement.2017.01.062
9. Ma, C., Chen, H., Zhang, W., Liang, J.: Determining calibration interval of measuring instrument based on gray markov prediction model. Yi Qi Yi Biao Xue Bao/Chinese Journal of Scientific Instrument **34**(6 SUPPL.), 1–5 (2013)
10. Natalinova, N.M., Rozhkova, O.V., Moldovanova, E.A., Ilina, N.L., Demin, N.S.: Filtering for stochastic systems in the case of continuous observation channels with memory of arbitrary multiplicity and anomalous noise. 2016 International Siberian Conference on Control and Communications, SIBCON 2016—Proceedings, 2016, 7491866 (2016). doi: https://doi.org/10.1109/SIBCON.2016.7491866
11. Jinzhe, X., Jiulong, X.: Calibration interval optimization and calibration conclusion risk analysis on automatic test system. ICEMI 2017—Proceedings of IEEE 13th International Conference on Electronic Measurement and Instruments 2018-January, 236–241 (2017). doi: https://doi.org/10.1109/ICEMI.2017.8265776
12. Chimitova, E.V., Chetvertakova, E.S.: Goodness-of-fit testing for the degradation models in reliability analysis. 14th International Scientific-Technical Conference on Actual Problems of Electronic Instrument Engineering, APEIE 2018—Proceedings, 2018, 8546176 (2018). doi: https://doi.org/10.1109/APEIE.2018.8546176
13. Egorov, A., Kochneva, E., Larionova, A., Lyukhanov, E., Shender, S.E.: AMR complexes metrological characteristics control at calibration time intervals. IEEE 60th Annual International Scientific Conference on Power and Electrical Engineering of Riga Technical University, RTUCON 2019—Proceedings, 8982327 (2019). doi: https://doi.org/10.1109/RTUCON48111.2019.8982327
14. Chen, Z.L., Zhao, F., Zhang, X.: Combined forecast of calibration interval based on linear trend model and neural network. Appl. Mech. Mater. **635–637**, 662–665 (2014). doi: https://doi.org/10.4028/www.scientific.net/AMM.635-637.662
15. Chvosteková, M.: Multiple use confidence intervals for a univariate statistical calibration. Measur. Sci. Rev. **19**(6), 264–270 (2019). doi: https://doi.org/10.2478/msr-2019-0034

Innovations in the Field of Production and Process Safety

Priorities for Improving Safety of Housing Management Services

Alina Gorkunova, Ludmila Redko, Inna Plotnikova, and Marina Yanushevskaya

Abstract Management concepts are dynamic and respond to the business needs of enterprises that strive to ensure continuous operation by keeping their customers satisfied with products or services. Improved management is a consequence of the functions and methods used to increase its safety and efficiency. The paper discusses an approach to improving the quality of housing services provided to residents of apartment buildings on the example of a service company. Such companies aim to provide consumers with comfortable and safe living. The paper presents statistical analysis of consumer requiests in order to increase the safety of housing services. The reasons for the requests were analyzed, and the factors stimulating competition and forcing service companies to consider the problems of comfortable living were identified. It is proved that increased responsibility of employees, increased efficiency of management processes, and improved functioning of all the processes contribute to the management system improvement. The incoming requests were thoroughly analyzed to refine the activities of the company in order to improve the quality and customer satisfaction. The proposed measures will improve the management of companies operating in competitive conditions that require knowledge and skills for its appropriate management.

A. Gorkunova · L. Redko (✉) · I. Plotnikova
Division for Control and Diagnostics, National Tomsk Polytechnic University, 30, Lenin Ave, Tomsk, Russia
e-mail: laredko@tpu.ru

I. Plotnikova
e-mail: inna@tpu.ru

M. Yanushevskaya
Faculty of Innovation Technologies, Tomsk State University of Control Systems and Radioe-lectronics, 40, Lenin Avenue, Tomsk, Russia 634058
e-mail: ymn@2i.tusur.ru

E. Lysenko et al. (eds.), *Recent Developments in the Field of Non-Destructive Testing, Safety and Materials Science*, Studies in Systems, Decision and Control 433,
https://doi.org/10.1007/978-3-030-99060-2_4

1 Introduction

Globalization of the world economy and fierce competition in local and international markets force enterprises striving to survive and develop to actively seek additional competitive advantages [1]. The competitiveness of any enterprise, regardless of its size, organizational and legal form and field of activity, depends primarily on the safety and quality of products and services, and on the balance in price and quality. Housing services can contribute to an increase in the indicators of economic activity [2, 3].

Services provided to the population to ensure comfortable living in houses can be divided into two groups: housing services and utility services. Utility resources include cold water, hot water, electricity, gas, thermal energy, heat carrier in the form of hot water in open heat supply systems (hot water supply), household fuel gas, solid fuel for stove heating used to provide public services and consumed when maintaining common property in an apartment building. The resources also imply wastewater discharged through centralized networks of engineering and technical support. Removal of solid municipal waste is a public service. After the utility resources of proper quality have been supplied to a residential building, the company that manages this house is responsible for the quality of these resources. Housing services are aimed at comfortable and safe living of consumers. The works in the sphere of housing services include:

- works for proper maintenance of load-bearing structures (foundations, walls, columns and pillars, floors and roofs, beams, cross-beams, ladders, load-bearing elements of roofs) and non-load-bearing structures (partitions, interior decoration, floors) of apartment buildings);
- works for proper maintenance of equipment and engineering systems that are part of the common property in an apartment building;
- works and services for maintenance of other common property in an apartment building.

For consumers, not only the availability of the listed services, but the quality of their provision is important [4, 5]. Requirements for the quality of utility services and housing services are stated in the regulatory documents.

Utility services meet clearly stated requirements and regulatory meanings. Quality control of these services is not complicated. It is sufficient to measure the voltage in the circuit or to take water sample and analyze it, compare the obtained results with the normative ones and make an appropriate decision [6].

Housing services are different. Quality control of housing services is subjective since no clear indicators are specified. One can take a scraping from the wall or floor of common facilities (CF), carry out a laboratory analysis for the degree of pollution, etc., but these data cannot be compared with the requirements, since there are no standard values for the sanitary state of the halls. The only thing that needs to be observed is the frequency. In the event of consumer dissatisfaction therefore, the service company must be ready to provide supporting documents to prove the services were actually provided. In comparison to cleaning, this document should be the Act of Acceptance of the Work Performed. Its reads as follows: date of completion of work; full name of the specialist; signature of the specialist; number of the apartment that accepted the work (consumer); consumer signature. Fulfillment of other services can be confirmed by various service order-tasks, schedules, acts of survey, service acceptance acts, records in the dispatcher's log, recordings of conversations, the history of actions of the passport office program, and others.

If the complaints do not stop or repeat, customer satisfaction should be assessed to improve the quality of the services provided [3]. The service quality is not only compliance with the requirements of standards, but also compliance with consumer expectations. That is, it is necessary to know expectations of the consumer to satisfy him with the quality of the services provided. Customer satisfaction can be assessed by questionnaires, telephone surveys or surveys on the company's website [4]. To improve customer satisfaction, calls of customers to the emergency dispatch service can be analyzed.

The main goal of the company is to ensure favorable and safe living conditions for citizens, proper maintenance of facilities, provision of utility services in line with the requirements of technical regulations, standards, rules, state, sanitary and epidemiological rules and regulations, hygiene standards and other legal acts.

The main services are as follows:

- maintenance of buildings, structures, equipment and communications;
- current and capital repairs of common property;
- round-the-clock provision of emergency dispatch service to consumers;
- elimination of accidents and malfunctions;
- current work with residents (settlement and cash services, passport office services).

The main services of third-party companies involved are as follows:

- utility services;
- removal of municipal solid waste (MSW);
- maintenance of fire alarms and executive devices for the smoke exhaust system in apartment buildings;
- maintenance of video surveillance and barriers;
- comprehensive maintenance of elevator facilities in apartment buildings
- sanitary cleaning of common facilities and adjacent territory of apartment buildings.

The service company is in charge of apartment residential buildings, the adjacent territory of apartment residential buildings, sidewalks and footpaths, lawns, playgrounds, sports grounds, facilities for operation of apartment buildings, and meeting the needs of its residents and tenants of non-residential premises.

2 Research Method

The service company conducted a customer satisfaction assessment via a telephone survey. The interviewer asked the service consumer to provide information on what he wanted to improve in the service company. The consumers responded that when contacting the emergency dispatch service, their requests were not always timely fulfilled. Some consumers reported that they had to apply twice with the same issue, since the first request was not transferred to the contractor.

The company was subjected to an internal auditing. The internal audit revealed the problem with untimely execution of orders. The service company increased the number of serviced houses, which resulted in an increased number of requests from the customers. Handwritten registration of requests by the operator became inconvenient. It was decided to purchase a special program that automatically enters the data of the consumer upon repeated contact, instantly forwards requests to the performers, monitors the execution deadline, and it is also supplemented with a mobile application for employees. Since incoming requests from consumers – homeowners are entered in an electronic version, another advantage of the program is the ability to structure data. Namely, to identify what services consumers complain about most often, why and how their priority changes [5].

The requests received over six months were analyzed. During this period, 3,426 requests concerning various issues were received. Requests directly related to poor quality housing services were highlighted. It was done to attain reliability of information, since the emergency dispatch service receives requests not only with complaints about poor-quality services, but also such as: 'neighbor's apartment leaks', 'plumbing equipment leaks', 'power outage'. These requests are not associated with the quality of the provided housing services.

A total of 1,917 requests received were grouped into categories and analyzed. The requests were divided into 3 main categories – Plumbing Services, Electrical Services, and Improvement and Maintenance of Common Facilities. Further, these categories were considered separately.

The service company is responsible for several serviced areas with a different number of serviced houses, which have different number of stores, different commissioning dates and many other features. The number of requests directly depends on the number of people living in a given microdistrict. However, it is problematic to control the exact number of residents in the microdistrict, therefore, the service company considers the indicator Number of Requests per 100 Apartments for each microdistrict separately. Table 1 shows the calculation results for this indicator.

The values highlighted differ from those obtained for other microdistricts in a given month or not typical of this microdistrict. Thus, it is possible to compare the work of employees of the service company in different microdistricts and control the dynamics in each microdistrict separately.

Below is the analysis of requests for the category of Plumbing Services.

The table highlights the values that are different from the mean values. The obtained values show that there were more requests in microdistricts 2 and 3

Table 1 Number of requests per 100 apartments

Microdistrict	1	2	3	4	5
Number of apartments	734	797	1,154	1,954	1,954
Sanitary fittings					
October	0.5	0.5	1.6	1.7	2.0
November	0.4	0.4	1.0	2.4	3.0
December	0.7	0.3	1.0	1.6	1.0
January	0.8	1.3	0.3	0.4	0.6
February	0.4	0.4	0.5	0.6	0.6
March	0.7	0.1	0.6	0.5	0.9
April	0.8	0.6	0.3	1.4	1.5
Electrical facilities					
October	2.2	3.1	2.7	2.4	1.6
November	1.8	2.9	1.8	2.2	2.0
December	2.3	3.8	2.1	1.0	1.8
January	1.5	3.0	1.8	1.9	1.8
February	1.9	1.0	1.6	1.8	1.2
March	0.8	3.1	1.1	2.5	1.4
April	0.7	1.5	1.5	1.5	1.3
Common facilities					
October	0.1	1.3	0.8	1.4	1.1
November	0.4	0.8	0.5	2.0	2.6
December	0.4	0.8	1.1	1.3	1.9
January	1.2	1.1	1.9	1.9	2.0
February	1.1	0.5	1.2	1.2	1.9
March	0.1	1.5	0.8	1.5	0.8
April	0.3	0.8	1.0	1.4	0.6

compared to other microdistricts. In microdistricts 4 and 5, more requests were received compared to other microdistricts. In order to analyze the occurrence of requests registered in these periods, the reasons for requests were thoroughly analyzed.

Let's analyze using statistical methods [7]. The analysis of the data obtained showed that in microdistrict 2 in January 80% of requests were associated with one house and one reason, namely the problems with general utilities which were finally eliminated [8].

In microdistrict 4, an increased number of requests was registered in October, November and December. A Pareto chart is shown in Fig. 1.

During three months, most requests were associated with general utilities of the house. Every month there were complaints from residents of various microdistricts

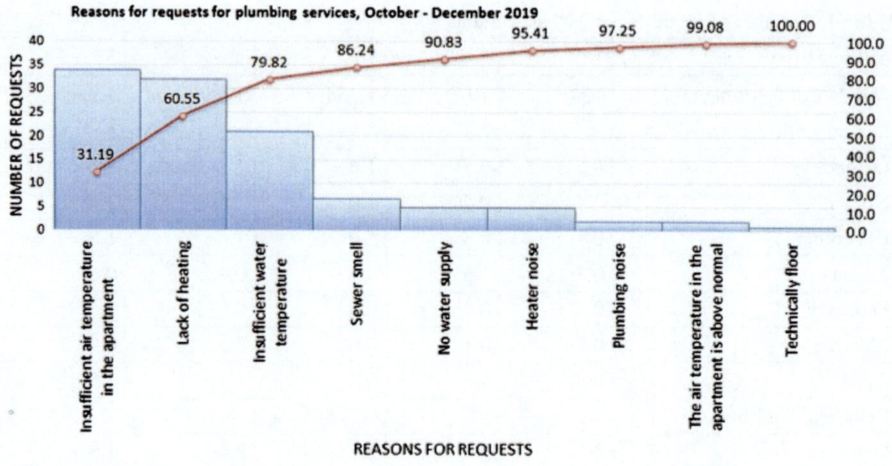

Fig.1 Reasons for requests for plumbing services in microdistrict 4

about low temperature in the apartment, about insufficiently warm water or about the lack of heating. At the beginning of the heating season, the problems were eliminated, and for the next three months no 'outbreak' of complaints was observed.

Over the next month, the number of requests increased again. The reasons for these requests are presented in Fig. 2.

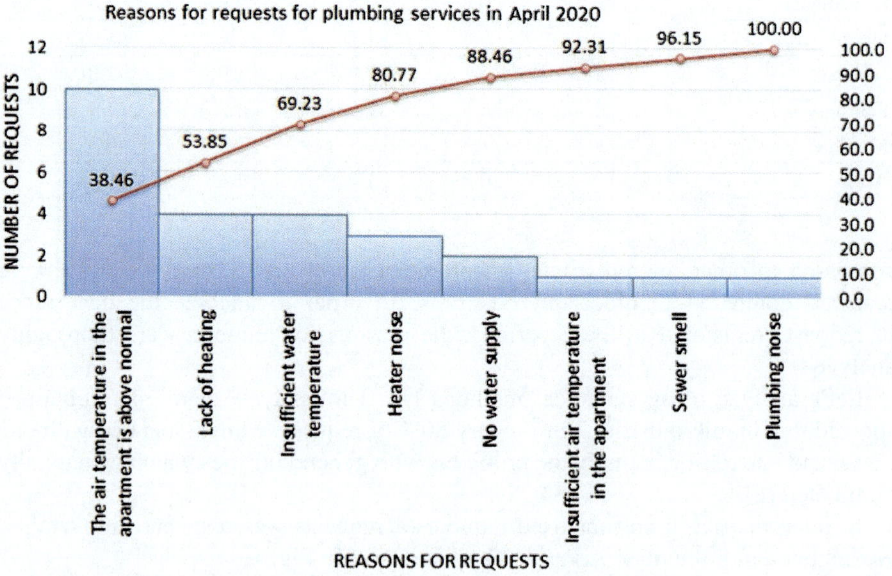

Fig. 2 Reasons for requests for plumbing services over a month

In this case, 80% of requests show other reasons, and the main reason is an increased air temperature in the apartment. In April 2020, the weather was abnormally hot, and air temperature was on average 7 ∊C higher than that in the previous five years. The number of sunny days attained 16, and the number of days fair with some clouds was 12. Since most people are usually at home in the evening only, they cannot feel high air temperature in the apartment. However, the previous mounth showed abnormally warm weather, the air temperature was 7 °C higher than that in the current month (data over five years). It can be assumed that during this period people used various devices such as computers and TVs more often, which also generate heat. Thus, the abnormally hot weather and the use of electrical devices contributed to the temperature rise in the apartment during the daytime. The service company had no right to turn off heating and had to supply the minimum amount of heat to consumers.

A similar situation could be observed in microdistrict 5. The reasons for requests registered in October and November were low air temperature in the apartment, lack of heating, sewage smell and heater noise. These reasons can be found in 80% of the total number of requests. It is worth noting that the majority of requests are associated with the first two reasons. In April, the priority of requests changed significantly. A total of 80% of requests included complaints about low temperature of water, sewage smell and high air temperature in the apartment.

Microdistrict 5 is currently being built, and the number of residents increases daily. New residents of the microdistrict do not always know that they are able to solve some problems with comfortable living on their own. For example, sewage smell in most cases is due to loose joining of sanitary facilities. In addition, the lack of heating may be associated not only with the problems with general utilities but also with the turned-off radiator or unscrewed control valve. As a result, people may experience low air temperature in the apartment.

Next, analyze the requests for electrical services.

The most significant number of requests per 100 apartments was registered in microdistrict 2. The Pareto diagram in Fig. 3 was used to study the most frequent requests to the service company in the category of Electrical Services.

Among the requests received within three months, 80% of requests are related to intercom and front entrance lighting.

A large number of requests concerning non-working intercoms are mainly those related to the broken intercom in the apartment. The reason for this is a deliberate restriction of consumers of housing services, namely intercom services. Restrictions are mostly placed on those consumers who consistently refuse to pay for housing services. This is permissible, since the consumer is not restricted from using a common intercom. This method is one of the few effective methods for dealing with non-payers [9, 10].

The second most common reason for requests for electrical services is front entrance lighting. To solve this problem, preventive measures are regularly taken, that is, daily inspections of the entrances by employees of the service company for burned out bulbs [11]. However, such requests are inevitable, since the consumer can more quickly reveal such a problem than employees of the service company.

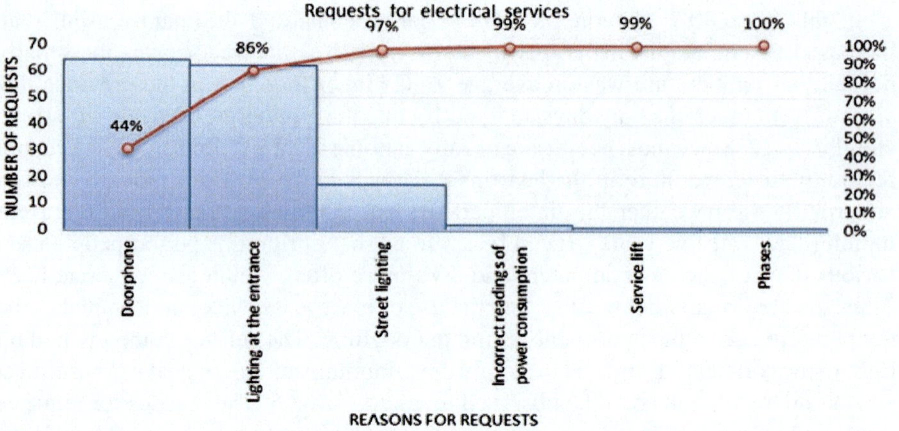

Fig. 3 Requests for electrical services in microdistrict 2

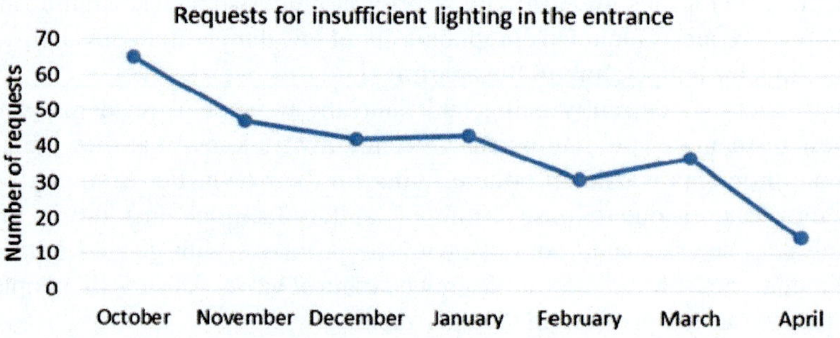

Fig. 4 Complaints about insufficient entrance lighting

Figure 4 shows changes in the number of complaints about insufficient entrance lighting over seven months.

The reduced number of requests could be affected by two factors: installation of a software product and increased daylight hours.

Due to the software product, consumers' requests are transmitted, and information on these requests cannot be accidentally lost [12, 13]. An increase in daylight hours leads to the fact that consumers do not pay attention to insufficient artificial entrance lighting which is compensated for natural lighting.

Below is the analysis of requests registered over six months for the category of Improvement and Maintenance of Common Facilities.

The number of requests registered in November in microdistrict 5 is abnormal, and all other values remain stable. Figure 5 presents the reasons for consumers' requests that month.

About 80% of requests received from consumers include the following reasons: door, ventilation, cleaning the entrance, cleaning the territory, handle. One of five

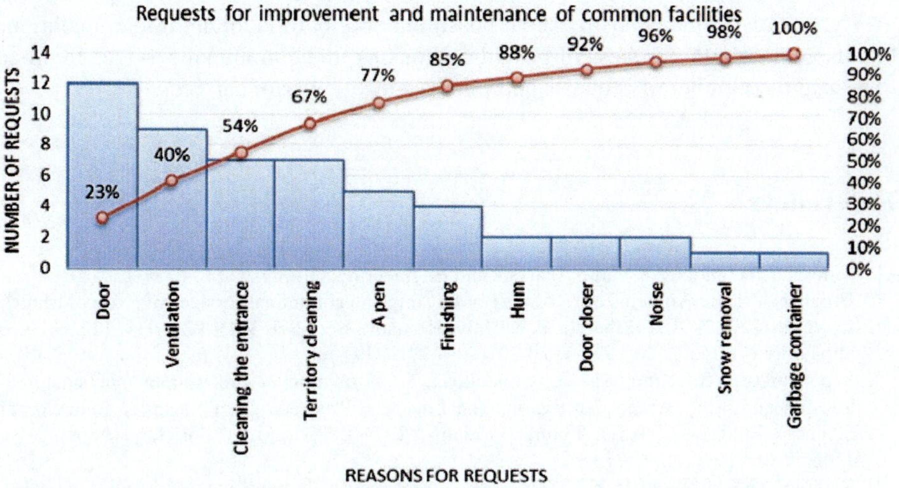

Fig. 5 Reasons for requests for improvement and maintenance of common facilities in November 2019, residents of microdistrict 2

reasons is not typical—ventilation. Other reasons are typical since they are visible and consumers pay attention to them immediately. The reasons for these requests are associated with cold, which is inherent in November. The requests for ventilation were found to come from residents of the same house and entrance. The comments indicate that the smoke exhaust system was turned on, and cold air entered the apartments through open ventilation grilles.

3 Conclusions

Since the regulatory requirements for housing services do not indicate the quality of the services provided but only state their frequency, the indicator Number of Requests per 100 Apartments can be used by the service company as one of the methods for quantitative assessment of the quality of the services provided. That is, in the future, it will be possible to define a normal range of values of this indicator for a particular microdistrict. The months with the increased number of requests can be forecasted, as well as affecting factors and measures to be taken so that the indicator does not go beyond the acceptable range of values. The analysis of requests received by the service company revealed factors that can affect the number of customers' requests. These are weather conditions (ambient temperature, precipitation);duration of daylight hours; number of non-payers; age of the house; untimely replenishment of consumables. In addition, the analysis enables adjustment of the procurement.

Thus, statistical analysis makes it possible not only to facilitate implementation, but also enhance the process of providing housing and communal services in order to reduce the number of requests and improve the quality of services.

References

1. Miles, L. Techniques of Value Analysis and Engineering. Hill, N. Y. Mc-Craw (2014).
2. Urozhaeva, T. P.: Problems of reforming of housing and communal services of cities of irkutsk region in the 1990s. J. Siberian Fed. Univ. Humanit. Soc. Sci. **9**(4), 923–939. (2016). doi: https://doi.org/10.17516/1997-1370-2016-9-4-932-939
3. Sapozhnikova, S., Kramlikh, O., Chudakova, S.: Formation of functioning mechanism of housing and utilities sector of region. The European Proceedings of Social & Behavioural Sciences EpSBS– Groznyi: Future Academy, 2774–2782. (2019). doi: https://doi.org/10.15405/epsbs.2019.12.04.373
4. Rampini, G.S., Berssaneti, F.T., Saut, A.M.: Insertion of risk management in quality management systems with the advent of ISO 9001:2015: Descriptive and content analyzes. Springer Proceed. Math. Statis. **281**, 209–221 (2018). https://doi.org/10.1007/978-3-030-14973-4_20
5. Jimenez, A., Medina-Merodio, J.: An intelligent framework for the evaluation of compliance with the requirements of ISO 9001:2015. Sustainability **12**(13), 5471–5477 (2020). https://doi.org/10.3390/su12135471
6. Obuhov, S., Plotnikov, I., Surkov, M., Sumarokova, L., Savkin, K.: Development of the structural and functional design of the laboratory bench for experimental research diesel generator sets on variable speed. Matec. Web. Conferences. **91**, 01036 (2017). https://doi.org/10.1051/matecconf/20179101036
7. Plotnikova, I., Redko, L., Chicherina, N.: The analysis of the process of complaints consideration using statistical methods of control. IOP. Conf. Ser. Mater. Sci. Eng. **516**(1), 101–105 (2019). https://doi.org/10.1088/1757-899X/516/1/012013
8. Buckley, J.J.: Fuzzy statistics: regression and prediction. Soft. Comput. **9**(10), 769–775 (2005). https://doi.org/10.1007/s00500-004-0453-9
9. Hungshu, M., Hsien, C.W.: Monitoring imprecise fraction of nonconforming items using control charts. J. Appl. Stat. **37**(8), 1283–1297 (2010)
10. Marimuthu, G., Ramesh, Dr.G.: On moderate fuzzy analytic hierarchy process pairwise comparison model. Int. J. Sci. Res. 2319–7064 (2013)
11. Ezrahovich, A.Y., Vladimirtsev, A.V., Livshitz, I.I.: Risk-based thinking of ISO 9001:2015—The new methods, approaches and tools of risk management. Proceedings of the 2017 Quality Management, Transport and Information Security, Information Technologies, 506–511 (2017). doi: https://doi.org/10.1109/ITMQIS.2017.8085872
12. Evtushenko, Y.I., Lesovik, R.V., Mityakina, N.A.: New approaches to creating quality management systems. Int. J. Appl. Eng. Res. **10**(24), 45176–45180 (2015)
13. Melnikov, V., Dorofeev, V., Gavrilov, M., Medvedev, M.: Energy-saving solutions in the construction and housing and utilities sectors of Russia. IOP Conference Series: Materials Science and Engineering **698** (2020). https://doi.org/10.1088/1757-899X/896/1/012121

Potentiometric Sensor for the Ion Speciation in the Industrial Waters

Anna Vtorushina, Ekaterina Larionova, Eleonora Romanenko, and Sergey Romanenko

Abstract The research provides the key concept for construction of new type of the potentiometric sensor to automate the measurement procedure of the ion species in the industrial waters and environmental samples to serve prospective screening purposes. In this work the possibility of automation of determination of ion species in the industrial waters such as the chloride and nitrate ions concentration is considered with help of combined potentiometric sensors. The design of the measuring equipment supposes its installation into flowing system. Measurement of the ion concentration is implemented by the combined sensors with two ion-selective electrodes used as working and reference electrodes. This work studies the possibility to control the ion-selective electrode calibration and the workability by use of the system of electrogeneration of the required ion in the measured medium. This study shows the possibility of developing a potentiometric sensor that has the advantages of high simplicity, designability, stability of potential and can be easily installed on the automated system to monitor the ion species in the industrial waters. The performance of the proposed sensor, such as stability, the effect of stirring and concrete admixtures was investigated. The potentiometric sensor can be used to monitor chloride and nitrate ions in the industrial waters and wastewaters within the 11–3540 and 6.2–6200 mg/dm^3 range. Accuracy of the proposed technique is verified by the added-found method. Recovery errors are within the acceptable range (4–22%) for the screening purposes. The results are in good agreement with the classic ionometry

A. Vtorushina (✉) · E. Larionova · E. Romanenko · S. Romanenko
National Research Tomsk Polytechnic University, 30 Lenin Avenue, Tomsk, Russia 634050
e-mail: anl@tpu.ru

E. Larionova
e-mail: evl@tpu.ru

E. Romanenko
e-mail: ulk@tpu.ru

S. Romanenko
e-mail: svr@tpu.ru

S. Romanenko
Siberian State Medical University, 2 Moskovsky trakt, Tomsk, Russia 634055

E. Lysenko et al. (eds.), *Recent Developments in the Field of Non-Destructive Testing, Safety and Materials Science*, Studies in Systems, Decision and Control 433,
https://doi.org/10.1007/978-3-030-99060-2_5

method. The proposed potentiometric system is to be constructed taking into account the effect of temperature, the shift of the potential, the effect of stirring and the presence of interfering substances.

1 Introduction

Industrial processes emit wastewaters during one or more stages of production. Wastewaters can be arisen from the preparation of technological solutions, application ones for etching materials and coating them, washing parts and baths after dumping waste solutions and treating the room. The main substances discharged by wastewaters exceeding the maximum permissible concentration (MPC) are iron, copper, zinc, nitrite, nitrate and ammonium, suspended substances (dust, metal scales, etc.), alkalis, acids, emulsions and petroleum products [1–4].

Currently, the legislative framework in the field of environmental protection is undergoing significant changes. In accordance with the Order of the Government of the Russian Federation of March 19, 2014, № 398-r "On a complex of measures aimed at refusing to use outdated and inefficient technologies, switching to the principles of the best available technologies and adoption of modern technologies", a transition from standardization based on the use of MPC standards to the adoption of the best available technologies (BAT) and technological regulation in the field of environmental protection is planned. The use of BAT implies the modernization of existing production facilities and the reduction of the negative impact on the environment (environmental risks). The amendment to Federal Law «About environmental protection» №7 from 10.01.2002 comes into force from January 1, 2018. This amendment regulates that for the first category facilities the stationary sources should be equipped with automatic equipment for measuring and recording the volume or mass of emissions and discharges.

According to the changes in the legislative framework of the Russian Federation in the field of environmental protection the factories need to decrease the level of negative impact produced on water objects. It can be realized by modernization of the cleaning water equipment [5, 6], adoption of new technologies of water supply and automated control of wastewaters volume and composition in situ. Consequently, design and development of analytical complexes for monitoring the qualitative and quantitative parameters of the wastewaters is an up-to-date problem.

Nowadays monitoring process of wastewaters parameters before and after treatment actions is generally carried out on-site in laboratories. Sample preparation, carrying out an analysis and results treatment requires time resources and highly qualified personnel. Therefore, one of the main requirements to the modern monitoring systems is the possibility of automation of the process of measuring the water parameters [7–9]. It should also be noted that in environmental monitoring it is often necessary to determine the ionic form, rather than the total content of the element in the wastewater [10–15]. This also imposes restrictions on the choice of the analytical method.

There are different methods applied for the element speciation [16–22]. However, automation of some methods is limited due to different reasons, for example necessity of sample preparation, the chemical pretreatment stages, unconventional and high cost equipment and etc. In our opinion the most promising method for automation of speciation is the ionometry method. Ionometry is the simple express method provided by low-cost equipment. It does not require the multistage sample preparation, highly qualified personnel, unconventional equipment or results treatment. Moreover, the duration of analysis does not to exceed one or two minutes [23–27].

The list of controlled parameters in water objects is rather wide and regulated by standards. In this work the possibility of automation of potentiometric determination of concentration of such water pollutants as chloride and nitrate ions is considered. The sources of ions emissions to the water objects can be operating and human wastes.

Therefore, the goal of this work is to develop the automated potentiometric system for the chloride and nitrate ions determination. The construction of ionometry system could be realized in the flow condition. In order to achieve the goal the following problem is to be solved: (i) to develop the construction of the ionometry sensor and choose the method of calibration; (ii) to develop the technique to estimate the workability.

2 Materials and Methods

2.1 Reagents

A two electrode system was used as two commercial working and reference ion-selective electrodes. Potentiometric measurements were enabled by the use of the in-house built potentiometer connected to the computer. All chemicals used in experiment were AR grade without additional purification. All solutions were made with double-fold distilled water. The 11–3540 mg/dm^3 standard solutions of chloride ions and 6.2–6200 mg/dm^3 standard solutions of nitrate ions were prepared.

2.2 Sensor Design

The equilibrium potential of ion-selective electrode is traditionally measured with respect to the silver chloride electrode [28–30]. However, in the case of determining the concentration of chloride ions it is necessary to use the silver chloride electrode providing a salt bridge. However, the limitations of the silver chloride electrode include the instability of potential caused by the gradual dissolution of AgCl from the electrode surface and high polarization ability. This significantly complicates

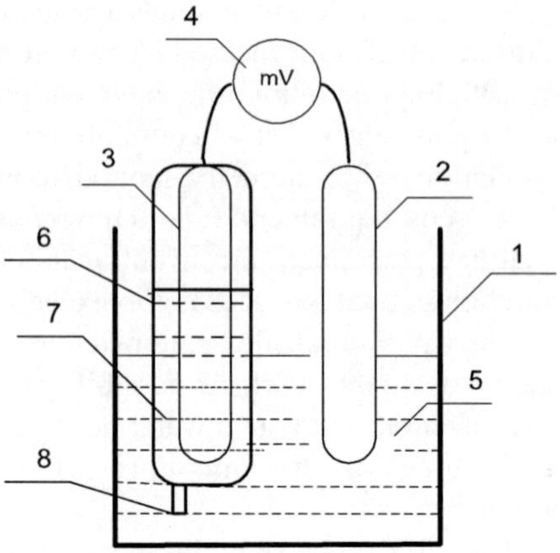

Fig. 1 Schematic representation of a potentiometric cell containing two ion-selective electrodes: 1—potentiometric cell, 2—ion-selective electrode as working electrode, 3—ion-selective electrode as reference electrode, 4—potentiometer, 5—sample, 6—body of reference electrode, 7—solution of target ion with a low concentration, 8—ceramic membrane

the process of automation to determine the content of the components in the water samples.

In this paper we propose the potentiometric sensor as the system of two ion-selective combined electrodes as a working electrode and a reference electrode. The reference ion-selective electrode is immersed in the solution containing the target ion with low concentration. The scheme of the proposed system is shown in Fig. 1.

Ion selective electrodes can go unserviceable due to mechanical contaminations or decrease in the membrane selectivity. The system of hydraulic cleaning of the measured system is proposed to eliminate mechanical contamination.

2.3 *Control of the Sensor Workability*

In this work the control of electrode workability is proposed by changing the autocalibration curve slope. The autocalibration of the potentiometric system is performed by loss of the known amount of chloride ions. Further to that, two electrodes (silver and graphite) were added to the system. The direct current flow passes through the cell during a certain time period. The amount of electricity passed through the cell corresponds to the weight of the resulting precipitate AgCl. Therefore, the decrease of chloride ions can be calculated by the known electrolysis time and the current intensity as follows:

$$m = \frac{Mq}{nF} = \frac{Mit}{nF}, \tag{1}$$

where M is the molar mass; q is the quantity of electricity; n is the amount of a substance; F is the Faraday constant; i is the amount of direct current; t is the flow-through direct current time. Therefore the autocalibration curve is plotted as dependence of the potential on the concentration of chloride ions. Potential is measured by the proposed sensor of two ion-selective electrodes. Then the workability of electrodes is estimated by change in the slope of the autocalibration curve comparing the calibration plot from the real and the model systems.

3 Results and Discussion

3.1 Sensor Calibration

The concentration of chloride ions can be determined by two ways: (i) by decrease of chloride ions; (ii) by the calibration curve of the system of two ion-selective electrodes. The concentration of nitrate ions can be estimated using the second way.

Using the fist way to determine the chloride ions in the wastewaters, the equilibrium potential E_x of ion-selective electrode should be recorded. Using the Nernst equation the concentration can be calculated as:

$$C_{Cl^-} = \frac{E^\circ - E_x}{b}, \tag{2}$$

where E° is the standard potential, E_x is the equilibrium potential, b is the autocalibration curve slope. The amount of b is obtained by the technique discussed in the section above. The amount of E° is determined as follows. It is necessary to plot the dependence of the equilibrium potential on the quantity of electricity (Fig. 2). The standard potential is estimated at the point where concentration of silver and chloride ions is equal:

Fig. 2 Potential plotted versus the quantity of electrisity

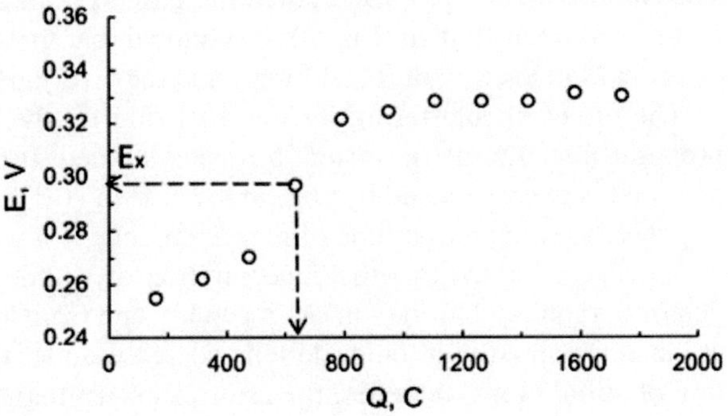

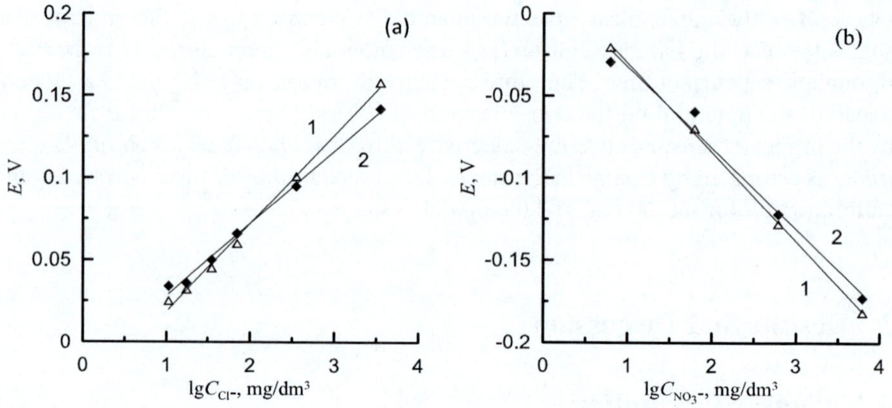

Fig. 3 Calibration plots of chloride **a** and nitrate ions **b** and stability of calibration plots: 1—initial calibration plot, 2—calibration plot after 14 days of potentiometric system working

$$C(Ag^{+}) = C(Cl^{-}) = \sqrt{K_{SP}} = \sqrt{1.78 \times 10^{-10}} \quad (3)$$

According to the Nernst equation the standard potential is $E_x = E^{\circ}$ at $C(Ag^{+}) = C(Cl^{-})$.

Using the second way to determine chloride and nitrate ions in the wastewaters the calibration plots obtained by the proposed potentiometric sensor can be used. We obtained the calibration plots of chloride ions from 11 to 3540 mg/ dm^3 and nitrate ions from 6.2 to 6200 mg/dm^3 (Fig. 3). The appropriate correlation coefficients (within the range 0.994 – 0.992) indicate the linearity in the whole range of concentrations. The intercepts on the ordinates in the graphs are negligible.

3.2 *Effect of Variables*

The stability of the proposed potentiometric system during 14 days in the flowing system has been studied (Fig. 3). It is shown that the stability of the potentiometric system is satisfactory during 14 days and the error does not exceed 10%.

The effect of interfering ions and stirring on the value of the potential of the proposed potentiometric system has been studied. It is shown that the value of the potential is overestimated by 16% in the case of stirring (Fig. 4). There is also a drift of potential at the initial time of measurements.

We also obtained that the concentration of sulfide ions significantly affects the determination of chloride ion. As it can be seen from Table 1, the values of potential at the determination of chloride ions decrease under the increase of the concentration of sulfide ions, therefore the error of determination of chloride ions increases. Significant effect of interfering components on the determination of nitrate ions has not been detected.

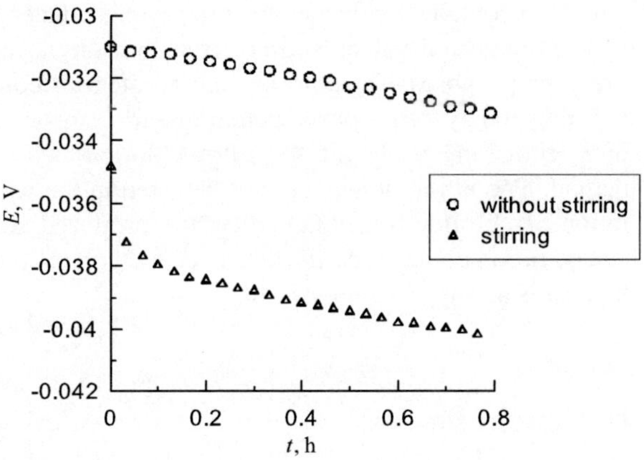

Fig. 4 Influence of stirring on potential at 62 mg/dm^3 nitrate concentrations

Table 1 Effect of sulfide ions concentration on the values of potential, V on the determination of chloride ions

Concentration of S^{2-}, [mg/dm^3]	Concentration of Cl^-, [mg/dm^3]		
	5	50	500
0	0.255 ± 0.003	0.184 ± 0.002	0.141 ± 0.003
5	0.253 ± 0.003	0.181 ± 0.004	0.136 ± 0.005
50	0.215 ± 0.004	0.166 ± 0.006	0.133 ± 0.003
500	0.163 ± 0.005	0.144 ± 0.003	0.126 ± 0.004

The temperature of industrial wastewaters can considerably vary. The effect of temperature can be considered by the Nernst equation.

3.3 Analytical Applications

The accuracy of the method was evaluated by the added-found method. The analytical results along with the error of the analysis are presented in Table 2. As shown in

Table 2 Analysis of chloride and nitrate ions in the water samples by the added-found method (N = 3)

Chloride ions			Nitrate ions		
Added C, [mg/dm^3]	Found C, [mg/dm^3]	Δ, %	Added C, [mg/dm^3]	Found C, [mg/dm^3]	Δ, %
17.7	16 ± 2	8	31	30 ± 2	4.2
177	159 ± 4	10	310	291 ± 19	5.8
1770	1381 ± 177	22	3100	2976 ± 124	4

Table 2 the obtained chloride and nitrate ions concentrations are in good agreement with the expected values and errors within the acceptable range (4–22%) for the screening purposes. The chloride and nitrate ions content of the wastewater sample was analyzed by the proposed potentiometric sensor. The potentiometric system with ion selective electrode and the silver chloride electrode was used as the reference method. The nitrate ions were not detected in the wastewater sample. The concentration of chloride ion obtained by the proposed sensor is 30 ± 2 mg/dm^3. The concentration obtained by the reference method is 25 ± 2 mg/dm^3. As it can be seen, the results are in good agreement.

4 Conclusion

In this work the potentiometric sensor for the analysis of chloride and nitrate ions by ionometry is proposed. We suggest the appropriate construction of electrode system based on two ion-selective electrodes. The workability of electrodes is verified by changing the slope of the autocalibration curve. The autocalibration curve is directly obtained by generation of chloride ions decrease in the system.

It should be emphasized that the calibration plots of the proposed potentiometric system should be constructed taking into account the effect of temperature, the shift of the potential, the effect of stirring and the presence of interfering substances.

In the wastewaters the chloride concentration can be vary from 10 to 400 mg/ml and nitrate ions concentration can be vary from 10 to 100 mg/ml; and concentration can change depending on the source of wastewater. The proposed potentiometric sensor of chloride and nitrate ions allows controlling the concentration of the ions within sufficient concentration range (chloride ions from 11 to 3540 mg/ml and nitrate ions from 6.2 to 6200 mg/ml). Moreover the proposed system can be used to monitor the concentration of ions at the level of maximum permissible concentrations of chloride and nitrate ions for screening purposes.

It is important to note that the proposed potentiometric sensor of two ion-selective electrodes can be realized to determine a wide range of ions in the wastewaters and environmental samples. A portable, submersible and automated analyzer can be designed by the proposed potentiometric sensor.

References

1. Aleksandrin, A.P., Egorshev, A.A., Katsereva, O.V., Mynin, V.N., Terpugov, G.V.: Purification and regeneration of technological solutions and waste of machine-building factories with application of ceramic membranes. Tyazheloe Mashinostroenie **11**, 22–24 (2001)
2. Porozhnyuk, L.A., Lupandina, N.S., Porozhnyuk. E.V.: Research of mechanisms of copper ions removal out of sewage of machine-building productions by combined reagent based on industrial wastes. Solid State Phenomena **265**, 851–855. (2017). doi: https://doi.org/10.4028/www.scientific.net/SSP.265.851.

3. Ding, B., Wang, J., Tao, S., Ding, Y., Zhang, L., Gao, N., Li, G., Shi, H., Li, W., Ge, S.: Fabrication of multi-functional porous microspheres in a modularfashion for the detection, adsorption, and removal of pollutants in wastewater. J. Colloid. Interface. Sci. **522**, 1–9 (2018). https://doi.org/10.1016/j.jcis.2018.03.060
4. Maksimov, D.B., Zakharov, A.V., Malts, I.E., Khomchenko, O.A., Solovev, E.M.: Production of cathode copper by electric extraction at kola mining and metallurgical company. Tsvetnye. Metally. **10**, 65–68 (2013)
5. Jachuła, J., Kołodyńska, D., Hubicki, Z.: Methylglycinediacetic Acid as a New Complexing Agent for Removal of Heavy Metal Ions from Industrial Wastewater. Solvent Extr. Ion Exch. **30**(2), 181–196 (2012). https://doi.org/10.1080/07366299.2011.581088
6. Seruga, P., Krzywonos, M., Pyżanowska, J., Urbanowska, A., Pawlak-Kruczek, H., Niedźwiecki, L.: Removal of Ammonia from the Municipal Waste Treatment Effluents using Natural Minerals. Molecules **24**(20), 3633 (2019). https://doi.org/10.3390/molecules24203633
7. Shannon, M., Bohn, P.W., Elimelech, M., Georgiadis, J.G., Mariñas, B.J., Mayes, A.: M: Science and technology for water purification in the coming decades. Nature **452**(7185), 301–310 (2008). https://doi.org/10.1038/nature06599
8. Laxmeshwar, L.S., Jadhav, M.S., Akki, J.F., Raikar, P., Kumar, J., Prakash, O., Mahakud, R., Raikar, U.S.: Quantification of chloride and iron in sugar factory effluent using long period fiber grating chemical sensor. Sens. Actuators, B Chem. **258**, 850–856 (2018). https://doi.org/10.1016/j.snb.2017.11.139
9. Tang, W., Ping, J., Fan, K., Wang, Y., Luo, X., Ying, Y., Wu, J., Zhou, Q.: All-solid-state nitrate-selective electrode and its application in drinking water. Electrochim. Acta **81**, 186–190 (2012). https://doi.org/10.1016/j.electacta.2012.07.073
10. Rani, S., Sharma, N., Malik, A.K.: A review on recent applications of high-performance liquid chromatography in metal determination and speciation analysis. Crit. Rev. Anal. Chem. **47**(6), 524–537 (2017). https://doi.org/10.1080/10408347.2017.1343659
11. Kiss, T., Enyedy, E.A., Jakusch, T.: Development of the application of speciation in chemistry. Coord. Chem. Rev. **352**, 401–423 (2017). https://doi.org/10.1016/j.ccr.2016.12.016
12. Alahi, Md.E.E., Mukhopadhyay, S.C., Burkitt, L.: Imprinted polymer coated impedimetric nitrate sensor for real-time water quality monitoring. Sens. Actuators, B Chem. **259**, 753–761 (2018). https://doi.org/10.1016/j.snb.2017.12.104
13. Mahmud, M.A.P., Ejeian, F., Azadi, S., Myers, M., Pejcic, B., Abbassi, R., Razmjou, A., Asadnia, M.: Recent progress in sensing nitrate, nitrite, phosphate, and ammonium in aquatic environment. Chemosphere **259** (2020). https://doi.org/10.1016/j.chemosphere.2020.127492
14. Vahl, K., Kahlert, H., Scholz, F.: Rapid automatic determination of calcium and magnesium in aqueous solutions by FIA using potentiometric detection. Electroanalysis **22**(19), 2172–2178 (2010). https://doi.org/10.1002/elan.201000146
15. Zolotov, Y.A.: Analytical Chemistry in Russia. Anal. Chem. **88**(17), 8348–8354 (2016). https://doi.org/10.1021/acs.analchem.5b04279
16. Zolotov, Y.A.: Russian Contributions to Analytical Chemistry. Springer International Publishing, Switzerland (2018)
17. Apyari, V.V., Gorbunova, M.O., Shevchenko, A.V., Furletov, A.A., Volkov, P.A., Garshev, A.V., Dmitrienko, S.G., Zolotov, Y.A.: Towards highly selective detection using metal nanoparticles: A case of silver triangular nanoplates and chlorine. Talanta **176**, 406–411 (2018). https://doi.org/10.1016/j.talanta.2017.08.056
18. Sohail, M., Roland De Marco, Lamb, K., Bakker, E.: Thin layer coulometric determination of nitrate in fresh waters. Analytica. Chimica. Acta. **744**, 39–44. (2012). doi: https://doi.org/10.1016/j.aca.2012.07.026
19. Myers, M., Khir, F.L.M., Podolska, A., Umana-Membreno, G.A., Nener, B., Baker, M., Parish, G.: Nitrate ion detection using AlGaN/GaN heterostructure-based devices without a reference electrode. Sens. Actuators, B Chem. **181**, 301–305 (2013). https://doi.org/10.1016/j.snb.2013.02.006
20. Baciu, A., Manea, F., Pop, A., Pode, R., Schoonman, J.: Simultaneous voltammetric detection of ammonium and nitrite from groundwater at silver-electrodecorated carbon nanotube electrode. Process Saf. Environ. Prot. **108**, 18–25 (2017). https://doi.org/10.1016/j.psep.2016.05.006

21. Yang, L., Wang, J., Wang, S., Liao, Y., Li, Y.: A new method to improve the sensitivity of nitrate concentration measurement in seawater based on dispersion turning point. Optik. **205**,(2020). https://doi.org/10.1016/j.ijleo.2020.164202
22. Scholz, F.: Electroanalytical methods. Guide to Experiments and Applications. Springer Berlin Heidelberg, Berlin (2010)
23. Mikhelson, K.N.: Ion-Selective Electrodes. Springer International Publishing, Switzerland (2013). doi: https://doi.org/10.1007/978-3-642-36886-8_8
24. Crespo, G.A.: Recent advances in ion-selective membrane electrodes for in situ environmental water analysis. Electrochim. Acta. **245**, 1023–1034 (2017). https://doi.org/10.1016/j.electacta.2017.05.159
25. Mandler, D., Scholz, F.: Electroanalytical methods. Guide to experiments and applications, 2nd ed. Analytical and Bioanalytical Chemistry **398** (7–8), 2771–2772. (2010). doi: https://doi.org/10.1007/s00216-010-4195-5
26. Pięk, M., Piech, R., Paczosa-Bator, B.: All-solid-state nitrate selective electrode with graphene/tetrathiafulvalene nanocomposite as high redox and double layer capacitance solid contact. Electrochim. Acta **210**, 407–414 (2013). https://doi.org/10.1016/j.electacta.2016.05.170
27. Blaz, T., Baś, B., Kupis, J., Migdalski, J., Lewenstam, A.: Multielectrode potentiometry in a one-drop sample. Electrochim. Acta **210**, 407–414 (2013). https://doi.org/10.1016/j.elecom.2013.05.021
28. Zuliani, C., Diamond, D.: Opportunities and challenges of using ion-selective electrodes in environmental monitoring and wearable sensors. Electrochim. Acta **84**, 29–34 (2012). https://doi.org/10.1016/j.electacta.2012.04.147
29. Hu, J., Stein, A., Buhlmann, P.: Rational design of all-solid-state ion-selective electrodes and reference electrodes. Trends Anal. Chem. **76**, 102–114 (2015). https://doi.org/10.1016/j.trac.2015.11.004
30. Cuartero, M., Bakker, E.: Environmental water analysis with membrane electrodes. Curr. Opin. Electrochem. **3**(1), 97–105 (2017). https://doi.org/10.1016/j.coelec.2017.06.010

Testing of Liquid Media In-Processes by Conductometry

Galina Vavilova, Anna Vtorushina, and Elena Liukiiu

Abstract In various industries, it is necessary to control the parameters of liquid media used in technological processes. The paper considers the indicator of the total degree of mineralization of the controlled media estimated by the conductometric method. The degree of mineralization was determined based on experimental calibration dependences of electrical conductivity on temperature and concentration of potassium chloride solutions. Experimental and theoretical data were compared, and the constant of the conductometric sensor for the system used was calculated. The described method can be used for any conductometric system to present the measurement results in the required form. The use of temperature corrections to eliminate the error is demonstrated. Taking into account the range of changes in the working concentrations of impurities and temperatures of liquid media for a particular technological process, the resulting calibration dependencies can be adjusted. The proposed method will make it possible to create an automated system for continuous monitoring of the total degree of mineralization of liquid media at different stages of various production processes.

1 Introduction

Water is an integral part of human life. The composition of water affects human health, the quality of products, and quality of production processes in general etc. Water objects are the main source for drinking and industrial water supply.

G. Vavilova (✉) · A. Vtorushina · E. Liukiiu
Division for Testing and Diagnostics, National Research Tomsk Polytechnic University, Tomsk 634050, Russia
e-mail: wgw@tpu.ru

A. Vtorushina
e-mail: anl@tpu.ru

E. Liukiiu
e-mail: esl18@tpu.ru

E. Lysenko et al. (eds.), *Recent Developments in the Field of Non-Destructive Testing, Safety and Materials Science*, Studies in Systems, Decision and Control 433,
https://doi.org/10.1007/978-3-030-99060-2_6

As a result of anthropogenic impact, natural water is polluted with various harmful substances, which leads to deterioration in its quality [1–3].

The main sources of water object pollution are the deposition of pollutants from the atmosphere, surface runoff from contaminated areas, and waste water from municipal and industrial enterprises [4, 5]. Water objects are the main source for drinking and industrial water supply [6, 7].

The spread of some mass diseases, such as brucellosis, salmonellosis, cholera, etc., is associated with the chemical composition of the water used for drinking [8–10]. Therefore, drinking water undergoes special treatment before it reaches the consumer. Water pre-treatment is also required at industrial enterprises. Special attention is paid to water treatment at food industry enterprises [11], breweries [12], in agriculture [13, 14], and nuclear and thermal power engineering [14].

For production processes, each industry sets its own requirements for the quality of water used in the technological process [4, 11, 12, 14, 15].

The control of the parameters of liquid media is carried out at different stages of production: incoming test, after water treatment procedure, at separate stages of the technological process for timely detection of deviations in parameters of liquid media from requirements of regulatory documentation.

The quality of liquid media is determined by various chemical, physical and biological characteristics. One of the chemical characteristics is the indicator of total mineralization (total salinity) [16].

Most often, total mineralization is characterized by the presence of inorganic salts (mainly bicarbonates, chlorides and sulfates of calcium, magnesium, potassium and sodium) and a small amount of water soluble organic substances [17].

The presence of salts of organic and inorganic origin in water affects electrical properties of water, in particular, the value of electrical conductivity. Electrical conductivity is a numerical expression of the ability of an aqueous solution to conduct electric current [18]. The electrical conductivity of water depends on its composition and degree of mineralization (concentration of dissolved substances), as well as on the temperature of the solution [16, 18].

One of the methods to determine electrical conductivity is the conductometric method [19, 20]. Conductometry is a type of chemical analysis based on measuring the electrical conductivity of solutions depending on the concentration, nature of the solute and temperature of the solution [18, 20].

This work considers the possibility of controlling the degree of mineralization of liquid media at different stages of the production process by the value of its electrical conductivity.

2 Building Calibration Dependencies

To control the degree of mineralization of various liquid media in any technological process, it is necessary to analyze the effect of concentration and temperature of the solution on the value of electrical conductivity. The dependence of the electrical

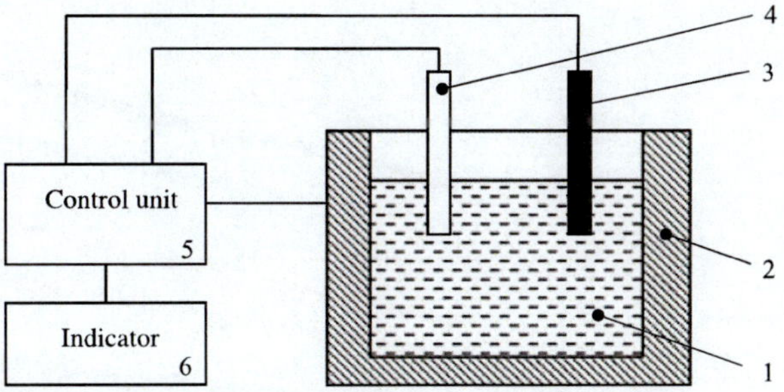

Fig. 1 Experiment block diagram

conductivity on the concentration of the solution is determined by the conductometric method. The conductometric method is easy to implement, it does not require large economic and time costs, and shows high accuracy.

The first part of the work is aimed at obtaining calibration dependencies. In the experiment, solutions of potassium chloride (KCl) with known concentrations of 0.0025; 0.005; 0.01; 0.02; 0.05; 0.1; 0.5; 1 mol/l were used [21].

The conductometric measuring cell consists of two electrodes located at a certain distance and immersed in the analyzed solution. The ratio of the electrode geometric dimensions (the surface area of the electrode and the distance between the electrodes) is invariable and is referred to as the constant of the measuring cell C [18, 22]. The cell constant depends on the shape of the vessel and the volume of the solution [18, 20]. The conductometric measuring cell is structurally represented by a conductometric sensor, with electrodes placed inside to perform primary processing of measuring signals.

The block diagram of the experiment is shown in Fig. 1. The analyzed solution 1 was placed in thermostat 2 to exclude the impact of the environment. During the experiment, the solution temperature is monitored using temperature sensor 3, and the solution conductivity is measured with conductometric sensor 4 [23].

In the automated mode, control unit 5 controls heating of the thermostat, the speed of stirring the solution, measurement of the required parameters, in particular the temperature and conductivity of the analyzed solution, and displays the required data on indicator screen 6.

In the experiment, the analyzed KCl solutions are heated in the range from +24 to +45 °C. The heating uniformity was achieved by constant automatic stirring. The electrical conductivity of the solution was recorded by changing the temperature with a step of 0.5 °C. The experimental results in the form of graphs of the dependence of the electrical conductivity L (*mS*) on temperature for KCl solutions of different concentrations are shown in Fig. 2.

The analysis of the dependences showed that at increased temperature, the value of electrical conductivity increases as the concentration of salt in the solution grows. The concentration growth increases the effect of temperature. The results obtained do not contradict previously published data.

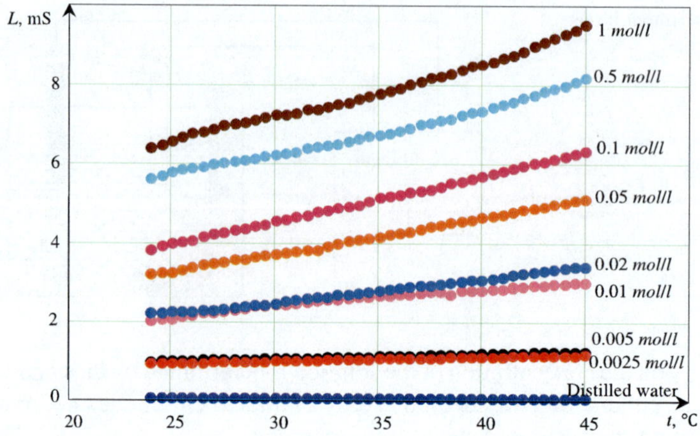

Fig. 2 Graph of the dependence of electrical conductivity on temperature changes for KCl solutions of different concentrations

The electrical conductivity of distilled water was measured in the same temperature range. From the graph (Fig. 2), it can be seen that the value of electrical conductivity practically does not change, since distilled water contains a small amount of electrically conductive particles [24, 25].

Theoretical relationships are usually given in units of electrical conductivity (*S*/*m*). In this work, the values of electrical conductivity were measured in units of *mS*. The comparison of the obtained values should be performed using specific electrical conductivity values.

The use of specific electrical conductivity values will eliminate the influence of the parameters of the measuring installation (the shape and volume of the vessel used, an external environmental impact) on electrical conductivity measurement results. The transition to specific electrical conductivity values can be carried out using the constant of the conductometric measuring cell C [22].

The value of the measuring cell constant C is difficult to calculate with sufficient accuracy; therefore, in practice it is determined using standard solutions of a given concentration, for which the values of specific electrical conductivity are known [21]. The relationship between the value of electrical conductivity and that of specific electrical conductivity is expressed by the equation [18, 20]:

$$C = \frac{l}{S} = \frac{R}{\sigma} = \frac{1}{L \cdot \sigma}, \quad (1)$$

where C is measuring cell constant (m^{-1});

σ is electrical conductivity (*S*/*m*);

$L = 1/R$ is electrical conductivity (*mS*), the reciprocal of the resistance R of the solution inside the conductometric measuring cell.

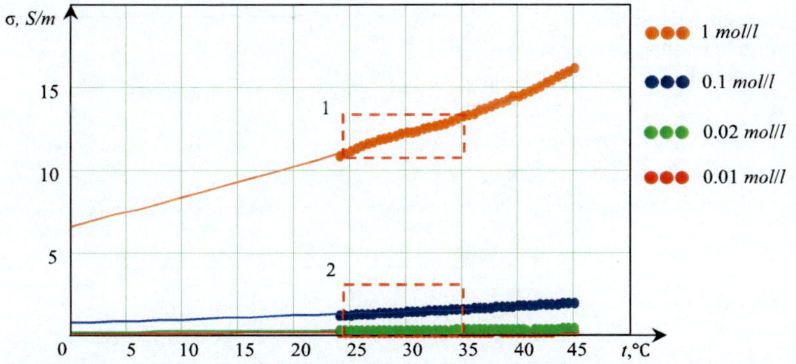

Fig. 3 Comparison of theoretical and experimental dependences of the specific electrical conductivity of KCl solutions

In [21], theoretical values of the specific electrical conductivity of KCl standard solutions with concentrations of 0.01 are presented; 0.02; 0.1 and 1 mol/l in the temperature range from 0 to +35 °C (Fig. 3).

A comparison was made between the theoretical values of the specific electrical conductivity of the KCl standard solutions and the values of the electrical conductivity *L* (mS) obtained experimentally at the same temperature (from 24 to 35 °C).

The comparison results were used to calculate the value of the measuring cell constant C in accordance with Eq. (1).

The values of the measuring cell constant are used to turn from the measured values of electrical conductivity (*mS*) to specific values (*S/m*).

Figure 3 shows the theoretical and experimental dependences of the specific conductivity of KCl solutions for concentrations of 0.01; 0.02; 0.1 and 1 mol/l. The dots on the graph indicate the experimental values in the temperature range from 24 to 45 °C, recalculated into the specific values of electrical conductivity for the considered concentration according to Eq. (1). Solid lines show the theoretical values of the specific electrical conductivity in the temperature range from 0 to 35 °C.

When considering the area of change in electrical conductivity in the range from 25 to 35 °C (Fig. 3, region 1, 2), it can be seen that the recalculated experimental values correlate well with the theoretical ones. Therefore, the experimental values of electrical conductivity (*mS*) can be used to construct calibration dependencies in the required concentration range.

The research results show the dependence of the measuring cell constant *C* on the concentration of solutions; therefore, it is necessary to obtain the function describing this dependence. The numerical values of the measuring cell constant C for concentrations 0.01; 0.02; 0.1 and 1 mol/l are obtained by Eq. (1) and are shown in Fig. 4 with red dots.

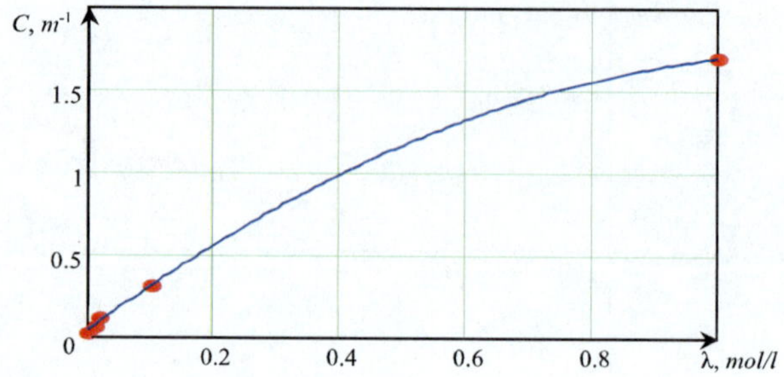

Fig. 4 Dependences of the vessel constant on the concentration of KCl in solution

The formula for the function of the dependence of the measuring cell constant on the concentration was obtained by regression analysis. The graph of the dependence is shown in Fig. 4 with a blue solid line.

The resulting dependence makes it possible to determine the measuring cell constant for solutions with concentrations, the specific conductivity values of which are not presented in the reference data [21]. The numerical values of the measuring cell constant *C* can be used to expand the temperature range when determining the specific conductivity of solutions of different concentrations.

The experimental values of electrical conductivity were expanded in the temperature range from 0 to 45 °C with regard to the obtained values of the measuring cell constant. Figure 5 shows the calibration dependences of the change in electrical conductivity on temperature for solutions of different concentrations.

The obtained calibration dependences can be used to control the degree of mineralization of water by the value of its electrical conductivity during production control [26, 27].

The proposed calibration makes it possible to estimate the overall degree of mineralization in terms of the concentration of a certain substance.

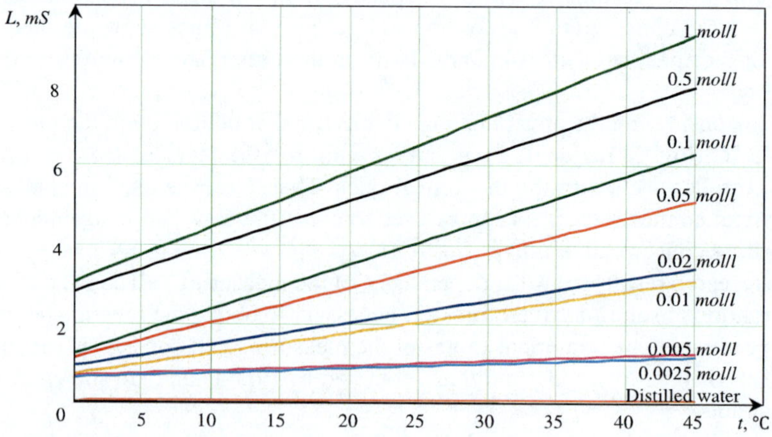

Fig. 5 Calibration dependencies

If required, calibration dependences can be obtained experimentally on the basis of other solutions. In this case, the total degree of mineralization is estimated in terms of a certain substance.

3 Practical Application

If temperature is known, the obtained calibration dependences can be used to estimate the total degree of mineralization of any water sample in terms of the *KCl* concentration by the value of electrical conductivity. The presented calibration is given in the values of electrical conductivity (*mS*), but if required, it can be recalculated into specific values to unify the proposed calibration (using another electrode, vessel, etc.) [20, 26, 28, 29].

The proposed calibration can be used as the basis for software to record electrical conductivity in an automated mode. The software will quickly recalculate the values of electrical conductivity into the value of the *KCl* concentration (total mineralization) [16, 26].

This system will simplify the process of monitoring water parameters in technological processes of any production [11–15].

The obtained calibration was tested on drinking water samples and snow samples taken in different parts of the city. Snow samples were pre-melted in the laboratory in a natural way. All the samples were filtered from suspended particles.

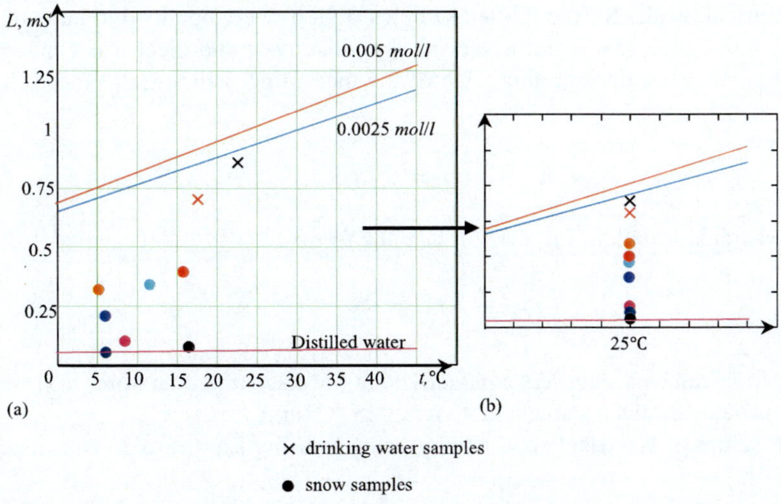

Fig. 6 Practical application of calibration dependencies: **a** Results of the measurement of electrical conductivity for water samples; **b** Values of electrical conductivity reduced to 25 °C

Figure 6a shows the values of the electrical conductivity of the studied samples of water and snow. Calibration dependences allow us to estimate the overall degree of mineralization of samples in terms of the KCl concentration.

It should be noted that during electrical conductivity measurement, water samples showed different temperatures. Since the temperature affects the numerical value of electrical conductivity, it is necessary to bring the obtained values to a uniform temperature to compare the results obtained.

To eliminate the temperature error, the conductometric analysis should be carried out at similar temperature, which is quite difficult to implement in real conditions. In any technological process, maintaining a uniform temperature of liquid media is very difficult and energy-consuming, and sometimes impossible.

The temperature error can be eliminated by introducing a temperature correction. It was established [18, 26] that the dependence of the specific electrical conductivity of a solution on temperature can be expressed by a simplified equation applicable for practical purposes:

$$\sigma_t = \sigma_T[1 + \alpha(t - T)] \quad (2)$$

where t is temperature to conduct research (°C);

σ_t is value of specific electrical conductivity at the current temperature (S/m);

α is temperature coefficient of conductivity of the solution, (°C^{-1});

T is uniform temperature to which the values should be reduced (°C). It is typically 25 °C [18, 26];

σ_T is conductivity value at 25 °C (S/m).

Empirical results for the studied samples (Fig. 6) were obtained in units of electrical conductivity (S). Specific electrical conductivity and electrical conductivity are interconnected through the value of the measuring cell constant (according to Eq. (1)):

$$\sigma = C \cdot L, \quad (3)$$

After transformation, Eq. (2) will take the form:

$$L_t = L_T[1 + \alpha(t - T)], \quad (4)$$

where L_t is value of electrical conductivity at the current temperature (mS);

L_T is value of electrical conductivity at 25 °C (mS).

The values of electrical conductivity, reduced using Eq. (4) to 25 °C, are shown in Fig. 6b.

The conductivity values reduced to a uniform temperature make it possible to compare the degree of mineralization for different water samples. Analysis of the data plotted in the Fig. 6 shows that the total mineralization of water samples does not exceed the concentration of 0.0025 mol/l in terms of KCl, which indicates a low content of electrically conductive impurities in these samples and complies with the requirements of documents (limiting value ≈ 0.0133 M (1000 mg/l)) [7, 24].

The mineralization of the snow samples is close to the mineralization of distilled water, which indicates that the samples contain a small amount of electrically conductive impurities.

4 Conclusion

The development of conductometry in the modern world makes it possible to determine numerous physico-chemical characteristics of solutions [30–34]. Conductometry has found wide application in the technological control of liquid-phase processes (water purification, water treatment, synthesis of sulfuric and phosphoric acids, ammonia fertilizers and a number of other chemical processes in the analysis of biological fluids) [35–37].

This paper shows the possibility of controlling the overall degree of mineralization of liquid media in various industries at different stages of the technological process. A method for controlling the total degree of mineralization of liquid media using calibration dependencies is proposed. The calibration dependences are constructed on the basis of experimental dependences of electrical conductivity on temperature and concentration for KCl solutions using the conductometric method.

The method of conductometry is quite simple in hardware design and expressive [38, 39]. The described advantages of the method allow it to be used to solve problems of controlling the total degree of mineralization of various liquid media in an automated mode.

The modern development of computer systems enables automation of physico-chemical analysis [40–42].

The use of information technology makes it possible to recalculate the value of electrical conductivity into a numerical value of the degree of total mineralization of water samples in various industries. This paper shows the possibility of recalculation using a model of calibration characteristics based on regression data processing.

Based on regression data processing, models of calibration characteristics of the dependence of electrical conductivity on temperature for samples with a known concentration of KCl were obtained. The obtained models can be employed to develop software used to determine the degree of total mineralization of samples of liquid media at different stages in various industries in an automated mode.

Further research will be aimed at development and testing of software to yield the result in units of the total degree of mineralization.

References

1. Lovinskaya, A., Kolumbayeva, S., Begimbetova, D., Suvorova, M., Bekmagambetova, N., Abilev, S.: Toxic and genotoxic activity of river waters of the Kazakhstan. Acta Ecol. Sinica **41**(6), 499–511 (2021). https://doi.org/10.1016/j.chnaes.2021.01.011
2. Perron, M.A.C., Frances, R.P.: Water quality effects on dragonfly and damselfly nymph communities: a comparison of urban and natural ponds. Environ. Pollut. **263** (2020). https://doi.org/10.1016/j.envpol.2020.114472
3. Jaffrezic-Renault, N., Dzyadevych, S.V.: Conductometric microbiosensors for environmental monitoring. Sensors **8**(4), 2569–2588 (2008). https://doi.org/10.3390/s8042569
4. Liu, Y., Wang, P., Gojenko, B., Yu, J., Wei, L., Luo, D., Xiao, T.: A review of water pollution arising from agriculture and mining activities in Central Asia: Facts, causes and effects. Environ. Pollut. **291**, 118209 (2021). https://doi.org/10.1016/j.envpol.2021.118209
5. Aboelkheir, M.G., Pal, K., Cardoso, V.A., Celestino, R., Yoshikawa, N.K., Resendea, M.M.: Influence of concrete mixer washing waste water on the chemical and mechanical properties of mortars. J. Mol. Struct. **1232**,(2021). https://doi.org/10.1016/j.molstruc.2021.130003
6. Hafeez, A., Shamair, Z., Shezad, N., Javed, F., Fazal, T., Rehman, S., FahadRehman, F.: Solar powered decentralized water systems: a cleaner solution of the industrial wastewater treatment and clean drinking water supply challenges. J. Clean. Prod. **289** (2021). https://doi.org/10.1016/j.jclepro.2020.125717
7. Sewak, R., Bena, D., Sewak, P., Sh., Mahajana: Ground-tested principles of designing a robust drinking water treatment system. Water Secur. **14** (2021). https://doi.org/10.1016/j.wasec.2021.100103
8. Xu, Y., Li, Y., Scott, K., Lindh, Ch.H., Jakobsson, K., Fletcher, T., Andersson, E.M.: Inflammatory bowel disease and biomarkers of gut inflammation and permeability in a community with high exposure to perfluoroalkyl substances through drinking water. Environ. Res. **181**, 108923 (2019). https://doi.org/10.1016/j.envres.2019.108923
9. Adamowicz, W., Dupont, D., Krupnickc, A., Zhang, J.: Valuation of cancer and microbial disease risk reductions in municipal drinking water: An analysis of risk context using multiple valuation methods. J. Environ. Econ. Manage. **61**(2), 213–226 (2011). https://doi.org/10.1016/j.jeem.2010.10.003
10. Shamsollahi, H.R., Ghoochani, M., Sadeghi, K., Jaafari, J., Masoud M., Sillanpää, M., Alimohammadi, M.: Evaluation of the physical and chemical characteristics of water on the removal efficiency of rotavirus in drinking water treatment plants and change in induced health risk. Process Saf. Environ. Prot. **130**, 6–13 (2019). https://doi.org/10.1016/j.psep.2019.07.014
11. Tekerlekopoulou, A.G., Economou, Ch.N., Tatoulis, T.I., Akratos, C.S., Vayenas, D.V.: Wastewater treatment and water reuse in the food industry. The Interact. Food Ind. Environ. **8**, 245–280 (2020). https://doi.org/10.1016/B978-0-12-816449-5.00008-4
12. Georgiadis, G.P., Elekidis, A.P., Georgiadis, M.C.: Optimal production planning and scheduling in breweries. Food Bioprod. Process. **125**, 204–221 (2021). https://doi.org/10.1016/j.fbp.2020.11.008
13. Siddiqiac, A., Wescoat Jr., Ja. L., Muhammad, A.: Socio-hydrological assessment of water security in canal irrigation systems: a conjoint quantitative analysis of equity and reliability. Water Secur. **4–5**, 44–55 (2018). https://doi.org/10.1016/j.wasec.2018.11.001
14. Feng, Y.-X., Yu, X.-Z., Zhang, H.: A modelling study of a buffer zone in abating heavy metal contamination from a gold mine of Hainan Province in nearby agricultural area. J. Environ. Manage. **287**, 112299 (2021). https://doi.org/10.1016/j.jenvman.2021.112299
15. Pawar, B., Park, S., Hu, P., Wang, Q.: Applications of resilience engineering principles in different fields with a focus on industrial systems: A literature review. J. Loss Prev. Process Ind. **69** (2021). https://doi.org/10.1016/j.jlp.2020.104366
16. Devesa, R., Dietrich, A.M.: Guidance for optimizing drinking water taste by adjusting mineralization as measured by total dissolved solids (TDS). Desalination **439**, 147–154 (2018). https://doi.org/10.1016/j.desal.2018.04.017

17. Azoulay, A., Garzon, P., Eisenberg, M.J.: Comparison of the mineral content of tap water and bottled waters. J. Gen. Intern. Med. **16**(3), 168–175 (2001). https://doi.org/10.1111/j.1525-1497.2001.04189.x
18. Boyd, C.E.: Electrical conductivity of water, part 1. Retrieved from https://www.globalseafood.org/advocate/electrical-conductivity-water-part-1/
19. Moldoveanu, S.C., David, V.: Short overviews of analytical techniques not containing an independent separation step. Sel. HPLC Method Chem. Anal. 31–53 (2017) https://doi.org/10.1016/B978-0-12-803684-6.00002-0
20. Christy, F.A., Shrivastav, P.S.: Conductometric studies on cation-crown ether complexes: a review. Crit. Rev. Anal. Chem. **41**(3), 236–269 (2011). https://doi.org/10.1080/10408347.2011.589284
21. Rafdel, A.A., Ponomareva, A.M.: Brief Reference Book of Physico-Chemical Values. ARIS, Moscow (2010)
22. Surzhikov, A.P., Lysenko, E.N., Vasendina, E.A., Sokolovskii, A.N., Vlasov, V.A., Pritulov, A.M.: Thermogravimetric investigation of the effect of annealing conditions on the soft ferrite phase homogeneity. J. Therm. Anal. Calorim. **104**(2), 613–617 (2011). https://doi.org/10.1007/s10973-010-1074-4
23. Surzhikov, A.P., Frangulyan, T.S., Ghyngazov, S.A., Vasil'ev, I.P., Chernyavskii, A.V.: Sintering of zirconia ceramics by intense high-energy electron beam. Ceram. Int. **42**(12), 13888–13892 (2016). https://doi.org/10.1016/j.ceramint.2016.05.198
24. ASTM D1193–06 Standard Specification for Reagent Water (2018)
25. HOST R 58144–2018 Distilled water. Specifications (2018) Standartinform, Moskow, Russia
26. Yu., Satoh, Kakiuchi, H.: Calibration method to address influences of temperature and electrical conductivity for a low-cost soil water content sensor in the agricultural field. Agric. Water Manage. **255** (2021). https://doi.org/10.1016/j.agwat.2021.107015
27. Vavilova, G.V., Gol'dshtein, A.E.: Instrument for in-process control of capacitance per unit length of an electrical wire. Meas. Tech. **61**(3), 278–283 (2018). https://doi.org/10.1007/s11018-018-1421-6
28. Ponziani, M., Slob, E.C., Ngan-Tillard, D.J.M.: Experimental validation of a model relating water content to the electrical conductivity of peat. Eng. Geol. **129–130**, 48–55 (2012). https://doi.org/10.1016/j.enggeo.2012.01.011
29. Gavrilin, A., Moyzes, B., Zharkevich, O.: Constructive and processing methods of reducing vibration level of the metalworking machinery elements. J. Vibroeng. **17**(7), 3495–3504 (2015)
30. Kabir-ud-Din, Al-Ahmadi, M.D.A., Naqvi, A.Z., Mohd, A.: Conductometric study of antidepressant drug-cationic surfactant mixed micelles in aqueous solution. Colloids Surf. B: Biointerfaces **64**(1), 65–69 (2008). https://doi.org/10.1016/j.colsurfb.2008.01.005
31. Khayatian, G., Shariati, S., Shamsipur, M.: Conductance study of the thermodynamics of binding of some macrocyclic polyethers with Tl^+ Ion in dimethylformamide-acetonitrile mixtures. J. Incl. Phenom. **45**(1–2), 117–121 (2003). https://doi.org/10.1023/A:1023024206317
32. Jiang, Y., Dong, X., Li, Y., Liang, Y., Zhang, M.: An environmentally-benign flow-batch system for headspace single-drop microextraction and on-drop conductometric detecting ammonium. Talanta **2241** (2021). https://doi.org/10.1016/j.talanta.2020.121849
33. Jaikang, P., Grudpan, K., Kanyanee, T.: Conductometric determination of ammonium ion with a mobile drop. Talanta **132**, 884–888 (2015). https://doi.org/10.1016/j.talanta.2014.10.046
34. Kakhki, R.M.Z., Heydari, S.: A simple conductometric method for trace level determination of brilliant green in water based on β-cyclodextrin and silver nitrate and determination of their thermodynamic parameters. Arab. J. Chem. **7**(6), 1086–1090 (2014). https://doi.org/10.1016/j.arabjc.2011.09.016
35. Jacobs, P., Suls, J., Sansen, W.: Performance of a planar differential-conductivity sensor for urea. Sens. Actuators: B. Chem. **20**(2–3), 193–198 (1994). https://doi.org/10.1016/0925-4005(94)01180-X
36. Muhammad-Tahir, Z., Alocilja, E.C.: A conductometric biosensor for biosecurity. Biosens. Bioelectron. **18**(5–6), 813–819 (2003). https://doi.org/10.1016/S0956-5663(03)00020-4

37. Makahleh, A., Saad, B., Siang, G.H., Saleh, M.I., Osman, H., Salleh, B.: Determination of underivatized long chain fatty acids using RP-HPLC with capacitively coupled contactless conductivity detection. Talanta **81**(1–2), 20–24 (2010). https://doi.org/10.1016/j.talanta.2009.11.030
38. Smith, A.C., Wynn, P.M., Barker, P.A., Leng, M.J.: Drip water electrical conductivity as an indicator of cave ventilation at the event scale. Sci. Total Environ. **532**, 517–527 (2015). https://doi.org/10.1016/j.scitotenv.2015.06.037
39. Cleskey, R.B., Nordstrom, D.K., Ryan, J.N., Ball, J.W.: A new method of calculating electrical conductivity with applications to natural waters. Geochimica et Cosmochimica Acta **77**, 369–382 (2012). https://doi.org/10.1016/j.gca.2011.10.031
40. El Ichi, S., Leon, F., Vossier, L., Marchandin, H., Errachid, A., Coste, J., Fournier-Wirth, C.: Microconductometric immunosensor for label-free and sensitive detection of Gram-negative bacteria. Biosens. Bioelectron. **54**, 378–384 (2014). https://doi.org/10.1016/j.bios.2013.11.016
41. Hnaiein, M., Hassen, W.M., Abdelghani, A., Fournier-Wirth, C., Coste, J., Bessueille, F., Jaffrezic-Renault, N.: A conductometric immunosensor based on functionalized magnetite nanoparticles for E. coli detection. Electrochem. Commun. **10**(8), 1152–1154 (2008). https://doi.org/10.1016/j.elecom.2008.04.009.
42. Tiecco, M., Corte, L., Roscini, L., Colabella, C., Germani, R., Cardinali, G.: A novel, rapid and automated conductometric method to evaluate surfactant-cells interactions by means of critical micellar concentration analysis. Chem. Biol. Interact. **218**, 20–27 (2014). https://doi.org/10.1016/j.cbi.2014.04.012

Improving the Surface of Titanium Alloys with Wave Cutters

Georgy Korovin, Aleksey Gavrilin, Sergey Petrushin, Georgy Odnokopylov, and Dmitry Ermakov

Abstract The article is devoted to the consideration of issues on improving the surface of titanium alloys using breakwaters. It is shown in the work that the processing of the titanium alloy used is a rather difficult task, since work in an aggressive environment imposes strict requirements on materials, namely: no reaction to external influences without loss of specified properties (solutions of diethanolamine, hydrogen sulfide), as well as a high specific strength, non-magnetic. The work investigates wave-shaped end mills made of high-speed steel for machining titanium alloys. The problem of calculating the geometry of wavy cutters is considered. It was established, on the basis of testing and measuring the vibration activity of the Part - Tool - Device - Machine tool (PTDM) system, that the roughness of the machined surface of the titanium alloy is less in wavy cutters compared to standard cutters made according to GOST 23,248–78. The stronger the titanium alloy, the less roughness when machining with wave cutter.

G. Korovin
Industrial Company LLC (MION), 46/6 Vershinin Street, Tomsk, Russia 634034

A. Gavrilin
Division for Mechanical Engineering, National Research Tomsk Polytechnic University, 30 Lenin avenue, Tomsk, Russia 634050
e-mail: tom-gawral@list.ru

S. Petrushin
Department of metal-cutting machines and tools, Kuzbass State Technical University, 28 Vesennya street, Kemerovo, Russia 650000
e-mail: victory_28@mail.ru

G. Odnokopylov
Division for Electric Drives and Equipment, National Research Tomsk Polytechnic University, 30 Lenin avenue, Tomsk, Russia 634050
e-mail: UniversityOGIz@yandex.ru

D. Ermakov (✉)
Division for Testing and Diagnostics, National Research Tomsk Polytechnic University, 30 Lenin avenue, Tomsk, Russia 634050
e-mail: ermakdmit@yandex.ru

E. Lysenko et al. (eds.), *Recent Developments in the Field of Non-Destructive Testing, Safety and Materials Science*, Studies in Systems, Decision and Control 433,
https://doi.org/10.1007/978-3-030-99060-2_7

1 Introduction

With the entry of Russian oil and gas companies to the shelf, the content of titanium alloys in oil and gas equipment has increased dramatically. Of all the labor intensity of machining titanium alloys, 70% is milling. Central CB "Nefteapparatury", FSUE "Prometey" together with the Verkhnaya Salda Metallurgical Plant (VSMPO-AVISMA) have created new "marine" titanium alloys and titanium with a high level of strength PT17, PT-3 V, PT-5 V, PT-37. For shipbuilding, turbine blades with a diameter of 1500 mm with σ_B>1000 MPa, for a steam generator, for MPSP "Prirazlomnaya"- a pipeline bundle made of titanium alloys, fine cleaning hulls, etc.

Therefore, the task of creating an effective tool for machining titanium alloys with specified geometric and physical properties is the most urgent today.

The world's largest manufacturers of end mills for machining titanium alloys: Dauphin Precision Tool (USA), Melin Tool Company (USA), Hanita Metals Works (Israel), Minicat International Inc. (Canada) and others [1–9].

The industrial company "MION", Tomsk has mastered the serial production of wave-like cutters from high-speed steel with diameters from 20 to 50 mm for processing difficult-to-machine and titanium alloys [10]. Wave cutters are available with 1D to 4D working lengths, with Weldon shanks, 7:24 Morse taper # 30, 40, 50.

The main differences of these mills from the above presented foreign end mills:

1. These cutters can be used for both rough and fine milling, since the wavy cutting edge is located on the cylindrical generatrix of the cutter, and the front surface of the wave tooth is part of the generating line passing through the cutter axis, perpendicular to the cutter axis or at an angle. When the generating line rotates around the cutter axis and simultaneously moves along this axis according to a certain law. The locus of points created by such a movement will be the surface of the helicoid;
2. The "wave" of each next tooth is shifted relative to the previous one by half a step, which makes it possible for the cutter to work in antiphase, and to each next tooth, along a less work-hardened layer, and effectively break up the general energy center of self-oscillations. All this leads to a decrease in the specific load per unit length of the cutting edge, a decrease in vibration activity and an increase in the durability of the cutter [11–13].
3. The clearance angle in the working plane is structurally made constant throughout the cylindrical part of the wave tooth, which makes it possible to regrind the cutter along the rear edge on a universal grinding machine.

When working with a wave cutter, the scheme of "different-angle cutting" is used, the advantage of which increases when two adjacent teeth have the most different angle of inclination of the tooth at each point of the cutting edge along the entire length of the working part.

2 Research Method

The paper considers the problem of calculating the geometric parameters of the wave-shaped cutter blade (using the example of calculating the main rake angle), measuring the vibration activity of the technological system and the roughness of the machined surface of titanium alloy BT22 with wave cutters and cutters made in accordance with GOST 23,248–78 "End mills for processing hard-to-machine and titanium alloys on systems with CNC" at the specified processing modes.

The difference between a wavy cutter and a standard cutter made in accordance with GOST 23,248–78 is that on the cylindrical working part, the front and rear angle of the cutting wedge at each point of the cutting edge has a different value due to the variable angle of inclination of the blade (ω_i). The change in the angle of the cutting wedge is affected by the main front (γH) and main back angle αH changing at each point of the wave tooth. Consider the problem of calculating the main rake angle of the blade (γH) at an arbitrary point A [14, 15].

The initial data for the calculation are (Fig. 1):

- the angle of inclination of the wave ω^A at point A;
- working rake angle $\left(\gamma_r^A\right)$ at point A, located in the plane coinciding with the cutting speed vector and as a point on the second order tool path according to the formula:

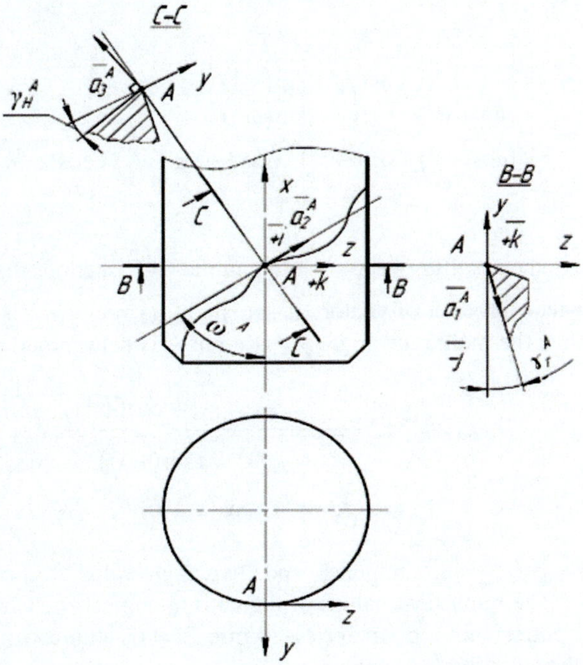

Fig. 1 Scheme for calculating the geometry of a wave milling cutter

$$\gamma_r^A = \arcsin\left(\frac{r\cos\left(\omega^A\right)}{R}\right); \tag{1}$$

where r is the second order tool radius; R is the radius of the cutter being produced.

It is necessary to define the main rake angle in a static coordinate system—γ_H^A.

Consider an arbitrary point A on the cutting edge of a wave tooth in an orthonormal rectangular coordinate system (Fig. 1).

Unit direction vector $\overrightarrow{a_1^A}$ tangent to the front surface of the tooth in section B-B at point A:

$$\overrightarrow{a_1^A} = -\cos\left(\gamma_r^A\right)\cdot\bar{j} + \sin\left(\gamma_r^A\right)\cdot\bar{k}; \tag{2}$$

The unit direction vector tangent to the wave part of the tooth at point A will be equal to:

$$\overrightarrow{a_2^A} = \cos\left(\omega^A\right)\cdot\bar{i} + \sin\left(\omega^A\right)\cdot\vec{k} \tag{3}$$

Then the normal $\overrightarrow{a_3^A}$ restored to the front surface of the cutting edge of the tooth in the C–C section is perpendicular to the vector $\overrightarrow{a_2^A}$, is determined by the vector product, and using a third-order matrix, we find the projection coordinates of the new vector:

$$\begin{aligned}\overrightarrow{a_3^A} &= \begin{vmatrix} \bar{i} & \bar{j} & \bar{k} \\ 0 & \cos\left(\gamma_r^A\right) & \sin\left(\gamma_r^A\right) \\ \cos\left(\omega^A\right) & 0 & \sin\left(\omega^A\right) \end{vmatrix} \\ &= \left(\cos\left(\gamma_r^A\right)\cdot\sin\left(\omega^A\right)\right)\cdot\bar{i} - \left(-\sin\left(\gamma_r^A\right)\cdot\cos\left(\omega^A\right)\right)\cdot\bar{j} + \left(-\cos\left(\gamma_r^A\right)\cdot\cos\left(\omega^A\right)\right)\cdot\bar{k};\end{aligned} \tag{4}$$

The resulting vector $\overrightarrow{a_3^A}$ lies in the C–C section, that is, the first condition of the vector product of vectors is satisfied, the vector $\overrightarrow{a_3^A}$ is perpendicular to $\overrightarrow{a_1^A}$ and a_2^A, then the main rake angle γ_H^A at point A is determined by the formula:

$$\begin{aligned} tg\left(\gamma_H^A\right) &= \frac{\left(-\cos\left(\omega^A\right)\right)\cdot\sin\left(\gamma_r^A\right)}{\sqrt{\left(\cos\left(\gamma_r^A\right)\right)^2\cdot\left(\sin\left(\omega^A\right)\right)^2 + \left(\cos\left(\gamma_r^A\right)\right)^2\cdot\left(\cos\left(\omega^A\right)\right)^2}} \\ &= -tg\left(\gamma_r^A\right)\cdot\cos\left(\omega^A\right); \end{aligned} \tag{5}$$

The "-" sign indicates a positive angle value.

The numerical values of the cutting angles γ_H^A, γ_r^A, α_H^A and the wedge angle for a cutter with a diameter of 50 mm Z = 6, with a working length l_w = 50 mm are shown in Table 1.

Table 1 Numerical values of cutting angles γ_H^A, γ_r^A, α_H^A and wedge angle

ω^A	γ_r^A	γ_H^A	α_H^A	Wedge angle
35°	11.340	9.30	12.15	68.55
45°	9.770	6.94	14.00	69.05
57.5°	7.510	4.10	17.93	71.90
30°	11.996	10.43	11.51	68.06
15°	13.400	12.96	10.34	66.70
7.5°	13.770	13.64	10.08	66.28
25°	12.560	11.41	11.00	67.59

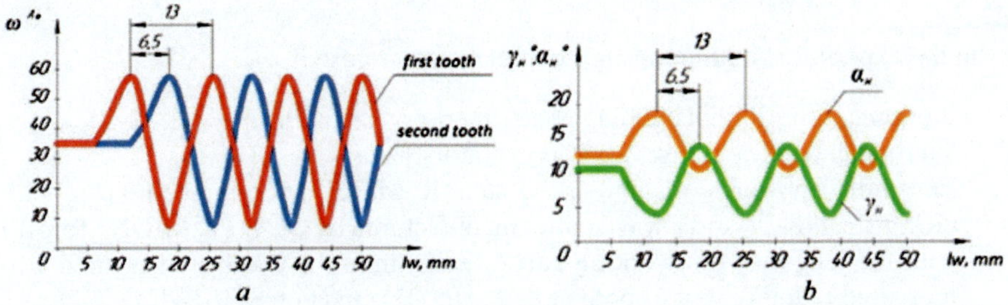

Fig. 2 Angle change data: **a** graph of the change in the angle of inclination ω^A for two adjacent teeth at the length of the working part of 50 mm, **b** graphs of the change in the main rake angle γ_n^A and the main clearance angle α_H^A at the length of the working part 50 mm for the first tooth

The clearance angle in the working plane α_r is constant in value over the entire length of the wave (working) part of the tooth and is equal to $\alpha_r = 10°$.

This fact makes it possible to regrind the cutter along the rear edge on a universal sharpening machine.

Figure 2a shows the graphs of the change in the angle of inclination ω^A for two adjacent teeth at the length of the working part of 50 mm.

Figure 2b shows the graphs of changes in the main rake angle γ_H^A and the main rear angle α_H^A over the length of the working part of 50 mm, depending on the change in the angle of inclination ω^A for two adjacent teeth.

3 Results and Considerations

The milling process is intermittent with a variable thickness of the cut layer. When machining structural steel parts, the fluctuation of the cutting force created by the chip formation process is not dominant [6, 16].

But it is shown that with an increase in the strength of titanium alloys, the lag of the cutting force from the thickness of the cut layer in phase increases [7, 17–21]. This means that the cutting force does the work going on to excite self-oscillations.

Work [8] shows that fluctuations in the cutting force arising in the processing of titanium alloys are directly related to the process of chip formation. Therefore, one cannot neglect the influence of the chip formation process on the dynamic component of the cutting force when machining low-plastic titanium all.

All known mechanisms of the initiation of self-oscillations are interconnected and, under normal cutting conditions, operate simultaneously in the technological system. However, there are no theories of vibration in metalworking that could comprehensively explain all the observed effects [9, 22–30]. In any case, with the growth of self-oscillations of the technological system, the roughness of the processed surface also grows.

In the experiment on measuring the roughness, we used:

1. Equipment: milling machining center model MCV-400.
2. Vibrodiagnostic complex "Vibroregistrator-F".
3. End mills Ø50 made of high-speed steel R6M5K5 made in accordance with GOST 23,248–78 and a wavy cutter manufactured by OOO PK MION (Tomsk) with the length of the working part $l_w = 50$ mm, the number of teeth Z = 6 measured radial runout of both cutters – 0.005 mm (Fig. 3) [31–42].
4. Laser scanning electron microscope model LEXT OLS4100.
5. Processed material: titanium alloy of high strength BT22 grade (forgings). Titanium was produced at the Verkhnaya Salda Metallurgical Plant (VSMPO-AVISMA).
6. Processing modes:
   - cutting speed $V = 10$ m / min, feed per tooth $S = 0.1$ mm;
   - cutting depth $t = 4$ mm, milling width $B = 24$ mm.
7. Cooling Liquid "Beazocut 5%" at a Pressure of 4 Atmospheres.

Figure 4 shows a diagram of the machining of a titanium alloy billet with the tested cutters.

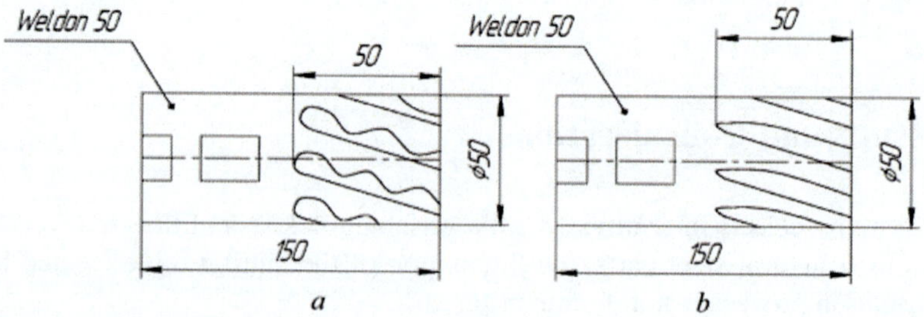

Fig. 3 Cutters: **a** wave milling cutter Ø50 mm, $Z = 6$; **b**) end mill cutter 50 mm, made in accordance with GOST 23,248–78, $Z = 6$

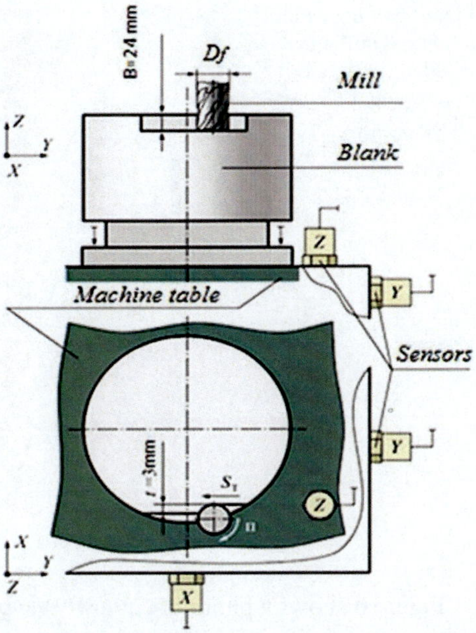

Fig. 4 Scheme of machining a titanium alloy billet

The Fig. 5 shows the location of the sensors on the workpiece when milling on the MCV-400 machine.

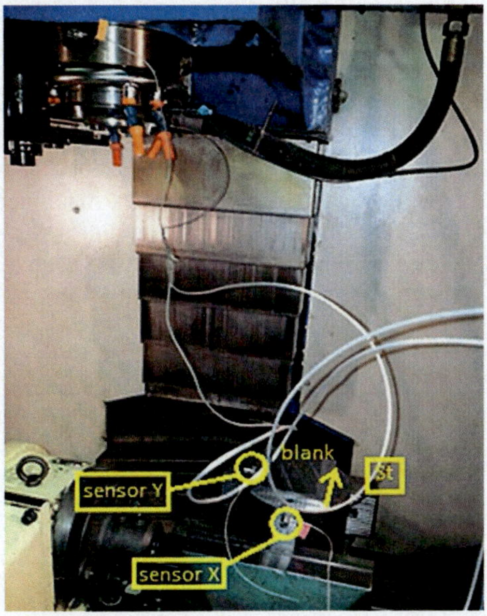

Fig. 5 The location of the sensors on the workpiece when milling on the MCV-400 machine

Fig. 6 Milling cutter: **a** "Wavy" milling cutter Ø20mm (left); **b** TK572 milling cutter Ø20mm GOST 23,248–78

Figure 6 shows a photo of cutters "Wavy" Ø50mm (left), cutters Ø50mm TK572 GOST 23,248–78, and "Wavy" milling cutter design [30].

The measured vibration activity of the technological system during processing of titanium alloy BT22 at specified processing modes is shown in Table 2.

The roughness of the machined surface is measured in the form of 3D parameters corresponding to ISO 25178 on a laser microscope model LEXT OLS4100. Due to the fact that this microscope has an estimate of the surface area, the roughness measurements S_Z and S_a were carried out with a high degree of reliability.

Figures 7, 8, 9, 10, 11, 12 show the roughness and 3D distribution of heights of the surface relief of the titanium alloy VT22.

The presented surface (Figs. 10, 11, 12), machined with a cutter Ø50 made in accordance with GOST 23,248–78 has a roughness $S_a = 1.86$ μm, $S_Z = 20$ μm.

The presented surface (Fig. 8), machined with a cutter made in accordance with GOST 23,248–78 Ø50 has a roughness $S_a = 1.86$ μm, $S_Z = 20$ μm.

The presented surface, machined with a wave milling cutter, has a roughness $S_a = 0.64$ μm, $S_Z = 18.6$ μm.

Table 2 The results of the measured vibration activity of the technological system during the processing of titanium alloy BT22 at the specified processing conditions

Cutter designation	Processed material	Root mean square vibration displacement along the axes (μm) and total			
		A_X	A_Y	A_Z	A_{sum}
Wave milling cutter Ø50	BT22	4	6	8	10.7
End mill cutter Ø50, made in accordance with GOST 23,248–78		15	20	8	26.25

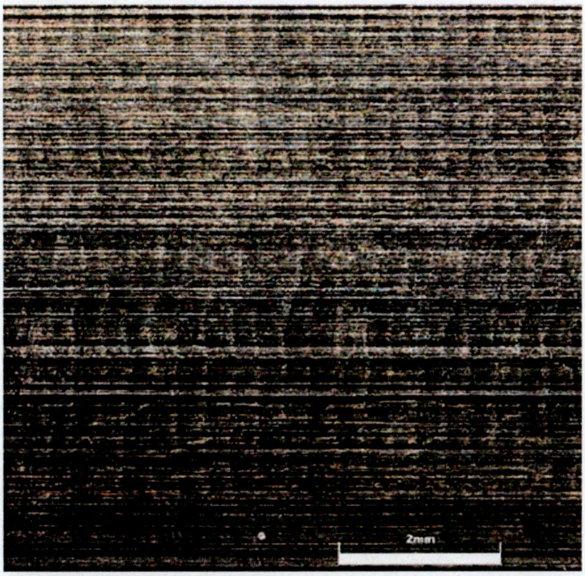

Fig. 7 Image of the BT22 alloy surface with a machined cutter Ø50 mm made in accordance with GOST 23,248–78

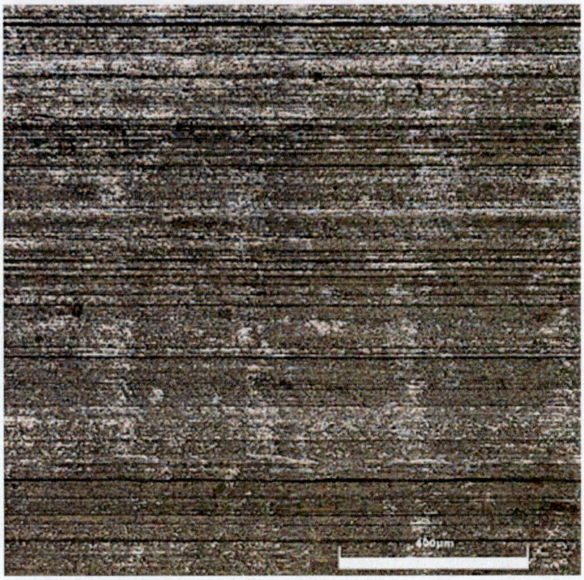

Fig. 8 An enlarged fragment of the surface

Analyzing the measurement results, it can be noted: surface roughness when machining high-strength titanium alloy VT22 with a wavy cutter in all roughness parameters ($S_a = 0.64$ μm, $S_Z = 18.6$ μm) is lower than that of a cutter made in accordance with GOST 23,248–78 ($S_a = 1.86$ μm, $S_Z = 20$ μm) by the S_a parameter by 2.9 times, by the S_Z parameter by 7.5%.

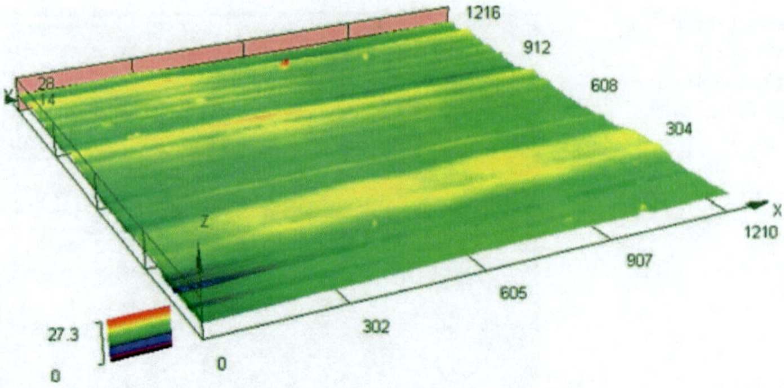

Fig. 9 3D distribution of surface relief heights

Fig. 10 Image of the surface of BT22 alloy processed by a wave milling cutter

4 Conclusion

The lower the vibration activity of the technological system, the lower the roughness of the processed surface. A change in the angle of inclination of the cutter tooth along the length of the working part leads to a change in the dynamic cutting angles, which leads to a change in the volume of plastic deformation [14] and, as a consequence, a change in the specific work of chip formation [11], and hence the acting forces, which are functions of the specific work.

The result of a change in the magnitude of the acting forces is the rotation of the resulting vector of forces at each point of the wave tooth [11, 14, 16]. As a result of the processing of high-strength titanium alloy BT22 with a wavy cutter, the quality

Fig. 11 An enlarged fragment of the surface

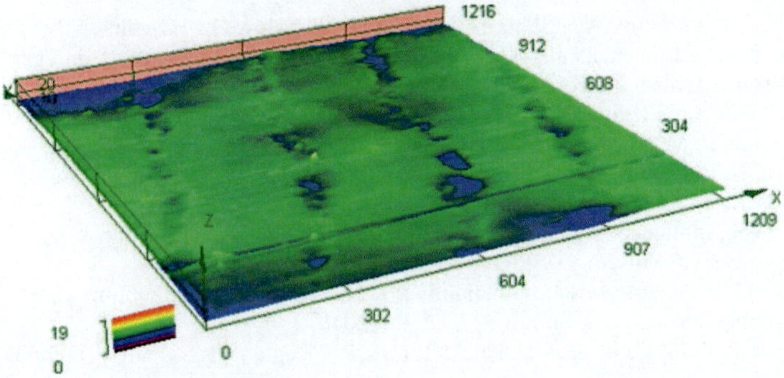

Fig. 12 3D distribution of the heights of the surface relief

indicators of the processed steel surface are significantly higher compared to the results of processing with a cutter made in accordance with GOST 23,248–78.

The constantly changing (wavy) volume of plastic deformation along the length of the wave tooth destroys a single focus of self-oscillations.

Production tests carried out at the production base of IC MION LLC showed that the service life of a "wave" cutter is more than 3 times higher than that of similar end mills for machining titanium alloys: Dauphin Precision Tool (USA), Melin Tool Company (USA), Hanita Metals Works (Israel), Minicat International Inc. (Canada), etc.

References

1. Vinokurova, I.M., Korotkov, V.V.: Titanium alloys using in order to improve the operational parameters of oil and gas equipment. Modern trends in the development of science and technologies **11**(3), 28–33 (2016)
2. Liu, D, Zhang, Y, Luo, M, Zhang D.: Investigation of tool wear and chip morphology in dry trochoidal milling of titanium alloy Ti-6Al-4V. Materials (Basel) **12**(12), 1937 (2019)
3. Li, Y., Wang, X., Yang, S., Hou, L., Wei, Y., Zhang, Z., Yang, X.: Investigation on wear behavior of cryogenically treated Ti-6Al-4V titanium alloy under dry and wet conditions. Materials (Basel) **12**(18), 2850 (2019)
4. Niu, Y., Jiao, F., Zhao, B., Gao, G.: Investigation of cutting force in longitudinal-torsional ultrasonic-assisted milling of Ti-6Al-4V. Materials (Basel) **12**(12), 1955 (2019)
5. Bari, P., Law, M., Vakhi, P.: Geometric models of non-standard toothed end mills. Int. J. Adv. Manuf. Technol. **111**(11–12), 3319–3342 (2020)
6. Lee, L.C., Lee, K.S., Gan, C.S.: On the correlation between dynamic cutting force and tool wear. Int. J. Mach. Tools Manuf **29**(3), 295–303 (1989)
7. Safavi, M., Balazinski, M., Mehmanparast, H.: Experimental characterization of tool wear morphology and cutting force profile in dry and wet turning of titanium metal matrix composites (Ti-MMCs). Metals **10**(11), 1459 (2020)
8. Komanduri, R., Von Turkovich, B.F.: New observations on the mechanisms of chip formation when machining titanium alloys. Wear **69**(2), 179–188 (1981)
9. Geng, D., Zhang, D., Li, Z., Liu, D.: Feasibility study of ultrasonic elliptical vibration-assisted reaming of carbon fiber reinforced plastics/titanium alloy stacks. Ultrasonics **75**, 80–90 (2017)
10. HSS end mills for titanium alloys (n.d.). Retrieved from http://pkmion.ru/catalog/dlya-obrabotki-titanovykh-splavov/710-dlya-obrabotki-titanovykh-splavov.
11. Korovin, G.I., Gavrilin, A.N., Petrushin, S.I., Odnokopylov, G.I., Kladiev, S.N.: Determining rational modes of mechanical processing of titanium body elements of oil and gas equipment by end-end machines (according to ss 23248–78) and wave-shaped cutters. Bulletin of the Tomsk Polytechnic University, Geo Assets Engineering **332**, 75–84 (2021)
12. Memarianpour, M., Niknam, S.A., Turenne, S., Balazinski, M.: Study of the effects of initial cutting conditions and transition period on ultimate tool life when machining Inconel 718. Materials **14**(3), 592 (2021)
13. Bobrov, V.F.: Influence of the angle of inclination of the main cutting edge of the tool on metal cutting processes. Moscow, USSR: Masigiz (1962)
14. Korovin, G.I., Petrushin, S.J., Gubaidulina, R.H.: Machining of titanium alloys with wave milling cutters. Mater. Sci. Forum **927**, 79–85 (2018)
15. Terekhina, L. I., Fiks, I. I. Vector algebra. Tomsk, RF: Regional Information Technology Center (1999)
16. Sahu, N.K., Andhare, A.B.: Modelling and multiobjective optimization for productivity improvement in high speed milling of Ti-6Al-4V using RSM and GA. J. Braz. Soc. Mech. Sci. Eng. **39**(12), 5069–5085 (2017)
17. Sikhimbaev, M.R., Sherov, K.T., Zharkevich, O.M., Sherov, A.K., Tkachyova, Y.O.: Experimental studies of stabilization of boring cutter form—building top oscillation. Journal of Vibroengineering **14**(2), 661–670 (2012)
18. Fedoseev, S.N., Valuev, D.V., Mamadaliev, R.A., Sokolov, P.N.: Application of integrated methods to improve the technological properties of steel. Key. Eng. Mater. **736**(7), 110–115 (2017)
19. Luk'yanov, A.V., Alejnikov, D.P., Vishchenko, M.V.: Study of spatial vibrations and forces in high performance milling. Transportation infrastructure of the Siberian region **2**, 598–603 (2017)
20. Bukreev, V.G., Lyapushkin, S.V., Edgulov, M.M.: Accuracy enhancement of automated system for multicomponent batching of bulk materials. MATEC Web of Conferences **37**, 01012 (2015)

21. Bukreev, V.G., Shandarova, E.B., Rulevskiy, V.M.: Power supply system model of remote processing equipment. Bulletin of the Tomsk Polytechnic University, Geo Assets Engineering **329**(4), 119–131 (2018)
22. Rulevskiy, V.M., Bukreev, V.G., Kuleshova, E.O. Shandarov, S.M., Vasilyeva, Yu. Z.: The power supply system model of the process submersible device with AC power transmission over the cable-rope. IOP Conference Series: Materials Science and Engineering **77**(1), 012098 (2017)
23. Vajda, I., Glazyrin, A., Ustinova, I., Bolovin, E.: Influence of design methods a discrete model of separately excited DC motor on parameters estimation. Acta Polytechnica Hungarica **15**(6), 219–233 (2018)
24. Korovin G.I., Zolotukhin D. V., Rozhkov P.S., Sherstobitov S.Y. End milling cutter for treatment of titanium alloys. Patent RU 2601525 C2 IPC B23C 5/10.
25. Zhongde Shi, C.E., Baron, L., Balazinski, M.: Machining of titanium metal matrix composites: progress overview. Materials **13**(21), 5011 (2020)
26. Srikanta, C., Mervyn, H.: Performance improvement studies for cutting tools with a perforated surface when turning titanium alloy. MATEC Web of Conferences **144**, 03003 (2018)
27. Çelik, Y.H., Yildiz, H., Özek, H.C.: Effect of cutting parameters on workpiece and tool properties during drilling of Ti-6Al-4V. Materials Testing **58**, 519–525 (2016)
28. Tehranizadeh, F., Koca, R., Budak, E.: Investigating effects of serration geometry on milling forces and chatter stability for their optimal selection. International Journal of Machine Tools and Manufacture **144**, 103425 (2019)
29. Kapranov, B.I., Vavilova, G.V., Volchkova, A.V., Kuznetsova, I.S.: Mathematical modeling of tomographic scanning of cylindrically shaped test objects. IOP Conference Series: Materials Science and Engineering **363**(1), (2018). https://doi.org/10.1088/1757-899X/363/1/012015
30. Bari, P., Law, M., Wahi, P.: Improved chip thickness model for serrated end milling. CIRP J. Manuf. Sci. Technol. **25**, 36–49 (2019)
31. Dombovari, Z., Altintas, Y., Stepan, G.: Mechanics and dynamics of serrated cylindrical and tapered end mills. Int. J. Mach. Tools. Manuf. **50**(6), 511–520 (2010)
32. Gavrilin, A.N., Moyzes, B.B., Cherkasov, A.I., Mel′nov, K., Zhang, X.: Mobile complex for rapid diagnosis of the technological system elements. MATEC Web of Conferences **79**, 01078 (2016)
33. Surzhikov, A.P., Lysenko, E.N., Vasendina, E.A., Sokolovskii, A.N., Vlasov, V.A., Pritulov, A.M.: Thermogravimetric investigation of the effect of annealing conditions on the soft ferrite phase homogeneity. J. Therm. Anal. Calorim. **104**(2), 613–617 (2011)
34. Surzhikov, A.P., Frangulyan, T.S., Ghyngazov, S.A., Vasil'ev, I.P., Chernyavskii, A.V.: Sintering of zirconia ceramics by intense high-energy electron beam. Ceramics Int. **42**(12), 13888–13892 (2016)
35. Vorotnikov, A.A., Klimov, D.D., Melnichenko, E.A., Poduraev, Y.V., Bazykyan, E.A.: Criteria for comparison of robot movement trajectories and manual movements of a doctor for performing maxillofacial surgeries. Int. J. Mech. Eng. Robot. Res. **7**(4), 361–366 (2018)
36. Borovskii, G.V., Shavva, M.A., Zakharevich, E.M., Grubyi, S.V., Maslov, A.R.: Ultraprecision machining of brittle optical materials. Russ. Eng. Res. **35**(12), 883–886 (2015)
37. Maslov, A.R., Tivirev, E.G.: Proceedings from XLIV academic space conference, New York. American Institute of Physics, USA (2021)
38. Lubimyi, N., Annenko, D., Chepchurov, M., Kostoev, Z.: The research of the temperature effect on a metal polymer during flat grinding of a combined metal polymer part. Australian J Mech. Eng. 1786654 (2020)
39. Tabekina, N.A., Chepchurov, M.S., Evtushenko, E.I., Dmitrievsky, B.S.: Solution of task related to control of swiss-type automatic lathe to get planes parallel to part axis. J. Physics: Conference Series **1015**(5), 052030 (2018)
40. Chetverikov, B.S., Chepchurov, M.S., Teterina, I.A.: Automation of component selection of ball-bearing support of drilling bit. Int. J. Adv. Manuf. Technol. **90**(1–4), 1059–1065 (2017)

41. Nizhegorodov, A., Gavrilin, A., Moyzes, B., Ditenberg, I., Zharkevich, O., Zhetessova, G., Muravyov, O., Bets, M.: Stand for dynamic tests of technical products in the mode of amplitude-frequency modulation with hydrostatic vibratory drive. J. Vibroengineering **18**(6), 3734–3742 (2016)
42. Gavrilin, A., Moyzes, B., Zharkevich, O.: Constructive and processing methods of reducing vibration level of the metalworking machinery elements. J. Vibroengineering **17**(7), 3495–3504 (2015)

Solid State Damper Based on Foam Aluminum to Reduce Vibration Activity of Electromechanical Devices

Dmitry Ermakov and Viktor Dmitriev

Abstract The operation of electromechanical devices is accompanied by vibration generated by them. To ensure the reliability, it is important to solve the problem of reducing the vibration activity of electromechanical devices. One of the ways to reduce vibration levels is damping. In this paper, the developed construction of a solid-state damper based on foam aluminum. The features of the technological process of manufacturing foam aluminum with open pores are considered. The damping properties of foam aluminum are described. An idealized deformation diagram of highly porous materials with designated working areas is also shown. An example of the effective use of a solid-state damper based on foam aluminum is considered. An analytical study of the dynamic displacements of an electromechanical device in a steady-state operating mode was carried out on the basis of the use of the developed solid-state damper.

1 Introduction

Based on the goals of operation, each spacecraft is equipped with special precision equipment, with the help of which the immediate target tasks are performed. The performance of the assigned tasks depends on the quality of the precision target equipment. However, the quality of work is influenced not only by the high technical level of the target equipment itself, but also by the dynamic phenomena in the equipment that is installed on the spacecraft and has a certain level of vibration activity.

Scientists and designers, when creating space technology, pay great attention to dynamic problems associated with vibroactivity existing on board the spacecraft [1]. Increased requirements for the quality of information received from high-precision

D. Ermakov (✉) · V. Dmitriev
Division for Testing and Diagnostics, National Research Tomsk Polytechnic University, 30 Lenin avenue, Tomsk 634050, Russia

V. Dmitriev
e-mail: dmitriev@tpu.ru

E. Lysenko et al. (eds.), *Recent Developments in the Field of Non-Destructive Testing, Safety and Materials Science*, Studies in Systems, Decision and Control 433,
https://doi.org/10.1007/978-3-030-99060-2_8

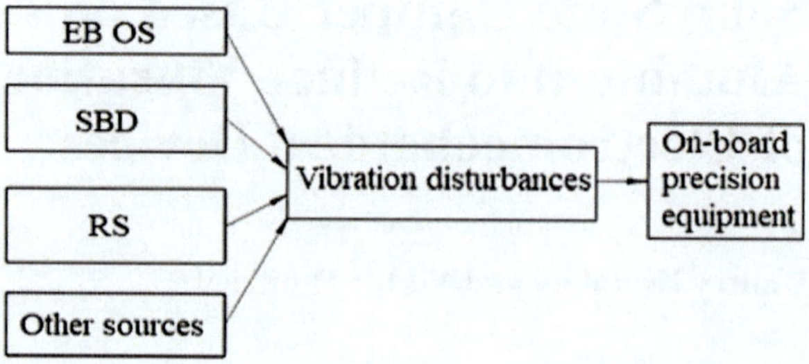

Fig. 1 The main sources of vibration activity on board the spacecraft

observation devices placed on a spacecraft create new problems in ensuring the dynamic regime under conditions of insufficient structural rigidity. Mechanical vibrations arising from vibration during controlled flight can lead to significant deviations from the permissible values of the technical characteristics of precision equipment.

Studying the materials of works [2, 3] by the authors on the topic of vibration disturbances from onboard sources on a spacecraft, two main directions for solving such problems can be distinguished. The first direction is based on limiting the intensity of work of electromechanical devices (ED—Electromechanical device), rotors and other active structural elements, which leads to a decrease in vibration disturbances on board the spacecraft. The second direction is based on the suppression of vibrations by means of vibration protection [4].

The spacecraft is a complex engineering structure with many components, each of which has its own dynamic characteristics [5]. On October 4, 1957, the first artificial Earth satellite, Sputnik-1, was launched, and from that time on, the complexity of the spacecraft design and the levels of tasks being solved began to grow in arithmetic, and in some areas—in geometric progression.

Thus, the mechanical activity of vibration sources located on the spacecraft has a significant negative impact on the operation of the target precision equipment. The main sources of vibration are: solar battery drives (SBD); executive bodies of the orientation system (EB OS); regulation systems (RS) and others (Fig. 1).

The trend forecast shows a further increase in the volume and mass of the payload, including due to a decrease in the mass of the service equipment, which includes the ED. At the heart of the identified trends (Fig. 2) are the results of the analytical studies carried out by the authors, as well as the general trends in improving the target equipment, as an integral part of the entire equipment of the spacecraft.

Analysis of the currently existing directions for improving the ED shows (Fig. 2) that such directions of development of space technology as increasing the accuracy of orientation of the spacecraft in space and minimizing the sources of vibration activity are of increasing importance.

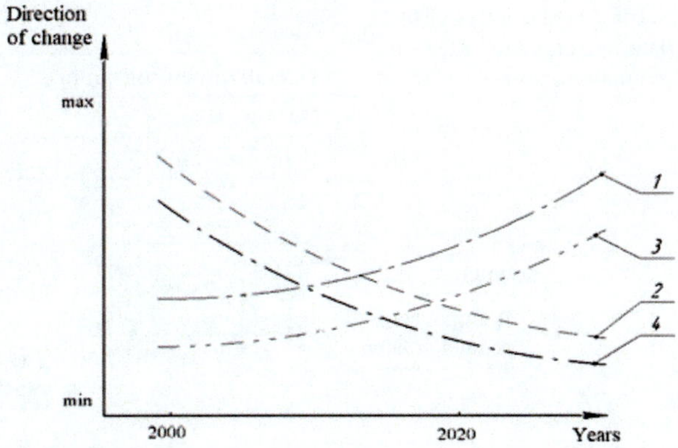

Fig. 2 Trends in the development of ED requirements on board the spacecraft: 1—the accuracy of the spacecraft orientation in space; 2—predicted level of influence of ED vibration activity; 3—technical level of optoelectronic equipment; 4—ED mass

2 Proposed Technical Solution

The sensitivity of the target equipment of the spacecraft is increasing all the time, and, accordingly, the requirements for reducing the level of vibration activity ED are increasing. One way to reduce the level is to minimize the ED vibration activity by means of damping.

Since it is not technically feasible to completely eliminate the transmission of vibration from ED on a spacecraft, it can only be reduced to values acceptable for the operation of the target precision equipment.

A technical solution to minimize vibration ED when using hysteresis damping is proposed—a solid-state damper based on foam aluminum (Fig. 3).

If we consider the technological process of manufacturing foam aluminum, then before its direct production, a polymeric porous material (template for the subsequent foam metal) with open pores with the desired relative density and pore size is created. For this, a sandy-clay mixture and polymer components are poured into the casting mold, then the mixture is thoroughly mixed. The casting mold with the molded mixture begins to be heated, the sandy-clay mixture hardens, while the mixture

Fig. 3 Fragment of the damping element of the vibration dampener of a solid-state damper based on aluminum foams

Table 1 Parameters of the damping element based on foam aluminum

Material	D16
Overall dimensions, mm	100 × 100 × 35
Density, g/sm^3	0.7
Pore size, mm	2 ÷ 6

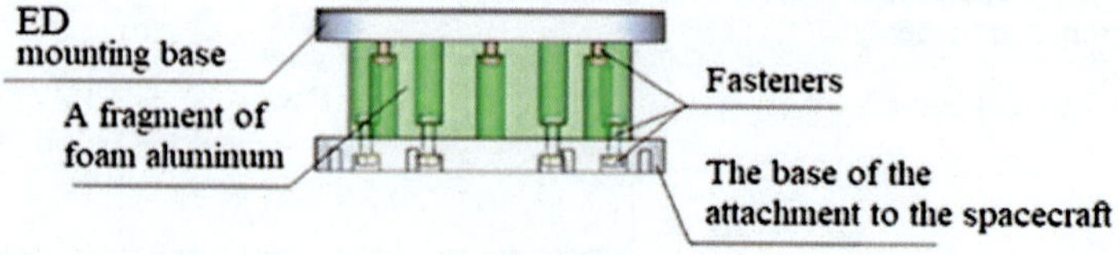

Fig. 4 Mounting scheme of a solid-state damper between an electromechanical device and a spacecraft

dehydrates and carbon dioxide is released. Next, molten metal is injected into the mold with the resulting mixture under pressure. After complete penetration of the melt into the mixture, the temperature is slightly increased. As a result, the polymer components melt, and their volume is occupied by molten metal. Next, the mold is cooled, the molded product is removed from it, and the sand is removed by vibration. This method resembles the investment casting method in traditional metal casting technologies and gives a structure with open pores and pore sizes of 2 … 6 mm in diameter (Table 1).

As noted in [6], the experimental data of tensile tests of porous samples indicate that the deformation value of the sample ε, at which the flow of aluminum foam from D16 material begins, practically remains constant in a wide range of porosity.

The attachment of the foam aluminum to the ED landing plane and the spacecraft is designed to provide kinematic decoupling between the ED and the spacecraft through a solid-state damping piece. The solid-state damper mounting diagram is shown in 3D models (Fig. 4).

Under the influence of an external load on the material (foam aluminum), it deforms, and, accordingly, the layers move (friction), which leads to energy dissipation. When developing the design of a solid-state damper, it is necessary to take into account the depth of the grooves for installing the screws. Since the different depth of the grooves determines the different rigidity of the foam aluminum fragment in the places of its attachment, this is what determines the different resistance to deformation under the influence of vibration.

The dissipation of energy in the material is due to the forces of inelastic resistance, which require energy to overcome. With the cyclic deformation of the material under loading and unloading, the dissipated energy can be shown in the form of a hysteresis characteristic (Fig. 5).

The damping properties of porous metals, which include foam aluminum, are manifested during plastic deformation and compaction, and the stress is less than the yield point of "solid" metals [7]. An idealized deformation diagram [7–9], highly

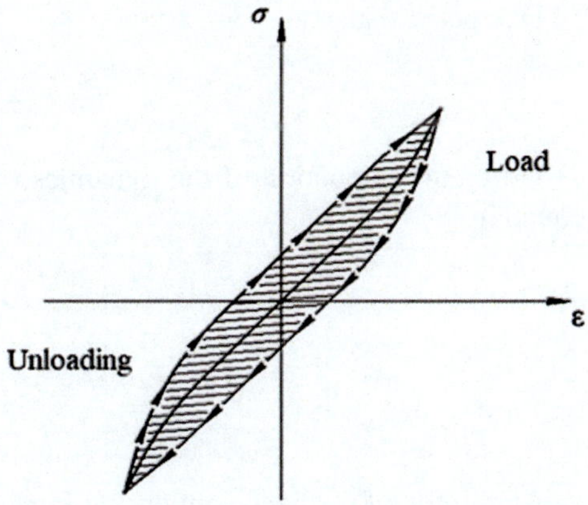

Fig. 5 Hysteresis characteristic: σ—voltage; ε—deformation

porous metals contains three main areas: elastic (I); sealing-plastic (II) and plastic (III). The area under the deformation diagram determines the specific absorption energy [7–14].

The peculiarity of the idealized deformation diagram [7–9], namely, region II, which limits the amplitude of the excitation force transmitted from the ED to the spacecraft, explains the choice of using this foam as a damping fragment of a solid-state damper [10]. The deformation area between ε_1 and ε_2 is limited by the dissipation of the power disturbance energy transmitted from the ED to the spacecraft.

3 Mathematical Description

Next, we will consider dynamic displacements at steady-state vibrations of a mechanical system with one degree of freedom (Fig. 6). Analytical studies of dynamic displacements are described based on studies of the application of hysteresis damping.

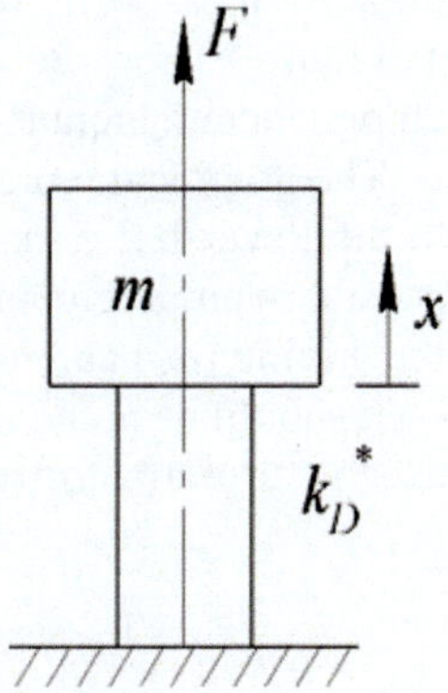

Fig. 6 Calculation scheme: k_D^*—elastic-dissipative element that takes into account stiffness and damping; F—forces that excite vibrations; x—emerging movements in the system; m—weight (ED)

Damper resistance

$$F_D = -k_D^* \dot{x} \tag{1}$$

Differential equation of the dynamics of a mechanical system and the reaction equation

$$m\ddot{x} + k_D^* \dot{x} + kx = F \sin \omega t$$

$$x = \frac{F}{k_D^*} \aleph \sin(\omega t - \varphi) \tag{2}$$

Amplitude of displacement during forced vibrations

$$A = \frac{h}{\sqrt{(1 - r^2)^2 + 4\xi^{*2}\omega^2}} \tag{3}$$

Steady-State Reaction of a Mechanical System

$$x = \frac{me\omega^2}{k_D^*} \aleph \sin(\omega t - \varphi) \tag{4}$$

or after mathematical transformations

$$\frac{m_1 x}{me} = r^2 \aleph \tag{5}$$

where: $\aleph = \frac{1}{\sqrt{(1-r^2)+(2\xi^* r)^2}}$—transmission ratio; $\xi* = f(k_D^*)$; $k_D^* = f(\eta)$.

In turn $\eta = \frac{D}{2\pi U}$—material loss factor; $D = \pi AF \sin \varphi$—the absorbed energy in the foam aluminum material; $U = \frac{1}{2} AF \sin \varphi$—dissipated energy during deformation of the foam aluminum material.

Expression (5) for the amplitude of the steady-state response was obtained in a dimensionless form, which is convenient for analyzing the results.

Figure 7 shows relationship (5) in graphical form, where the interdependence of all parameters included in this relationship is clearly visible.

The graphical representation (Fig. 7) shows that the best results are concentrated in the lower left corner of the graph, i.e. a solid-state damper with the maximum possible damping coefficient and the maximum possible ratio of the amplitudes of the driving (ω_e) and natural (ω_c) frequencies is required.

It should be noted that similar approaches to the consideration of dampers can be used in aircraft construction, shipbuilding, and instrument making [15–26].

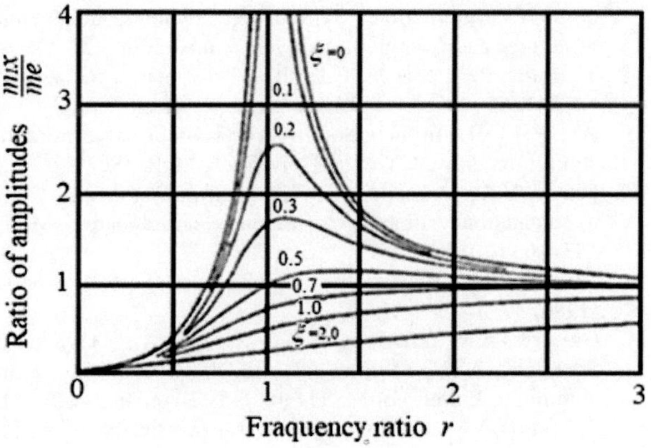

Fig. 7 The dependence of the steady-state reaction on the difference in frequencies r at different dimensionless coefficients ξ

4 Conclusion

The solution of technical problems of reducing the level of vibroactivity of the functional mechanisms of the spacecraft is an urgent task. The presence of vibroactivity in ED worsens their dynamics of operating conditions and, at the same time, degrades the operational characteristics of the target equipment installed on the spacecraft.

The basis for ensuring the effective functioning of the target precision equipment is the calculated and practical results in the design of the ED.

Since the existing trends dictate the steady minimization of the level of vibroactivity, this imposes mathematically justified requirements for methods and technical solutions in the process of their design.

The proposed technical solution allows minimizing the vibration activity of the ED functional units of the spacecraft and qualitatively improving the conditions for their reliable operation.

References

1. Taniwaki, S., Ohkami, Y.: Experimental and numerical analysis of reaction wheel disturbances. JSME Int. J. Ser. C Mech. Syst. Mach. Element. Manuf **46**(2), 519–526 (2003)
2. Nashif, A., Jones, D., Henderson, J.: Oscillation damping. Mir, Moscow, Russia (1988)
3. Nur, G.S., Ryan, R.S., Scofield, H.N., Seeme, J.L.: Dynamics of large space structures and their control. Aerosp. Eng. **6**, 129–147 (1985)
4. Telepnev, P.P., Kuznetsov, D.A.: Fundamentals of designing vibration protection for spacecraft: textbook. Ed. MSTU them. N.E. Bauman, Moscow (2019)
5. Taniwaki, S., Kanzawa, T., Ohkami, Y.: Sensing of micro noise caused by satellite fly-wheel using air floating disturbance detector. INTERMAC 2001 Joint Technical Conference (2001)
6. Butarovich, D.O., Smirnov, A.A., Ryabov, D.M.: Foam aluminum as an energy-absorbing material and its mechanical properties. Proc. High. Educ. Inst. Ser. Mech. Engi. **7**, 53–58 (2011)
7. Markov, V.A., Ovchinnikov, A.F., Pusev, V.I., Selivanov V.V.: On the main features of the mechanical and shock-absorbing properties of highly porous aluminum alloys. In: Materials

of the XVI International Symposium "Dynamic and Technological Problems of Mechanics of Structures and Continuous Media" named after A.G. Gorshkov, vol. 2, pp. 218–225 (2010)
8. Thorntor, P.H., Magee, C.L.: The deformation of aluminium foams. Met. Trans. A **6A**(6), 1253–1263 (1975)
9. Wei, P., Liu, L.: Influence of density on compressive properties and energy absorption of foamed aluminium alloy. J. Wuhan Univ. Tech. Mater. Sci. **22**(2), 225–228 (2007)
10. Petrov, Y.A., Makarov, V.P., Kolobov, A.Y., Aleshin V.F.: Spacecraft landing gear based on foam and honeycomb blocks. Science and Education. MSTU them. N.E. Bauman. Electron. **4**, 153–163 (2010)
11. Nikolsky, L.N., Keglin, B.G.: Rolling stock shock absorbers. Mashinostroenie. Moscow, Russia (1986)
12. Gusarov, A.V., Zharikov, V.A., Markov, A.F., Ovchinnikov, V.I., Pusev, V.V., Selivanov, Sochikov, A.N.: Mechanical and shock-absorbing properties of highly porous cellular aluminum. Bulletin of MSTU im. N.E. Bauman. Ser. Mech. Eng. **1**(74), 58–66 (2009)
13. Selivanov, V.V.: On the mechanical properties of highly porous aluminum alloys. Bulletin of the Nizhny Novgorod University. N.I. Lobachevsky **4**(4), 1760–1762 (2011)
14. Modheja, A., Zahrai, S.M.: Numerical study of visco-hyperelastic damper with high axial damping rubber subjected to harmonic loading. Structures **29**, 1550–1561 (2021)
15. Surzhikov, A.P., Peshev, V.V., Pritulov, A.M., Gyngazov, S.A.: Grain-boundary diffusion of oxygen in polycrystalline ferrites. Russ. Phys. J. **42**(5), 490–495 (1999). https://doi.org/10.1007/BF02508222
16. Gavrilin, A.N., Chuprin, E.A., Moyzes, B.B., Halabuzar, E.A.: Land-based sources of seismic signals. In: Proceedings of 2014 International Conference on Mechanical Engineering, Automation and Control Systems MEACS 2014, 6986947 (2014). https://doi.org/10.1109/MEACS.2014.6986947
17. Veismoradi, S., Mohamad, S., Yousef-beik, M., Zarnani, P., Quenneville, P.: Development and parametric study of a new self-centering rotational friction damper. Eng. Struct. **235**, 11209 (2021)
18. Amjadian, M., Agrawal, A.K., Silva, C.E., Dyke, S.J. : Experimental testing and validation of the dynamic model of a magneto-solid damper for vibration control. Mech. Syst. Signal Process. **166**, 108479 (2021)
19. Cavalagli, N., Agresta, A., Biscarini, C., Ubertini, F., Ubertini, S.: Enhanced energy dissipation through 3D printed bottom geometry in Tuned Sloshing Dampers. J. Fluid Struct. **106**, 103377 (2021)
20. Zaccardo, V.M., Buckner, G.D.: Active magnetic dampers for controlling lateral rotor vibration in high-speed rotating shafts. Mech. Sys. Signal Process. **152**, 107445 (2021)
21. Sui, Y., Yu, H.: Oscillation of damped second order quasilinear wave equations with mixed arguments. Appl. Math. Lett. **117**, 107060 (2021)
22. Zenelis, I., Wang, X.: A model-free sparse wide-area damping controller for inter-area oscillations. Int. J. Electr. Power Energy Syst. **136**, 107609 (2021)
23. Su, X., Kang, H., Guo, T.: Modelling and energy transfer in the coupled nonlinear response of a 1:1 internally resonant cable system with a tuned mass damper. Mech. Syst. Signal Process. **162**, 108058 (2021)
24. Sonowal, P., Pandey, K.M., Sharma, K.K.: Design and static analysis of landing gear shock absorber of commercial aircraft. Mater. Today Proc. **45**(7), 6712–6717 (2021)
25. Acanfora, V., Saputo, S., Russo, A., Riccio A.: A feasibility study on additive manufactured hybrid metal/composite shock absorbers. Compos. Struct. **268**, 113958. (2021)
26. Liu, M., Xu, J., Li, Q.: Design and experiment of piezoelectric-shape memory alloy composite shock absorber. Mater. Lett. **304**, 130538 (2021).

The Development of a Source for Impulse Impact on Coal Bed

Kirill Kuvshinov, Boris Moyzes, and Anatoly Nizhegorodov

Abstract Methane is a valuable raw material. The extraction of methane from coal beds is an urgent task. The solution of this problem allows degassing of coal beds—a decrease in the volume of methane in them. Also the solution of this problem allows increasing the safety of operation of coal mines by reducing the probability of methane explosions during development. The article reviews the development of a hydro-impulse source that allows creating microcracks in porous bed. In the article a schematic diagram of a hydraulic pulse source is presented. Also the mathematical models of the dynamics of valve opening and the process of filling the implosion chamber with liquid are presented in the article. The results of mathematical modeling are given. The effect of well pressure on the time of the chamber filling and the value of the generated impulse have been determined in the article. The research made it possible to develop the design of the hydro-impulse source and determine its parameters which characterize the impulse impact on the formation. The parameters of the impulse impact are the basis for the technology of impact on the coal bed by the developed source. During the research, the analysis of information sources in the field of research, the synthesis of constructive and technological solutions, a parametric and functional description of the source were carried out.

K. Kuvshinov
Division for Mechanical Engineering, National Research Tomsk Polytechnic University, 30 Lenin avenue, Tomsk, Russia 634050
e-mail: kuvshinov@tpu.ru

B. Moyzes (✉)
Division for Testing and Diagnostics, National Research Tomsk Polytechnic University, 30 Lenin avenue, Tomsk, Russia 634050
e-mail: mbb@tpu.ru

A. Nizhegorodov
Department of Construction and Road-Making Machines and Hydraulic Systems, Irkutsk National Research Technical University, 83 Lermontov Street, Irkutsk 664074, Russia

E. Lysenko et al. (eds.), *Recent Developments in the Field of Non-Destructive Testing, Safety and Materials Science*, Studies in Systems, Decision and Control 433,
https://doi.org/10.1007/978-3-030-99060-2_9

1 Introduction

In the process of coalification, plant residues on the Earth are biologically and thermally degraded [1–6]. Biodegradation is the result of the vital activity of bacteria, thermal destruction occurs as a result of an increase in temperature and pressure in the process of layer-by-layer deposition of plant substances. When the plant matter is converted to coal, gases are formed. The gases are either absorbed by the coal or accumulate in the interporous space in the coal bed.

The methane concentration in the mixture of natural gases of coal beds is 80…98%. Saturated coal-methane reservoirs can contain five times more gas than conventional gas reservoirs, when compared in size, temperature and pressure.

The rationale of the research is based on the fact that the regions of Kuzbass are rich only in coals suitable for methane extraction.

In world practice, various methods of extraction of methane from coal beds are used [1]. These methods are:

- degassing of coal beds to prevent methane from entering the mine atmosphere;
- methane extraction from closed mines;
- extraction from coal beds that are planned to be spent or which will not be extracted at all.
- There are mine and well methods of methane (extraction) [7–15].

The first method is used for direct coal mining. The purpose of the recovery is to reduce the volume of its release and reduce the explosion chance; therefore, methane is extracted in small quantities.

The commercial methane recovery is carried out by the second way, using the technologies to enhance gas recovery of coal-beds after the well is put into operation. When developing coal-bed methane deposits, the coal bed is exposed to pressure impulses followed by pumping out the produced water.

The coal bed is destroyed under the impact of a pressure pulse. The vertical cracks, connecting the wellbore with a system of natural fractures in the bed are formed. Then the liquid with methane dissolved in it is pumped out of the well. During the operation of the well, the volume of extracted methane is monitored by regular measurements. When the productivity of the extracted methane is decreased the, the coal-bed is re-affected.

2 Hydropulse Source for Impact on Coal Bed

The design of a hydraulic pulse source for impulse impact on coal bed consisting of an implosion chamber 1 and a valve 2, with windows 3 with an area S_0 (Fig. 1), has been developed at the Tomsk Polytechnic University.

To form the impact by a pressure pulse P_p, the source is lowered on the cable 4 in the well until the windows 3 are placed at the impact level. Then the initial

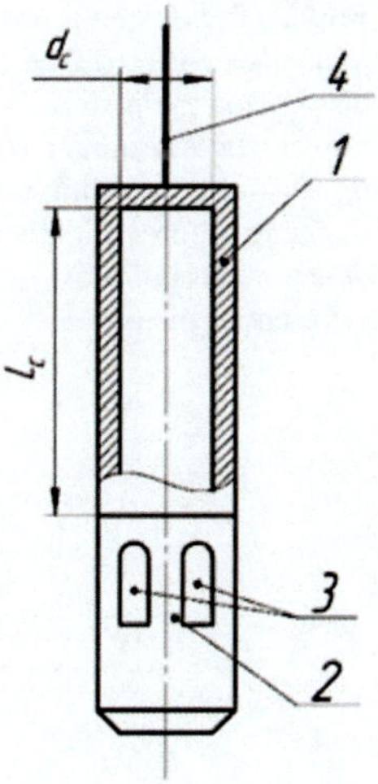

Fig.1 Schematic diagram of a hydraulic pulse. *Source* l_c—chamber length; d_c—inner diameter of the chamber.

pressure in the well P_0 is increased by supplying fluid to it. And when a certain pressure is reached, the source valve is triggered, connecting the space of chamber 1 with the well. After that, the fluid from the well directed into the implosion chamber, creating a sharp, short-term pressure drop at the level of the valve windows. Under that impact, cracks appear in the coal-bed, which open the pores of methane rocks. This intensify process of methane release [16–21].

The main parameters that determine the efficiency of the hydraulic pulse source are the valve opening time and time of the implosion chamber filling.

To study the dynamics of valve opening and chamber filling, let's create the mathematical models for each of these processes [22–36].

3 Mathematical Model of Valve Opening Dynamics

To create a mathematical model, let's consider the design scheme and the principle of operation of a valve with a differential piston (Fig. 2).

Under the action of the pressure P_0 in the well due to the difference in areas $S_1 = \frac{\pi}{4}\left(D_p^2 - d_h^2\right)$ and $S_2 = \frac{\pi \cdot D_h^2}{4}$ the differential piston of mass m will begin to move. Let's distinguish three phases of its work which correspond to the sections:

- $x_0 \ldots x_1$—acceleration of the piston with closed valve windows;
- $x_1 \ldots x_2$—opening of the valve windows;
- $x_2 \ldots x_3$—the differential piston deceleration after the valve windows are fully opened.

A mathematical model of a valve with a differential piston has been developed:

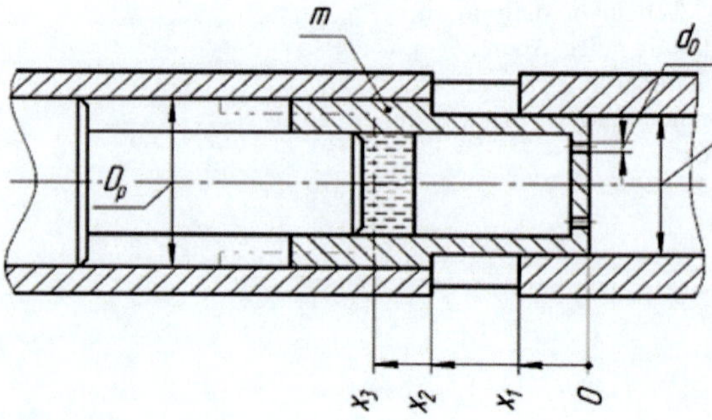

Fig. 2 The design schemes of a valve with a differential piston: x is the piston movement coordinate; d_0, d_h, D_p—the throttling holes diameters, holes and piston; m is the mass of the differential piston

$$m\frac{d^2x}{dt^2} + \alpha\frac{dx}{dt} = P_0 \cdot S$$
$$\alpha = \begin{cases} \alpha_1, x(t) \le x_2 \\ \alpha_2, x(t) > x_2 \end{cases}$$
$$S = \begin{cases} S_1 = \frac{\pi}{4}\left(D_p^2 - d_h^2\right), x(t) \le x_2 \\ S_2 = \frac{\pi \cdot D_p^2}{4}, x(t) > x_2 \end{cases} \tag{1}$$

where S is the area of the piston which is determined depending on the phase of work; α is the coefficient of viscous friction.

To solve the system (1) by the Runge–Kutta method, let's rewrite it in the following form:

$$\begin{cases} a(t) = \frac{dV(t)}{dt} = \frac{P_0 \cdot S - \infty \cdot V(t)}{m} \\ V(t) = \frac{dx(t)}{dt} \end{cases}, \tag{2}$$

where $V(t)$, $a(t)$ are valve opening speed and acceleration.

To carry out modeling in the Mathcad software product, an experiment plan was made. A fragment of the plan is given below (Table 1).

Initial values are accepted on condition.

$$\frac{dx(0)}{dt} = 0.$$

The modeling results (2) are presented in the form of the dependence graphs of the displacement x (t) and the speed $V(t)$ of valve opening on the valve response time (Figs. 3, 4 and 5). The displacement $x(t)$ and speed $V(t)$ equal to zero and correspond to the values $x = x_2$ и $V = V_2$.

Table 1 A fragment of the model experiment plan

Experience number	Well pressure at the moment of valve actuation P_0, [MPa]	Area, [cm^2]		Movement length, [mm]			Viscous friction coefficients, [N·s/m]		Piston weight, [kg]
		S_1	S_2	x_1	x_2	x_3	α_1	α_2	
1	30	2.2	6	3.5	5.5	6	0.003	0.03	0.1
2									0.15
3									0.2

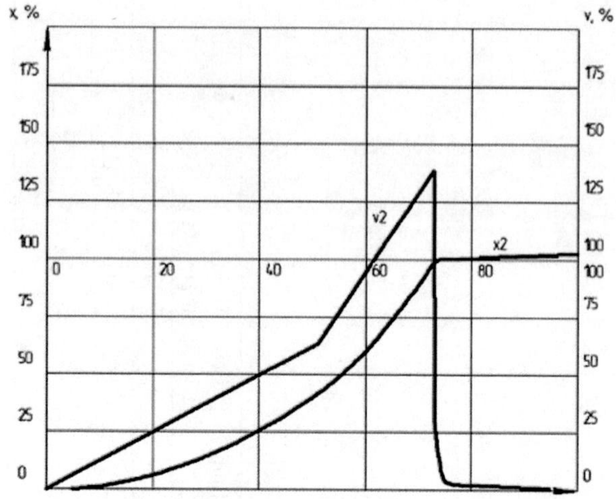

Fig. 3 Experiment No 1 results

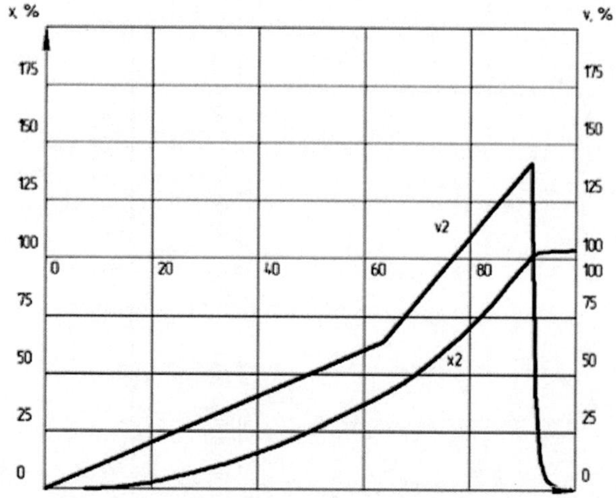

Fig. 4 Experiment No 2 results

Analyzing the obtained graphs, it can be concluded that the valve response time significantly depends on the mass of the differential piston. Therefore, when designing such valves, the piston mass should be reduced.

4 Mathematical Model of the Process of Filling the Implosion Chamber

To create a mathematical model of the filling process of the implosion chamber, let us consider the design scheme (Fig. 6).

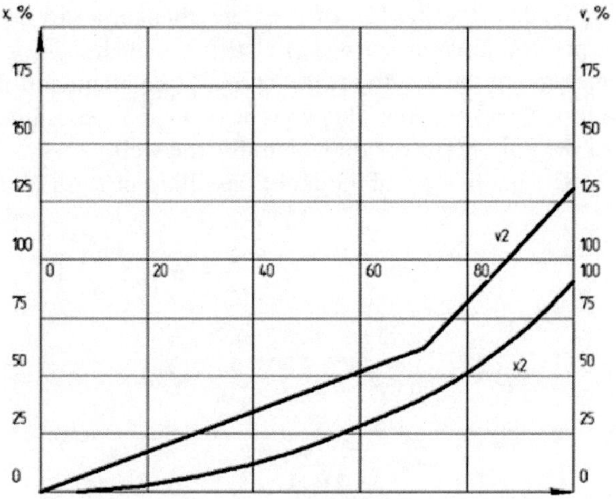

Fig. 5 Experiment No 3 results

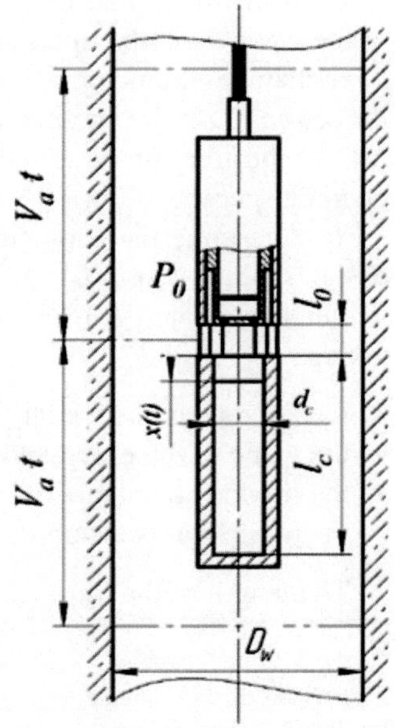

Fig. 6 The design scheme: $x(t)$ is the piston movement coordinate; V_a is a valve opening speed; l_0 is the initial position of the valve in the chamber; D_w—well diameter; P_0 is the operating pressure in the well; l_c, d_c are chamber length and diameter

In this case, let's make the assumption that the process of filling the chamber is isothermal and the fluid flow is continuous.

The features of the system:

- the dependence of the mass of the accelerated fluid which fills the chamber on the coordinate $x(t)$;
- change in the force which acts on this mass due to the pressure drop in the well.

To describe the law of pressure change, let us consider the fluid in the well as a pre-compressed spring. The stiffness of the spring changes over time due to an increase in the length of the "spring" (the volume of fluid involved in the process of filling the chamber). This volume change is determined by the speed of propagation of the fluid perturbation which fills the well.

The mathematical model of the filling process of the implosion chamber:

$$
\begin{aligned}
&(m_0 - m(x))\frac{d^2x}{dt^2} + \alpha(x)\left(\frac{dx}{dt}\right)^2 + C_a = P_0 \cdot S_c;\\
&m(x) = \rho \cdot x \cdot S_c;\\
&\alpha(x) = \rho \cdot \lambda \cdot \frac{x}{2d_c} \cdot S_c;\\
&C_a = \frac{S_c^2 \cdot E}{2V_a \cdot t \cdot S_c};
\end{aligned}
\tag{3}
$$

where m_0 is the initial mass of the fluid in the chamber; ρ is the density of the fluid in the well; P_0 is the operating pressure in the well; S_c is the area of the chamber; $d_\text{к}$ is chamber diameter; λ is the coefficient of hydraulic friction, x is the position of the piston, C_a is the coefficient taking into account the nature of the medium and the ratio of the areas of the chamber and the well, E is the impact energy, V_a is the valve opening speed.

To determine the geometric parameters of the implosion chamber, let's use the design scheme (Fig. 7).

Let's calculate the length of the implosion chamber and the speed of its filling, taking:

- well diameter and initial pressure in it—$d_c = 114$ mm and $P_0 = 10$ MPa [8];
- the value of the energy which acts on the coal-bed—$E = 5$ MJ [9];
- the implosion chamber diameter and the pressure in it—$d_c = 50$ mm;
- the implosion pressure in it $P_c = 0.1$ MPa.

The length of the implosion chamber is determined by the formula:

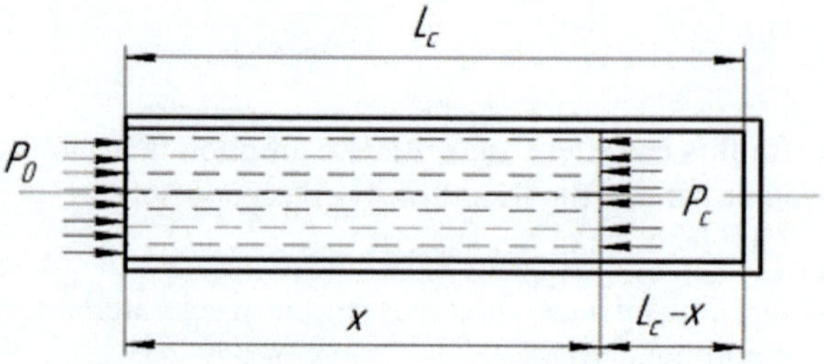

Fig. 7 The design scheme for determination the geometric parameters of the implosion chamber

$$l_{\mathrm{c}} = \frac{W}{S_{\mathrm{c}}} = \frac{E}{P_0} \cdot \frac{4}{\pi \cdot d_{\mathrm{c}}^2}, \tag{4}$$

where W is the chamber volume.

The speed of filling the chamber with liquid is determined through the change in pressure ΔP_{c} [9]:

$$V_a = \frac{\Delta P_{\mathrm{c}}}{\rho \cdot c} = \frac{P_0 - P_{\mathrm{c}}}{\rho \cdot c}; \tag{5}$$

where c is the speed of sound in air, 330 m/s; ρ is the water density, 1000 kg/m^3.

Taking into consideration that the pressure in the chamber is significantly lower than the pressure in the well $P_{\mathrm{c}} << P_0$, the following assumption can be made $\Delta P_{\mathrm{c}} \approx P_0$ and rewrite the Eq. (5)

$$V_a = \frac{P_0}{\rho \cdot c}.$$

Based on Eq. (4, 5), let's find the value of the chamber length and its filling speed:

$$l_{\mathrm{c}} = \frac{5 \cdot 10^6 \cdot 4}{10 \cdot 10^6 \cdot 3.14 \cdot 0.05^2} = 255\ \mathrm{m}.$$

The length of the chamber is unacceptable, we will reduce the impact energy on the formation to 10 kJ:

$$l_{\mathrm{c}} = \frac{10 \cdot 10^3 \cdot 4}{10 \cdot 10^6 \cdot 3.14 \cdot 0.05^2} = 0.5\ \mathrm{m};$$

$$V_a = \frac{P_0}{\rho \cdot c} = \frac{10^7}{1000 \cdot 330} \approx 30,3\ \mathrm{m/s}.$$

From the condition

$$P_{\mathrm{c}} \cdot W_0 = P_0 \cdot W_1,$$

let's determine the air volume W_0 in the chamber.

$$W_0 = l_{\mathrm{c}} \cdot S_{\mathrm{c}} = l_{\mathrm{c}} \cdot \frac{\pi \cdot d_{\mathrm{c}}^2}{4} = 0.5 \cdot \frac{3.14 \cdot 0.05^2}{4} = 10^{-3}\ \mathrm{m}^3;$$

let's determine the air volume W_1 in the chamber after filling the chamber with fluid

Table 2 Calculation results

Parameter	Value			
P_0, [MPa]	5	10	15	20
V_a, [m/s]	17	30	50	67
d_c, [mm]	50			
t, [s]	0.06	0.03	0.02	0.015
d_c, [mm]	80			
t, [s]	0.023	0.012	0.008	0.006

$$W_1 = \frac{P_c \cdot W_0}{P_0} = \frac{0,1 \cdot 10^{-3}}{10} = 10^{-5}\ \mathrm{m}^3.$$

Let's determine the value l_c–x:

$$l_c - x = \frac{W_1}{S_c} = \frac{10^{-5}}{20 \cdot 10^{-4}} = 5 \cdot 10^{-3}\ \mathrm{m}.$$

Determine the chamber filling time. Let's assume that the chamber is filled completely on the based of the condition L–x << L,

$$l_c = V_a \cdot t - \frac{a \cdot t^2}{2} = V_a \cdot t - \frac{V_a \cdot t^2}{2t} = \frac{V_a \cdot t}{2t}; t = \frac{2l_c}{V_a} = \frac{2 \cdot 0.5}{30.3} = 0.03\ \mathrm{s}.$$

Let us calculate the speed and time of the chamber filling at different initial pressures P_0 (5, 15, 20 MPa). The results are summarized in the table (Table 2).

Let us make the dependence graphs of the chamber filling time t on the pressure difference ΔP at different values of the implosion chamber diameter ($d_c = 50$ and 80 mm) and the initial pressure P_0 (5, 10, 15, 20 MPa) (Fig. 8).

The filling time of the chamber characterizes the impulse value which acts on the coal-bed. The larger the chamber diameter at a fixed volume, the shorter the filling time of the chamber and the larger the impulse.

5 Conclusion

The presented models, as well as diagrams, make it possible to determine the characteristics of the device, taking into account the chamber dimensions, the fluid type, the well depth and other parameters. Also they make it possible to set the required impulse energy or select the chamber dimensions and the mass of the piston. This will increase the flow rate of the well.

The approximating equations are determined in the research.

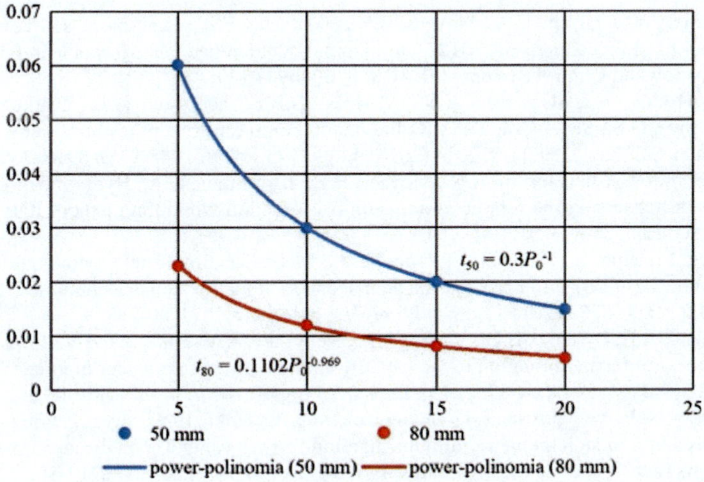

Fig. 8 Graphs of the dependence of the filling time of the chamber on the well pressure when the device valve is triggered

References

1. Trofimova, G.I., Trofimov, N.I., Bakushkina I.A., Cheremisina V.G.: Technologies of mining methane from coal seams of Kuzbass. Nauka i mir **1**(3), 80–84 (2017)
2. Li, W., Zhu, Y.M., Wang, G., Jiang, B.: Characterization of coalification jumps during high rank coal chemical structure evolution. Fuel **185**, 298–304 (2016)
3. Zhou, S., Liu, D., Cai, Y., Yao, Y.: Effects of the coalification jump on the petrophysical properties of lignite, subbituminous and high-volatile bituminous coals. Fuel **199**, 219–228 (2017)
4. Hu, S., Hao, G., Feng, G., Guo, H., Wu, D.: A method for improving the methane extraction concentrations of in-seam boreholes. Fuel **265**(1), 117006 (2020)
5. Liu, Y., Zhang, C., Zhang, Y.: Evolution mechanism of methane adsorption capacity in vitrinite-rich coal during coalification. J. Nat. Gas Sci. Eng. **96**, 104307 (2021)
6. Hao, C., Cheng, Y., Wang, L., Liu, H., Shang, Z.: A novel technology for enhancing coalbed methane extraction: Hydraulic cavitating assisted fracturing. J. Nat. Gas Sci. Eng. **72**, 103040 (2019)
7. Tagiev, S.M.: Materials of XI International Research and Practice Conference. Sheffield, UK (2015)
8. Vadetskij, Yu.V.: Driller's Handbook. Akademiya, Moscow (2008)
9. Ageev, N.P., Ageev, P.G., Desyatkin, A.S., Elsukov, G.A.: Seismic and geophysical studies of the results of plasma-pulse action on coal seams in order to extract methane. Gornaya promyshlennost' **5**(123), 70–73 (2015)
10. Butikov, E.I., Kondrat'ev, A.S.: Physics. Book 1. Mechanics. Nauka, Moscow (1994)
11. Bauman, V.A., Bykhovskij, I.I.: Vibration machines and processes in construction. Vysshaya shkola, Moscow (1977)
12. Nizhegorodov, A., Gavrilin, A., Moyzes, B., Ditenberg, I., Zharkevich, O., Zhetessova, G., Muravyov, O., Bets, M.: Stand for dynamic tests of technical products in the mode of amplitude-frequency modulation with hydrostatic vibratory drive. J. Vibroeng. **18**(6), 3734–3742 (2016). https://doi.org/10.21595/jve.2016.16994

13. Gavrilin, A.N., Chuprin, E.A., Moyzes, B.B., Halabuzar, E.A.: Land-based sources of seismic signals. In: Proceedings of 2014 International Conference on Mechanical Engineering, Automation and Control Systems MEACS, p. 6986947 (2014)
14. Nizhegorodov, A.I., Gavrilin, A.N., Moyzes, B.B., Cherkasov, A.I., Zharkevich O.M., Zhetessova, G.S., Savelyeva, N.A.: Radial-piston pump for drive of test machines. IOP Conf. Ser Mater. Sci. Engi. **289**(1), 012014 (2018). https://doi.org/10.1088/1757-899X/289/1/012014
15. Nizhegorodov, A.I., Gavrilin, A.N., Moyzes, B.B., Kuvshinov, K.A.: Hydraulic drive of vibration stand for testing the robotic systems units by random vibration method. IOP Conf. Ser. Mate. Sci. Eng. **516**(1), 012031 (2019)
16. Jingjing, L., Yijia, L., Xiaodong, T., Xiaojie, L.: Methane sulfonic acid ionic liquid extraction of gasoline desulfurization process and enhancement of microchannel technology. Chem. Eng. Sci. **242**(12), 116753 (2021). https://doi.org/10.1016/j.ces.2021.116753
17. Surzhikov, A.P., Lysenko, E.N., Vlasov, V.A., Malyshev, A.V., Nikolaev, E.V.: Investigation of the process of ferrite formation in the Li_2CO_3-ZnO-Fe_2O_3 system under high-energy actions. Russ. Phys. J. **56**(6), 681–685 (2013). https://doi.org/10.1007/s11182-013-0085-6
18. Surzhikov, A.P., Frangulyan, T.S., Ghyngazov, S.A., Lysenko, E.N.: Investigation of oxidation processes in non-stoichiometric lithium-titanium ferrites using TG analysis. J. Therm. Anal. Calorim. **102**(3), 883–887 (2010). https://doi.org/10.1007/s10973-010-0912-8
19. Fan, Y., Shu, L. Huo, Z. Hao, J., Li Y.: Numerical simulation of sectional hydraulic reaming for methane extraction from coal seams. J. Nat. Gas Sci. Eng. **95**, 104180 (2021)
20. Guo, H., Zhang, Y., Zhang, Y., Li, X., Li, Z., Liang, W., Ali, M.I.: Feasibility study of enhanced biogenic coalbed methane production by super-critical CO_2 extraction. Energy **214**(1), 118935 (2021)
21. Su, E., Liang, Y., Zou, Q., Xu, M., Sasmito, A.P.: Numerical analysis of permeability rebound and recovery during coalbed methane extraction: Implications for CO_2 injection methods. Process Saf. Environ. Prot. **149**, 93–104 (2021)
22. Liu, Z., Cheng, Y., Dong, J., Jiang, J., Wang, L., Li, W.: Master role conversion between diffusion and seepage on coalbed methane production: implications for adjusting suction pressure on extraction borehole. Fuel **223**, 373–384 (2018)
23. Fan, Z., Zhang, D., Fan, G. Zhang, L., Zhang, S., Liang, S., Guo, W.: Non-darcy thermal-hydraulic-mechanical damage model for enhancing coalbed methane extraction. J. Nat. Gas Sci. Eng. **93**, 104048 (2021)
24. Ge, Z., Zhong, J., Lu, Y., Cheng, L., Zheng, J., Zhou, Z., Cheng Y.: Directional distance prediction model of slotting–directional hydraulic fracturing (SDHF) for coalbed methane (CBM) extraction. J. Petrol. Sci. Eng. **183**, 106429 (2019)
25. Qu, Q.D., Xu, J.L., Xue, S.: Improving methane extraction ratio in highly gassy coal seam group by increasing longwall panel width. Proced. Earth Planet. Sci. **1**(1), 390–395 (2009)
26. Dong, J., Cheng, Y., Jin, K., Zhang, H., Liu, Q., Jiang, J., Hu, B.: Effects of diffusion and suction negative pressure on coalbed methane extraction and a new measure to increase the methane utilization rate. Fuel **197**, 70–81 (2017)
27. Englezos, P.: Extraction of methane hydrate energy by carbon dioxide injection-key challenges and a paradigm shift. Chin. J. Chem. Eng. **27**(9), 2044–2048 (2019)
28. Campuzano, R., González-Martínez, S.: Extraction of soluble substances from organic solid municipal waste to increase methane production. Biores. Technol. **178**, 247–253 (2017)
29. Zhao, P. Zhuo, R. Li, S. Shu, C. M., Jia, Y., Lin, H., … Xiao P.: Fractal characteristics of methane migration channels in inclined coal seams. Energy **225**, 120127 (2021)
30. Liu, Y., Zhou, F., Liu, L., Liu, C., Hu, S.: An experimental and numerical investigation on the deformation of overlying coal seams above double-seam extraction for controlling coal mine methane emissions. Int. J. Coal Geol. **87**(2), 139–149 (2011)
31. Zhang, Y., Zhang, X., Li, C., Liu, C., Wang, Z.: Methane moving law with long gas extraction holes in goaf. Proced. Eng. **26**, 357–365 (2011)
32. Li, H., Lin, B., Yang, W., Hong, Y., Wang, Z.: A fully coupled electromagnetic-thermal-mechanical model for coalbed methane extraction with microwave heating. J. Nat. Gas Sci. Eng. **46**, 830–844 (2017)

33. Bolson, N., Yutkin, M., Patzek, T.: Energy efficiency and sustainability assessment for methane harvesting from Lake Kivu. Energy **225**, 120215 (2021)
34. Yang, H., Bi, W., Zhang, Y., Yu, J., Yan, J., Lei, D., Ma, Z.: Effect of tectonic coal structure on methane adsorption. J. Environ. Chem. Eng. **9**(6), 106294 (2021)
35. Li, S., Xu, J.: Possibility analysis of simultaneous extraction of coal and coalbed methane in China. Fuel Energy Abstr. **43**, 246 (2020)
36. Dong, J., Cheng, Y., Pan, Z.: Comparison of transient and pseudo-steady diffusion of methane in coal and implications for coalbed methane control. J. Petrol. Sci. Eng. **184**, 106543 (2020)

Vermiculate Concentrate Pre-crushing for Firing in Electric Furnaces

Anatoly Nizhegorodov, Boris Moyzes, and Aleksey Gavrilin

Abstract The expanded vermiculite is widely used in many spheres of life. So, the great attention is paid to the vermiculite baking and to the technological preparation of concentrates. The preliminary preparation of concentrates includes not only grinding large particles to a size of about 5 mm, but also additional "narrow-band" fractioning of the original concentrates. Only granulometric homogeneous streams of vermiculite concentrates make it possible to carry out cost-effective firing without unnecessary energy costs, but also provide the highest quality of the expanded product at the exit from the furnaces. The need for grinding the row material is made by the severely expressed granulometric heterogeneity in all size groups of concentrates. The spread in the sizes of the nominal particle diameters can be fourfold, that is, for example, from 2 to 8 mm. The article reviews chopping units for grinding large-sized particles of vermiculite concentrates. The calculation of dynamic characteristics of the chopping process is carried out. The use of the dynamic characteristics to provide technological preparation of concentrates for thermal treatment of vermiculite in electric furnaces is being proved. The specificity of the designs of electric modular-furnaces and furnaces with movable base platforms is that due to the narrow slot-like firing space, high energy efficiency of such furnaces is achieved, which makes them competitive and safe, in contrast to furnaces operating on hydrocarbon fuel.

A. Nizhegorodov
Department of Construction and Road-Making Machines and Hydraulic Systems, Irkutsk National Research Technical University, 83 Lermontov Street, Irkutsk, Russia 664074

B. Moyzes (✉)
Division for Testing and Diagnostics, National Research Tomsk Polytechnic University, 30 Lenin avenue, Tomsk, Russia 634050
e-mail: mbb@tpu.ru

A. Gavrilin
Division for Mechanical Engineering, National Research Tomsk Polytechnic University, 30 Lenin avenue, Tomsk, Russia 634050

E. Lysenko et al. (eds.), *Recent Developments in the Field of Non-Destructive Testing, Safety and Materials Science*, Studies in Systems, Decision and Control 433,
https://doi.org/10.1007/978-3-030-99060-2_10

1 Introduction

The electric modular-firing furnaces [1] and furnaces with movable bases [2] for firing vermiculite differ from other types by a slot shaped firing chambers fitted in electrical modules. The gap between the base surface and the hanging wire heaters is usually from 5 to 10 mm and can be adjusted depending on the size group of the vermiculite concentrate to be fired. This is one of the main factors that determine the energy efficiency of such furnaces in comparison with combustion furnaces operating on hydrocarbon fuel [3–6].

However, concentrates of the same size group of any manufacturer are not uniform in granulometry. Let's consider, for example, KVC-4 concentrate. Specifications for its production [7] regulate the main fraction with variation of particle sizes from 2.5 to 5 mm within 80% by weight, that is, even inside the main fraction, the particle size changes by half. In this case, the remaining 20% contain particles of 2.0 and 8.0 mm, although there are not too many of them.

A quantitative estimate of the spread of particle sizes in any vermiculite concentrate is given by their size distribution, and it behaves in accordance with the law of normal distribution [8–16]. Therefore, if the concentrate flow is launched into the operating space of the firing modules without preliminary fractionation, small particles will be burned out, and large ones will not be expanded.

The small particles will lose strength and become brittle due to irreversible changes in the mineralogical structure [17], while in large ones the dehydration processes will not complete, they will not be well expanded and will not reach the minimum density.

Namely, these properties of expanded vermiculite that is to have sufficient strength in the massif and to be light provide its wide application in various fields of industrial, construction, environmental and other types of human activity [3, 4], etc. But producers of concentrates will not change their technological lines, therefore, consumers of raw materials must themselves provide the necessary preparation of typical vermiculite concentrates for firing in the electric furnaces mentioned above.

2 Additional Narrowband Fractionation

It is practical to subject the concentrates of vermiculite mica to additional, so-called narrow-band fractionation, starting with the second and fourth size groups: KBC-2 and KBC-4. In this case, if the separation into two fractions is carried out on wire sieves, then the diagonal of the sieve cell "in the light" should be determined by the formula [17–23]:

$$d = 0.73D_c, \tag{1}$$

where Dc is the conditional particle diameter that determines the size group (KVC-2, KVC-4, etc.) and is determined by the formula:

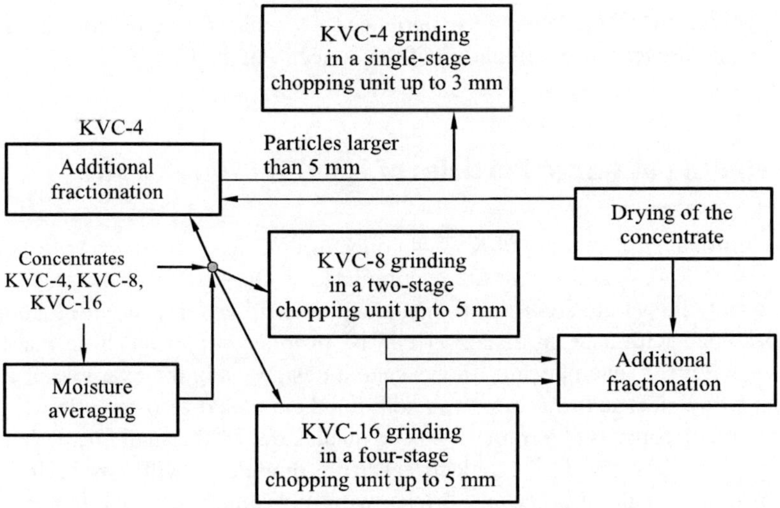

Fig. 1 Technological scheme for the preparation of vermiculite concentrates of various size groups before firing in electric furnaces

$$D_c = \sqrt{d_1 \cdot d_2},$$

d_1 and d_2 are two sizes of a flat particle taken in arbitrary directions.

It is sufficient for KBC-2 concentrate to be separated into two size groups on a sieve with a sieve cell diagonal determined by formula (1). An additional grinding in a chopping unit is not required, since there are no particles with a nominal diameter of more than 5 mm in it. With consideration of the formula (1), there are particles from 5 to 8 mm in the concentrate KBC-4 after additional fractionation and separation into two new size groups. They should not immediately be sent for firing to the furnace, but should be preliminarily subjected to grinding in a two-stage chipper (Fig. 1).

This group of concentrate is first fractionated (unlike KBK-8 and KBK-16) (Fig. 1) and then large particles are sent for the chopping unit. At the same time, in the process of additional fractionation, purified hot air is blown through closed drum sieves (drying of the concentrate).

It should also be noted that all manufacturers supply concentrates in big bags or soft containers (SC) made from polipropylene with polyethylene liners in open-top wagons. If the material is dry, with a moisture content of 5…6%, then it immediately is sent to preparation. In some cases, in spite of the packaging, rainwater or snow penetrates into the concentrate, which also, after thawing, overmoistens the raw material. But as a rule, it is 20…30% of the total. Therefore, before it is sent to preparation, it should be subjected to the operation of averaging moisture in a one and a half cube gravity mixer (Fig. 1). Due to the layered structure, the dry particles of vermiculite mica pull on themselves the water surplus and the most water-saturated particles of one SC. The production experience shows that after removing undesirable moisture, the process of fractionation of the concentrate is more efficient.

3 Grinding of Large Particles of KVK-4 Concentrate

After additional fractionation of KVC-4 concentrate (Fig. 1) particles with nominal diameters of 5, 6, 7 and 8 mm are accumulated at the exit from the sieve. If the production volumes are significant, then a separate firing furnace with a chopping unit combined with a drum dispenser can be installed for firing these massifs of particles. Figure 2 schematically shows such a design using the example of a three modular firing electric furnace with an additional, so-called zero-module.

The furnace consists of a frame 1, closed on all sides by thermally insulated walls and flaps (not shown in Fig. 1), electrical firing modules 2 with a refractory base 3, for example, made of refractory bricks, wire nichrome suspended heaters 4 and thermal covers 5, which closes the slot-like firing space.

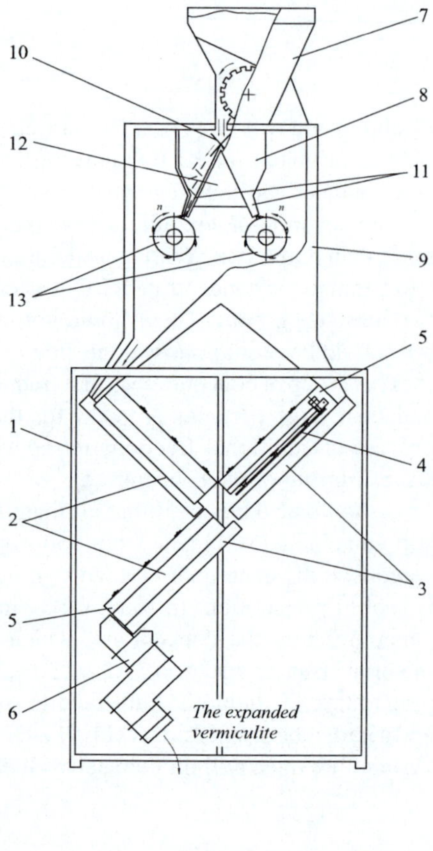

Fig. 2 Three-modular electric furnace with an additional zero-module and a built-in chopping unit for firing large particles of KVC-4 concentrates

The upper firing modules are connected in series and the third module is located from below, therefore, the flows of expanded vermiculite leaving the upper modules are summed up and enter it. This is the furnace with series-parallel interface of modules (SPI-oven) [24–26].

The lower electrical module is tightly coupled with the zero-module 6. The zero-module 6 makes it possible to use the thermal exergy accumulated by vermiculite in electrical modules and thermal radiation of small vermiculite particles to complete the mechanical transformation (structure formation) of vermiculite in the absence of an external heat source. The use of zero modules can reduce energy consumption and energy consumption of firing by 10…20%, depending on the place of its installation in the furnace is shown in the research.

A drum dispenser 7 is installed on the furnace frame, the drum feeds the large grains of KVK-4 concentrate with nominal diameters of 5, 6, 7 and 8 mm, obtained after its additional fractionation into a chopping unit 8, combined in a single housing 9 with a dispenser. Since the furnace under consideration has two parallel-coupled upper modules 5, then a flow divider 10 is installed in the dispenser. The flow divider is made in the form of two crossed and connected plates with alternating slots, which pro-vide an equal separation of the flow of large particles into two parts passing through the trays 11 to the upper modules. The firing process begins in those upper modules and continues in the third module that sums up the flows of expanded vermiculite, and ends in the zero module.

Due to the ribs 12, the trays form a rigid structure that resists shock loads from the blades of the high-speed rotors 13 which rotate synchronously in opposite directions with equal angular velocities n. When the rotor passing through the angle γ which corresponds to the arc L between adjacent blades (Fig. 3) a flat concentrate particle should have time to leave the tray by a value of z, for a given unit equals to 3 mm.

If particles with $D_c = 5$ or $D_c = 6$ mm are chopped off, their fragments (2 and 3 mm) fall out of the tray and sent to one of the upper modules 5 (Fig. 2).

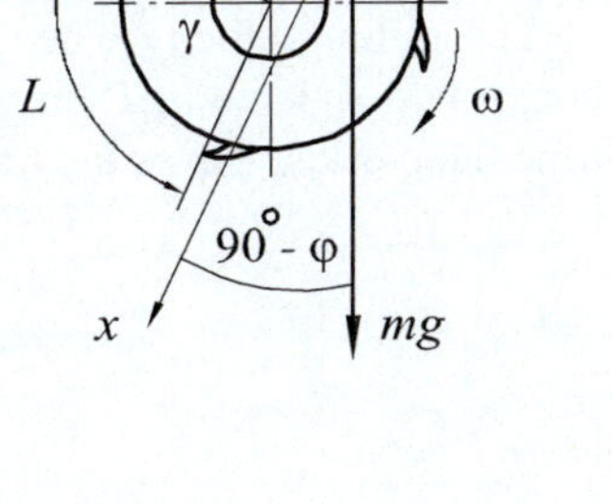

Fig. 3 Scheme for calculating of the technical characteristics of the chopping unit

If a particle with $D_c = 7$ mm is chopped off, its fragment (4 mm) will fall out of the tray under its own weight or will be pushed out from there with the next blade without chopping off, since the gap in the tray is twice the thickness of the average particle from the range of nominal diameters of 5…8 mm.

If a particle with $D_c = 8$ mm is chopped off, its fragment (5 mm) will be chopped off by the next rotor blade, fall out of the tray and sent to one of the upper modules 5 (Fig. 2).

The feeding of large particles of KBC-4 concentrate by the dispenser drum 7 (Fig. 2) is selected to exclude more than two particles from entering one place of the tray at once, in order to avoid their jamming.

Therefore, the mass productivity of the dispenser should be 10…15% less than the capacity of the built-in chopping unit.

4 The Calculation of Parameters and Values of the Chopping Unit Operating Process

Let's consider the design scheme shown in Fig. 3. The x axis is directed along the inner surface of the tray, where a mica particle with a center of gravity in the point a protrudes.

The particle is affected by gravity mg, normal reaction N, and friction force F.

To determine the exit time of a particle of mass m by the value z, let's compose the differential equation of its motion:

$$m\frac{d^2x}{dt^2} = m \cdot g \cdot \sin\varphi - f \cdot m \cdot g \cdot \cos\varphi, \tag{2}$$

where f is the coefficient of sliding friction of a vermiculite particle on steel, approximately equals to 0.68 [27–29]), φ is the angle between the line of gravity and the y axis.

Reducing Eq. (2) by m and dividing the variables, after double integration over time, we obtain the dependence:

$$x = \frac{1}{2}(g \cdot \sin\varphi - f \cdot g \cdot \cos\varphi)t^2 + x_0. \tag{3}$$

The initial velocity when a particle falls out is zero, and in Eq. (3) the initial coordinate x_0 is zero. Then, solving Eq. (3) with consideration to t, we obtain the time the particle leaves the tray until it is chopped off.

$$t = \sqrt{\frac{2x}{(\sin\varphi - f \cdot \cos\varphi)g}}. \tag{4}$$

In this case, x is equal to z and equals to 0.003 m, and let the angle φ be equal to 70°. Solving Eq. (4) with given initial data, we obtain $t = 0.0294$ s.

During the same time, the rotor will turn through the angle γ, and each of its blades will move along the arc L (Fig. 3).

Let's set the design parameters of the rotors of the chopping unit of the furnace:

- the outer diameter of the rotors along the blades—$D = 0.2$ m;
- the width of the blades and the rotors themselves—$B = 1.0$ m;
- the height of the blades (approximately equal to z)—$h = 0.0028$ m;
- the angle of inclination of the trays to the vertical is 20°.

The angular speed of rotation of the rotor blades n can be calculated on the base of the fact that the time of one complete revolution t_{cr} is equal to 4t. It will be equal (1/s):

$$n = \frac{1}{4t} = \frac{1}{4 \cdot 0.0294} = 8.5.$$

Then the angular velocity of the rotor knives will be equal to (rad/s):

$$\omega = 2 \cdot \pi \cdot n = 2 \cdot 3.14 \cdot 8.5 = 53.4$$

and the linear peripheral speed of knives can be determined by the formula (m/s):

$$\upsilon_k = \frac{D \cdot \omega}{2} = \frac{0.2 \cdot 53.4}{2} = 5.3.$$

For further modeling of the operating process of the chopper, we will make assumptions. Let all the concentrate particles entering the chopper have the same thickness $s = 0.0005$ m (0.5 mm) and have a square shape with sides equal to D_c (5, 6, 7 and 8 mm). Let us take the side of the square e equal to the arithmetic mean of the sum of the dimensions indicated in parentheses, that is, 6.5 mm or 0.0065 m. These square particles in each tray are arranged in one row tightly to each other over the entire width of the rotor ($B = 1.0$ m).

Then, in one revolution of two rotors, particles with a total area will be chopped off (m²): $S = 2{\cdot}4{\cdot}V{\cdot}z = 2{\cdot}4{\cdot}1{\cdot}0.003 = 0.024$.

After the first chopping, the averaged particle will have a size of $z = 3.5$ mm, therefore, on the second revolution of the rotors, the total area of the dropped particles will be (m²): $S = 2{\cdot}4{\cdot}V{\cdot}z = 2{\cdot}4{\cdot}1{\cdot}0.0035 = 0.028$.

The average area of the two rotors of the chopping unit dropped out during the rotation will be $S_a = 0.026$ m². With a particle thickness $s = 0.0005$ m, their total volume per hour of operation of the unit will be (m³): $V = 0.026{\cdot}0.0005{\cdot}8.5{\cdot}3600 = 0.4$.

If we leave the model of square particles and consider the nominal diameter inscribed in the square, then the volume of round particles will be equal to 0.31 m³.

The true density of hydromicas is approximately equal to 2300 kg/m^3 [30], so the resulting volume can be converted to mass kg: $m = \rho \cdot V = 2300 \cdot 0.31 = 713$.

The resulting value determines the theoretical mass hourly productivity of the chopping unit: $P_{\mathrm{m}} = 0.713$ t/h.

With an expanding coefficient $k_{\mathrm{e}} \approx 8.5$ m^3/t [29] and such a concentrate supply, the volumetric productivity of a three-module electric furnace will be approximately (m^3/h): $P_V = k_{\mathrm{e}} \cdot P_{\mathrm{m}} = 8.5 \cdot 0.713 = 6.1$.

5 Results and Considerations

In contrast to the KVC-4 concentrate, large-sized KVC-8 and KVC-16 concentrates, in a dry state or after a preliminary operation of moisture averaging, are immediately subjected for grinding to a size of 5 mm. And only after that, there is additional fractionation on a drum sieve with a given set of sieve meshes, during which purified hot air is driven through closed drum sieves (drying the concentrate) (Fig. 1).

In the KBC-8 concentrate, most of the particles are grouped near the nominal diameter $D_{\mathrm{c}} = 8$ mm, which determines the size group of the concentrate. But it necessarily contains particles with a minimum $D_{\mathrm{c}} = 4$ mm and a maximum $D_{\mathrm{c}} = 16$ mm in size, therefore, the entire volume of this raw material is subjected to grinding in a special two-stage chopping unit shown in Fig. 4.

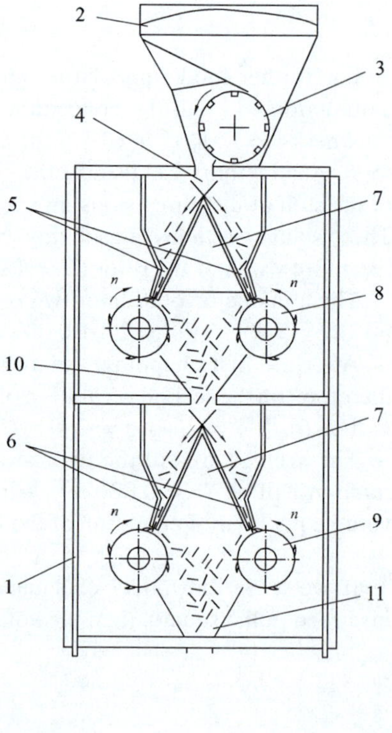

Fig. 4 Two-stage chopping unit for concentrates KVC-8

There is a drum dispenser 2 with a feed drum 3 on the frame 1. The rotation speed of feed drum 3 can be adjusted within the required limits, for example, using a frequency regulator. The batcher is equipped with a concentrate flow divider 4 into two parts, like the batcher of the three modular electric furnace shown in Fig. 2.

After separation, the concentrate particles enter the trays 5 of the first level and the trays 6 of the second level. These trays, together with the frame 1, due to the ribs 7, form a rigid structure that resists the shock loads from the blades of the high-speed rotors 8 and 9 of the first and second levels. Rotors of each level rotating synchronously towards each other with equal angular velocities n. During the rotation of the rotors through the angle γ (Fig. 3) corresponding to the arc L between adjacent blades, a flat mica particle should have time to leave the tray by the value z. For a special two-stage chopping unit, the size of the protruding part of a flat particle z should be 5 mm.

During the cutting process, the concentrate particles pass through trays 10 and 11, then they are collected in the bin and when accumulates is sent for additional fractionation (Fig. 1). In concentrate KVC-8, particles with nominal diameters from 4 to 16 mm are present in different amounts.

Particles with $D_c = 4$, 5 and 6 mm with a gap of $z = 5$ mm will simply pass through the chopping unit and fall out into the storage bin. Particles with $D_c = 7$, 8, 9, 10 and 11 mm will be chopped off by the blades of the first stage rotors with remainders of 2, 3, 4, 5 and 6 mm, respectively. These remainders will also freely pass through the unit into a storage bin located under the unit (not shown in Fig. 4).

Then there are particles with nominal diameters $D_c = 12$, 13, 14, 15 and 16 mm.

At the first stage, particles with $D_c = 12$ are divided into two particles with dimensions of 5 and a remainder of 7 mm, which in the second stage will be divided by blades into two more parts 5 and 2 mm.

Particles with $D_c = 13$ will be divided into two with dimensions 5 and a remainder of 8 mm, which in the second stage will be further divided into two more parts 5 and 3 mm, etc. Particles with $D_c = 14$ are divided into three with sizes of 5, 5 and 4 mm. Particles with $D_c = 15$ will be divided into three with sizes 5, 5 and 5 mm. And finally, the largest particles with $D_c = 16$ will be divided into three parts with sizes of 5, 5 and 6 mm.

So, we see that through the two-stage chopping unit significant amount of particles of 6 mm in size can be passed, which, under their own weight, will fall out of the trays or will be pushed out of them by other blades without chopping off, since the particle is fitted in the tray with a back-lash due to the slot in the tray is approximately twice the thickness of the average particle in the range of nominal diameters 4…16 mm. Therefore, after additional fractionating in the drum sieve, which passes only 5 mm particles, a small group of particles with a size of 5…6 mm will remain. As such particles accumulate in the chopper unit, the gap z should be readjusted to 3 or 3.5 mm.

The mass productivity of a two-stage unit is calculated in the same way as for the chopper of a three-module furnace shown in Fig. 2.

The grinding of concentrates of the size group KBC-16 should be carried out on a four-stage chopping unit. The four-stage chopping unit is designed as a set

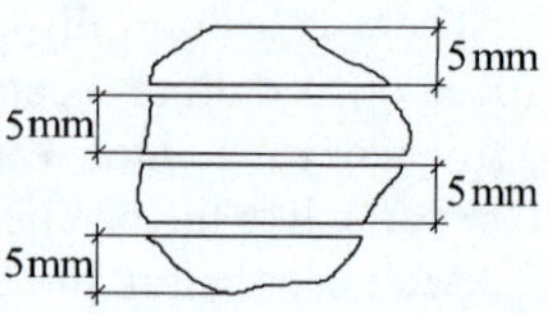

Fig. 5 Shapes of concentrate particles at the exit from the chopper unit

of sequentially vertically installed two-stage chopping units with a common drum dispenser. The capacity of such a unit will not exceed the capacity of the original two-stage unit [31, 32].

At the outlet of the grinder of large concentrates, the largest number of strip-shaped particles shown in Fig. 5. But despite the fact that one of the sizes of such particles exceeds 5 mm, the firing time of such particles will not exceed the firing time of rounded particles with $D_c = 5$ mm [32].

6 Conclusion

The particles of the raw material—vermiculite concentrate should not exceed 5 mm by their nominal diameters to ensure high energy efficiency of electric modular-furnaces and furnaces with movable base platforms for vermiculite firing. This is due to the fact that in the slot-like space for firing electrical modules, the gap between the base surface and the suspended heaters should be minimal, no more than 10 mm. This is one of the main factors that determine the energy efficiency of such furnaces.

Since the concentrates have different size groups, the most effective way to unify the particle size is to grind them in the chopping units reviewed in this article.

Despite the simplicity of the design of such machines, they have a high productivity. Thus, even a single-row chopping unit of a three-module electric furnace for firing of large particles of concentrates KVC-4 provides an hourly productivity of 0.713 t/h, and the furnace itself provides the production of expanded vermiculite up to 6.1 m^3/h.

It is obvious that new units for grinding large-sized concentrates are capable of providing pre-crushed vermiculite concentrates for firing in electric furnaces and have good prospects for application in the production of vermiculite.

References

1. Nizhegorodov, A.I.: Theory and practical use of modular-pouring electric furnaces for firing vermiculite. Refract. Ind. Ceram **56**(4), 361–365 (2015)
2. Bryanskikh, T.V., Kokourov, D.V.: Energy efficiency of electric furnaces with movable floor in firing of vermiculite concentrates of different size groups. Refract. Ind. Ceram **58**(4), 368–373 (2017)

3. Kogal, J.E., Trivedi, N.C., Barker, J.M., Krukowski, T.S.: Industrial Minerals and Rocks: Commodities, Markets and Uses. Society for Mining, Metallurgy, and Exploration Inc., Littleton, USA (2009)
4. Fuks, L., Herdzik-Koniecko, I.: Vermiculite as a potential component of the engineered barriers in low- and medium-level radioactive waste repositories. Appl. Clay Sci. **161**, 139–150 (2018)
5. Rashad, A.M.: Vermiculite as a construction material: a short guide for civil engineer. Constr. Build. Mater. **125**, 53–62 (2016)
6. Hombostel, C.: Construction Materials: Types, Uses, and Applications. Wiley, New York, USA (1991)
7. Panfilov, V.I.: Vermiculite fractionated Kovdorsky concentrate. Technical Conditions (TC 21–0281921–24–92). JSC "Kovdorsluda", Kovdor (1992)
8. Zvezdin, A.V., Bryanskikh, T.B.: Considering adaptation of electrical ovens with unit-type releasing to peculiarities of thermal energization of mineral raw materials. IOP Conf. Ser. Mater. Sci. Engi. **168**(1), 012003. (2017)
9. Wang, S., Gainey, L., Marinelli, J., Deer, B., Wang, T., Mackinnon, I., Yunfei, X.: Effects of vermiculite on in-situ thermal behaviour, microstructure, physical and mechanical properties of fired clay bricks. Constr. Build. Mater. **316**, 125828 (2022)
10. de Vargas Brião, G., da Silva, M.G., Vieira, M.G.A.: Expanded vermiculite as an alternative adsorbent for the dysprosium recovery. J. Taiwan Inst. Chem. Eng. **127**, 228–235 (2021)
11. Feng, J., Liu, M., Mo, W., Su., X.: Heating temperature effect on the hygroscopicity of expanded vermiculite. Ceram. Int. **47**(18), 25373–25380 (2021)
12. Koksal, F., Nazli, T., Benli, A., Gencel, O., Kaplan, G.: The effects of cement type and expanded vermiculite powder on the thermo-mechanical characteristics and durability of lightweight mortars at high temperature and RSM modelling. Case Stud. Constr. Mater. **15**, e00709 (2021)
13. Li, M., Shi, J.: Mechanical and thermal performance assessment of paraffin/expanded vermiculite-diatomite composite phase change materials integrated mortar: experimental and numerical approach. Sol. Energy **227**, 343–353 (2021)
14. Surzhikov, A.P., Lysenko, E.N., Vasendina, E.A., Sokolovskii, A.N., Vlasov, V.A., Pritulov, A.M.: Thermogravimetric investigation of the effect of annealing conditions on the soft ferrite phase homogeneity. J. Therm. Anal. Calorim. **104**(2), 613–617 (2011)
15. Surzhikov, A.P., Frangulyan, T.S., Ghyngazov, S.A., Vasil'ev, I.P., Chernyavskii, A.V.: Sintering of zirconia ceramics by intense high-energy electron beam. Ceram. Int. **42**(12), 13888–13892 (2016)
16. Feng, J., Liu, M., Fu, L., Ma, S., Yang, J., Mo, W., Su, X.: Study on the influence mechanism of Mg2+ modification on vermiculitethermal expansion based on molecular dynamics simulation. Ceram. Int. **46**(5), 6413–6417 (2020)
17. Nizhegorodov, A.I.: Optimal fractionation of vermiculite concentrates. Stroitel'nye materialy **5**, 98–99 (2008)
18. Balima, F., Nguyen, A.-N., Reinert, L., Le Floch, S., Pischedda, V., Duclaux, L., San-Miguel, A.: Effect of the temperature on the structural and textural properties of a compressed K-vermiculite. Chem. Eng. Sci. **134**, 555–562 (2015)
19. Chen, L., et al.: Preparation and characterization of the eco-friendly chitosan/vermiculite biocomposite with excellent removal capacity for cadmium and lead. Appl. Clay Sci. **159**, 74–82 (2018)
20. El Mouzdahir, Y., Elmchaouri, A., Mahboub, R., Gil, A., Korili, S.A.: Synthesis of nano-layered vermiculite of low density by thermal treatment. Powder Technol. **189**(1), 2–5 (2009)
21. Lozano-Lunar, A., Raposeiro Da Silva, P., De Brito, J., Fernández, J.M., Jiménez, J.R.: Safe use of electric arc furnace dust as secondary raw material in self-compacting mortars production. J. Clean. Prod. **211**, 1375–1388 (2019)
22. Santamaria, A., Faleschini, F., Giacomello, G., Brunelli, K., San José, J.-T., Pellegrino, C., Pasetto, M.: Dimensional stability of electric arc furnace slag in civil engineering applications. J. Clean. Prod. **205**, 599–609 (2018)
23. Xu, W., Zhang, J., Zhang, R.: Application of multi-model switching predictive functional control on the temperature system of an electric heating furnace. ISA Trans. **68**, 287–292 (2017)

24. Shyamal, S., Swartz, C.L.E.: Real-time energy management for electric arc furnace operation. J. Process Control **74**, 50–62 (2019)
25. Contreras-Serna, J., Rivera-Solorio, C.I., Herrera-García, M.A.: Study of heat transfer in a tubular-panel cooling system in the wall of an electric arc furnace. Appl. Therm. Eng. **148**, 43–56 (2019)
26. Nizhegorodov, A.I., Zvezdin, A.V.: Study of an electric furnace physical model for firing vermiculite with a "zero" module. Refract. Ind. Ceram **57**(3), 246–251 (2016)
27. Kutateladze, S.S.: Heat Transfer and Hydrodynamic Resistance: Reference Book. Energoatomizdat, Moscow (1990)
28. Kreith, F., Black, W.: Basic Heat Transfer. Harper & Row, New York, USA (1983)
29. Telegin, A.S., Shvydkii, V.S., Yaroshenko, Y.G.: Heat and Mass Transfer. ITsK Akademkniga, Moscow (2002)
30. Comparative properties of natural and synthetic micas (n.d.). http://cniga.com.ua/index.files/sluda.htm. Accessed 24 March 2020
31. Maqueda, C., Romero, A.S., Morillo, E., Pérez-Rodríguez, J.L.: Effect of grinding on the preparation of porous materials by acid-leached vermiculite. J. Phys. Chem. Solids **8**(5–6), 1220–1224 (2007)
32. Nizhegorodov, A.I., Zvezdin, A.V.: Transformation of vermiculite energy into mechanical transformation energy during firing in electric furnaces with "zero" module. Refract. Ind. Ceram **57**(3), 239–245 (2016)

Application of the System for Electrical Equipment Diagnostics and Its Analysis

Inna Plotnikova, Elena Sheveleva, and Rinat Narimanov

Abstract External factors, uncertainty and risk force the organization to seek for new management methods and concepts to guarantee its further development and strengthen its competitiveness. The paper suggests theoretical aspects of the system for monitoring and diagnosing electric motors, and describes the structure of technical diagnostics. The need for diagnostics and its application is substantiated; the principles of process control and the organization of control points are given. The cause-and-effect analysis reveals the causes that provoke risk factors. The paper provides a detailed discussion of the causes related to the equipment operation (downtime and depreciation of equipment, malfunction of measuring instruments and signal converters), to the diagnostic procedure (errors in measurement sequence, violation of technology, low accuracy of state recognition, insufficient initial data, incorrect choice of analysis method, low efficiency of information processing), and to human resources (qualification of personnel, involvement in other operations, time limits of operation performance, insufficient professional experience, lack of focus at work). Risk aftereffects are systematized using a decision tree. Recommendations are proposed to eliminate risks, which entail serious problems, such as engine breakdown or premature withdrawal from operation.

1 Introduction

New technologies force enterprises to constantly adapt to changing external conditions, while maintaining the overall direction of their business. Electrical engineering

I. Plotnikova (✉)
Tomsk Polytechnic University, 30, Lenin Avenue, Tomsk, Russia 634050
e-mail: inna@tpu.ru

E. Sheveleva · R. Narimanov
Tomsk State University, 34, Lenin Avenue, Tomsk, Russia 634050
e-mail: vasendina@tpu.ru

R. Narimanov
e-mail: ring_0@mail.ru

E. Lysenko et al. (eds.), *Recent Developments in the Field of Non-Destructive Testing, Safety and Materials Science*, Studies in Systems, Decision and Control 433,
https://doi.org/10.1007/978-3-030-99060-2_11

companies are no exception, as electric motors are used in all areas of industry to drive various mechanisms, such as pumps, fans, compressors, as well as in the household sphere [1]. The total energy consumption in electric motors is estimated to be about 80% of the total electricity produced. In most cases, these mechanisms operate under stressful conditions, which leads to damage to individual elements and the occurrence of emergency situations. Periodic diagnostics of the equipment allows to prevent such a development of the situation.

Diagnostics is defined as a set of measures that allow to inspect the equipment and establish signs of malfunction [2]. The diagnostic process is based on methods and means, based on the use of which it is possible to make a conclusion about the absence or presence of malfunctions. Monitoring and diagnostics of electric motors provide a full assessment of the condition of the engine. By searching and analyzing, first of all, internal equipment malfunctions are identified. With the help of visual inspection, all external defects are easily determined [3].

The object of diagnostics is the object whose condition needs to be determined, and the entire process of equipment research is called diagnostics. In order to increase the reliability of the equipment and achieve a qualitative assessment of the residual resource, it is necessary to set the main goal of diagnostics—to recognize the state of the technical system [4–6]. This goal is often achieved under conditions of a limited amount of initial data. Appropriate diagnostic systems have been developed for various technical systems, taking into account the requirements.

The experience of Russian and foreign enterprises has shown that the development and implementation of new methods for improving the diagnostic process makes it possible to identify and prevent the occurrence of various malfunctions, to ensure operational characteristics established by regulatory documentation and corresponding to consumer expectations [7–9].

2 Research Method

Judging by the annual analysis of the operation of electric motors, emergency failures in electric motors are found in the fourth part of the sample. Modern methods of equipment protection do not always provide a wide range of defect prevention that is necessary to improve the reliability of electric motors [10]. Failures of electrical devices can be caused not only by defects that are visible to the eye, but also by hidden inconsistencies (Fig. 1).

The Development Center of HSE University conducted an analysis of the dynamics of industrial production intensity indices. Let's consider the indicators of one of the manufacturing industries—the production of electrical equipment.

The intensity index of the electrical equipment production is at a high level. Every year more and more enterprises for the production of electrical equipment appear due to increased demand. The sphere of the electrical equipment operation is also intensively developing. Since every year a large amount of electrical equipment enters the enterprises, there is a need to establish a stable system for monitoring

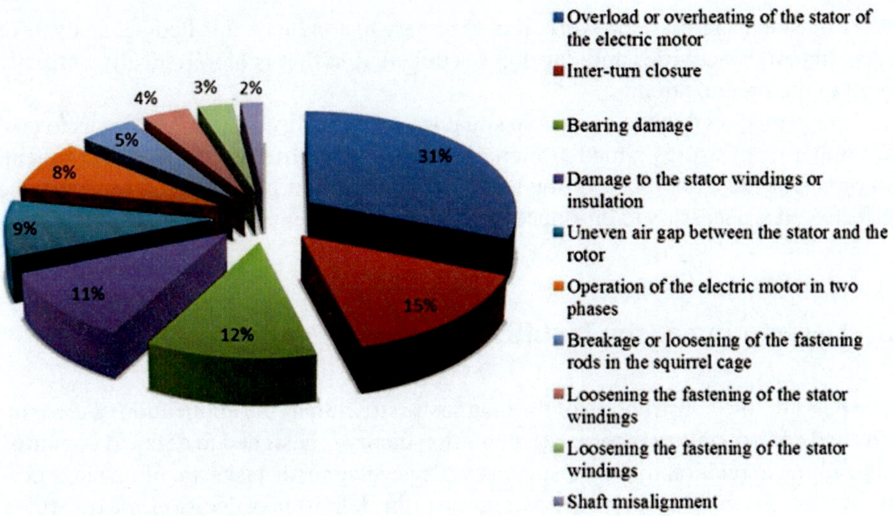

Fig. 1 Defects of asynchronous electric motors

and diagnostics of electric machines, in particular electric motors in operation. The market of large enterprises that are responsible for creating and maintaining a high-quality control and diagnostic process is as wide as it is closed.

Table 1 presents data for analyzing the dynamics of the intensity index in % by January 2014 [11].

One of the frequently observed facts is that the introduction of new management solutions in diagnostic departments occurs systematically when the system and management approaches are transformed throughout the enterprise as a whole. To

Table 1 Production of electrical equipment

	2014	2015	2016	2017	2018	2019	2020	2021
January	90.5	95.6	93.7	90.3	99.8	88.1	87.2	98.6
February	94.0	89.2	87.6	101.4	98.0	85.1	92.7	95.8
March	99.3	88.1	88.5	93.0	99.5	92.5	85.8	94.5
April	100.2	84.9	93.1	98.9	89.9	91.2	66.0	92.0
May	105.6	86.6	86.9	100.3	91.6	93.4	86.5	88.1
June's	106.9	86.4	85.8	96.3	89.1	100.2	85.7	82.3
July	102.7	85.7	94.2	95.6	93.2	91.4	89.0	88.4
August	101.2	89.8	86.9	92.7	99.8	90.8	90.7	83.5
September	94.7	93.3	93.3	90.9	92.4	97.0	89.1	86.0
October	93.4	95.7	98.0	92.5	93.0	91.5	89.4	
November	104.0	85.4	98.7	93.1	86.4	93.5	92.8	
December	97.8	91.6	89.3	94.5	101.9	89.2	92.0	

carry out such a large-scale work, it is necessary to conduct a full-fledged analysis of the company's activities and develop documentation that is hierarchically correctly built in the organization.

Due to the fact that the economic situation in the energy industry forces us to take a number of measures aimed at increasing the service life of electrical equipment. In order for the assessment of the technical condition of electrical equipment to be effective, it is necessary to introduce methods of instrumental control and diagnostics.

3 Description of the Installation

The basis for the construction of the diagnostic structure is the application of recognition and controllability theories. Recognition theory is designed to define recognition algorithms in relation to diagnostic tasks. These diagnostic tasks are often classification tasks. The main problem of recognition is the definition of decision-making rules. In the presented work, we are talking about diagnostics in the field of the theory of controllability, since the processes of obtaining information data about equipment, condition monitoring and troubleshooting, and initiating follow-up measures are being investigated.

Conventionally, the structure of technical diagnostics for any type and purpose of equipment is shown in Fig. 2.

In the theory of diagnostics, the fundamental criteria of analysis are diagnostic parameters, which determine the totality of all parameters necessary for analysis. The main rule of choosing diagnostic parameters is to fulfill the requirement of redundancy of information and reliability. At the same time, reliable information about operational parameters obtained in real conditions is investigated. Production conditions require the use of several diagnostic parameters at the same time to speed up the diagnostic process and return the equipment to the production cycle as soon as possible. As for the number of parameters, it is determined by how modern measuring

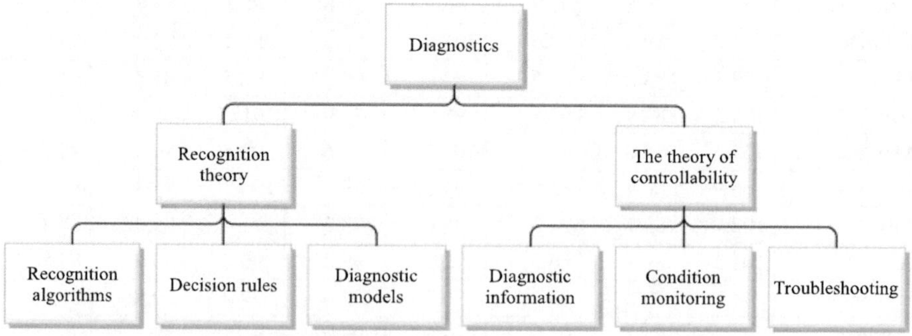

Fig. 2 Structure of technical diagnostics

instruments are used in the diagnostic process and whether they correspond to the degree of diagnostic methods development.

4 Results and Considerations

The process of monitoring and diagnostics of electric motors is a sequential process that requires the execution of a mandatory sequence of diagnostic steps. Although the directions of modern diagnostics of electrical equipment, depending on the goal pursued, may differ, the sequence of stages of the process is similar. At the moment, in accordance with the development of technical diagnostics, parametric, preventive diagnostics and fault diagnosis are distinguished. Of particular interest is troubleshooting, the purpose of which is to determine the magnitude and type of defect after the fact of the presence of a malfunction is registered.

The diagnostic results can be considered information about the condition of electrical equipment, information about detected defects and their characteristics (the scale of the area of distribution, the nature of the defect, the place of detection, causes of occurrence and possible consequences), as well as a forecast of the equipment condition.

It can be concluded that it is important to diagnose equipment defects in the early stages of the development of malfunctions, to increase and maintain the reliability level of electrical equipment, since its serviceable condition is the key to the successful execution of the production process as a whole.

The type of technical diagnostics depends on several conditions: the purpose of the object being diagnosed, the complexity of the object's design, the degree of criticality of the consequences of failures or the degree of danger of an emergency situation. Failure to comply with any of these conditions also causes risky situations that can provoke the onset of negative consequences. Before formulating the criteria by which the analysis of the condition of electrical equipment will be carried out, it is necessary to conduct a preliminary study, during which potential malfunctions are clarified. Also, this study analyzes the operational readiness and reliability of the machine, the list of the main components being diagnosed and the operating modes in which diagnostics can be carried out without taking the equipment out of the operating cycle [12].

The cycle of research on the need for diagnosis is shown in Fig. 3.

According to GOST 20,911-89 "Technical diagnostics. Terms and definitions", technical diagnostics is the definition of the technical condition of an object, and its tasks are the following: monitoring the technical condition, finding a place and determining the causes of a malfunction, as well as forecasting the technical condition [13]. Thus, control, i.e. checking the compliance of the object's parameter values with the requirements of technical documentation and determining the type of technical condition at the moment is included in the concept of diagnostics [14, 15]. From the point of view of the operations performed during the process, it is advisable to

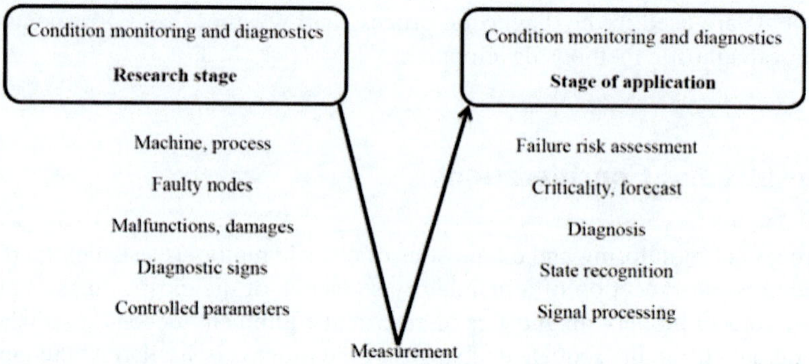

Fig. 3 The cycle of research on the need for diagnosis and its application

develop a unified organizational regulatory documentation for the monitoring and diagnostics system [15].

Let's consider the process of monitoring the condition and diagnosing electric motors by the efforts of the personnel of the electrical laboratory. The diagnostic process is carried out by competent personnel with extensive experience in this field, as well as novice specialists, therefore, high requirements are placed on qualifications and due to the specifics of the work (a deep store of theoretical knowledge and the ability to work with a large amount of information is required). Errors caused by the human factor are possible.

Let's turn to a more detailed description of the process of condition monitoring and diagnostics. At the measurement stage, in accordance with the diagnostic model of the equipment, parameters are measured using measuring instruments (measuring devices). Automated systems are also used to study the parameters, which facilitate rapid and often simultaneous measurements. At the stage of signal processing of measuring instruments, the necessary calculations are performed if the parameters in question are the result of indirect measurements. The parameters are monitored, compared with the normalized values in order to recognize the condition of the equipment, i.e. to determine whether it is serviceable or faulty and whether further diagnostics makes sense.

The laboratory staff checks electric asynchronous motors and DC motors with a capacity of up to 100 kW in an electrical diagnostic laboratory. The process map is presented in Table 2.

At the stage of diagnosis, based on the available deviations of the parameter values, the presence of a malfunction, its location, nature and causes of occurrence are investigated (Fig. 4). Determining the criticality of the condition allows you to determine how serious the malfunction is and how it affects the overall condition of the equipment. At the stage of condition forecasting, the study of the patterns of changes in the technical condition of the equipment with the development of the identified malfunction is carried out, the availability of information about previously performed measurements is of particular importance, on the basis of which it is possible to construct functions that determine the forecast.

Table 2 Process map

Management tools: Operation mode of the diagnostic department; Staff involvement		**Performance criteria**: Correctly measured parameters; Time spent on diagnosis; Availability of the equipment status report; The forecast of the equipment condition is formulated
Input data: Previous hardware status report; Technical regulations	**Process**: Diagnosing the condition of the electric motor	**Output data**: Electric motor status report; Recommendations for improving reliability
Responsibility: Employee of the diagnostic laboratory	**Resources:** Measuring instruments	**Process monitoring indicators:** Measured parameters; Repeated detection of the defect

At the stage of assessing the risk of failure, taking into account the information about forecasting, the probability of equipment failure due to the development of a malfunction is determined, as well as the severity of the consequences of the failure. All data obtained during the implementation of the operations of the condition monitoring and diagnostics of electrical equipment are used to develop measures to prevent the rapid onset of failure and the withdrawal of equipment from the operating cycle. Such measures include the appointment of current or major repairs carried out to restore the operability and maintain the serviceable condition of the equipment, or the revision of the schedule of control tests.

When carrying out the diagnostic process, it is necessary to ensure the necessary safe conditions at the workplace. The tests to which the equipment is subjected are carried out in accordance with the safety requirements, which every employee should be familiar with. All tests are carried out by the personnel of the electrotechnical laboratory according to the methodological instructions and state standard requirements.

5 Conclusion

Thus, when diagnosing and monitoring the operation of electric motors, it is necessary not only to apply some physical diagnostic method, but also to make measurements correctly, analyze the data obtained during diagnostics and maintain the equipment with which measurements are carried out, make decisions about replacing the electric motor in case of its failure. At each stage of the production process, it is important to comply with all prescribed requirements and rely on existing experience in assessing the condition, otherwise the diagnosis will be ineffective, and the assumption of

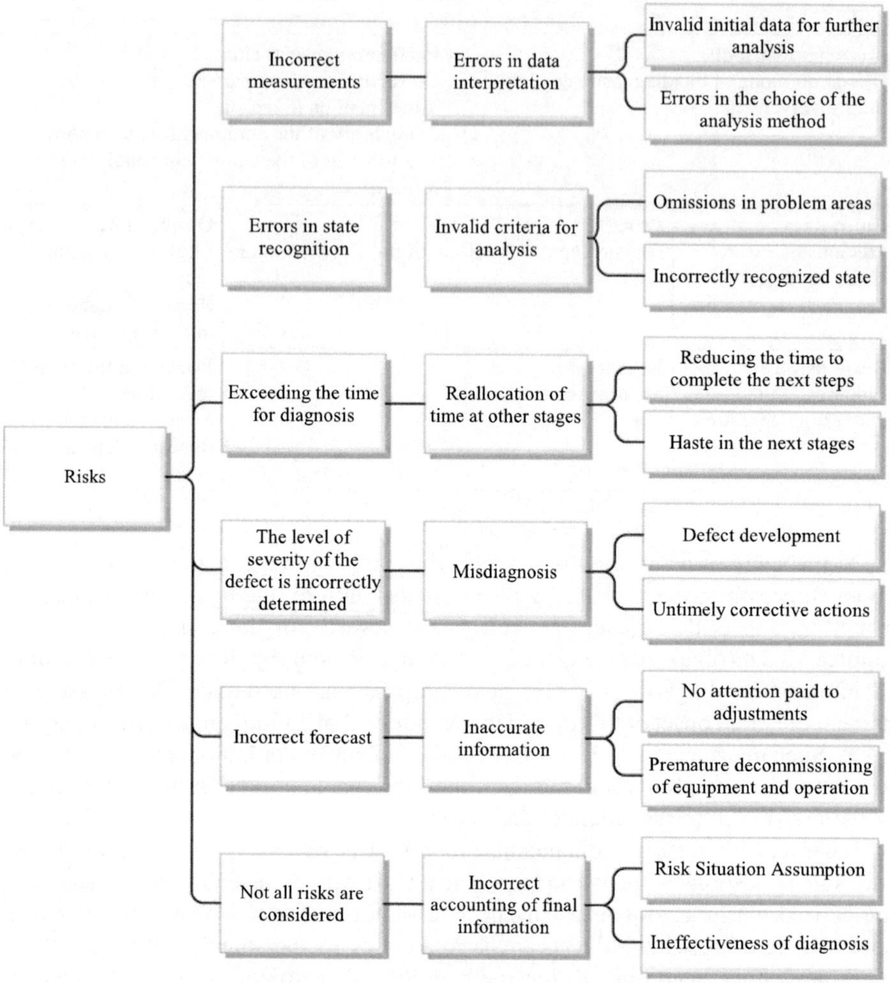

Fig.4 Risk consequence chart

consequences will lead to large financial costs. The proposed diagnostic process map, which is intended for organizing and carrying out work on fulfilling orders received from customers, as well as implementing the requirements of concluded contracts and developing new methods and tools for diagnosing electric motors, is aimed at continuous improvement of quality and results.

Acknowledgements This research was supported by TPU development program.

References

1. Chen, S., Lipo, T.A.: Bearing currents and shaft voltages of an induction motor under hard- and soft-switching inverter excitation. IEEE Trans. Ind. Appl. **34**(5), 1042–1048 (1998)
2. Faiz, J., Nejadi-Koti, H.: Eccentricity fault diagnosis indices for permanent magnet machines. State-of-the-art. IET Electr. Power Appl. **13**, 1241–1254 (2019)
3. Artigao, E., Honrubia-Escribano, A., Gómez-Lázaro, E.: In-service wind turbine DFIG diagnosis using current signature analysis. IEEE Trans. Industr. Electron. **67**, 2262–2271 (2020)
4. Antonino-Daviu, J.A., Quijano-Lopez, A., Rubbiolo, M., Climente-Alarcon, V.: Advanced analysis of motor currents for the diagnosis of the rotor condition in electric motors operating in mining facilities. IEEE Trans. Ind. Appl. **54**, 3934–3942 (2018)
5. Choi S., Haque M.S., Tarek M.T.B., Mulpuri V., Duan Y., Das, S., Garg, V., Mirafzal B.: Fault diagnosis techniques for permanent magnet ac machine and drives. A review of current state of the art. IEEE Trans. Transp. Electrif. **4**, 444–463 (2018)
6. Henao, H., Capolino, G.A., Fernandez-Cabanas, M., Filippetti, F., Bruzzese, C., Strangas, E., Pusca, R., Hedayati-Kia, S.: Trends in fault diagnosis for electrical machines. A review of diagnostic techniques. IEEE Trans. Ind. Electron. **8**, 31–42 (2014)
7. Faiz, J., Ojaghi, M.: Different indexes for eccentricity faults diagnosis in three-phase squirrel-cage induction motors. Mechatronics **19**, 2–13 (2009)
8. Salah, A.A., Dorrell, D.G., Guo, Y.A.: Review of the monitoring and damping unbalanced magnetic pull in induction machines due to rotor eccentricity. IEEE Trans. Ind. Appl. **55**, 2569–2580 (2019)
9. Aggarwal, A., Strangas, E.G.: Review of detection methods of static eccentricity for interior permanent magnet synchronous machine. Energies **12**, 4105 (2019)
10. Obuhov, S., Plotnikov, I., Surkov, M., Sumarokova, L., Savkin, K.: Development of the structural and functional design of the laboratory bench for experimental research diesel generator sets on variable speed. MATEC Web Conf. **91**, 01036 (2017). https://doi.org/10.1051/mateconf/20179101036
11. Intensity indices industrial production. https://www.hse.ru/pubs/share/direct/document/526679207.pdf
12. Vasendina E, Plotnikova I, Red'ko L, Zyablova N.: Study of types of defects in wood chipboard production. IOP Conf. Ser. Mater. Sci. Eng. **81.** 012070 (2015). https://doi.org/10.1088/1757-899X/81/1/012070
13. State standard 20911–89 "Technical diagnostics. Terms and definitions"
14. Melik-Shakhnazarov, V.A., Strelov, V.I., Sofiyanchuk, D.V., Bezbakh, I.Z.: New design of active vibroprotection devices. Techn. Phys. Lett. **38**(3), 283 (2012). https://doi.org/10.1134/S1063785012030303
15. William, T.T., Fenger, M.: Current signature analysis to detect induction motor faults. IEEE Ind. Appl. Mag. **7**, 23–29 (2000)

Indoor Greening for Volatile Organic Compounds Reduction

Ondřej Franek, Valeriya Frankova, Čeněk Jarský, and Igor Plotnikov

Abstract The article deals with the implementation of plants to reduce the concentration of volatile organic compounds (VOC) in the indoor environment of buildings, namely benzene, octane, toluene and trichlorethylene. Based on the simulation, which investigated the effect of 3 m^2 of implemented plants in an office space with a total air volume of 74.7 m^3, it was calculated that the plants have the potential to maintain benzene, toluene, octane and trichlorethylene concentrations and can replace part of the ventilated air. In the first state, the implemented greening was able to significantly reduce the requirement for the amount of ventilated air in the range of 96–100% because the implemented greenery is able to assimilate a similar amount of VOC as the expected imitation. In the second case, when the immediate reduction of the concentration to the level of 10% of the original value was required, the implemented greening did not significantly affect the amount of ventilated air, only in the range of 0.02–0.64%. However, when the required concentration corresponding to 10% of the initial value was reached, the plants again significantly reduced the requirement for ventilated air in the next 7 h of simulation, when they were able to assimilate similar amounts of VOC, octane, toluene and trichlorethylene exceeding the actual emission. Based on the simulation, the assumption was confirmed that the implementation of plants in the indoor environment can bring savings in the field of building ventilation, especially in the longer term, not with a short impact requirement for an immediate reduction in concentration. Based on the achieved results, it

---

O. Franek
Czech Technical University, Prague 160 00, Czech Republic
e-mail: ondrej.franek@fsv.cvut.cz

V. Frankova · Č. Jarský · I. Plotnikov (✉)
Tomsk Polytechnic University, 30, Lenin Avenue, Tomsk, Russia 634050
e-mail: igplot@tpu.ru

V. Frankova
e-mail: ondrej.franek@fsv.cvut.cz

Č. Jarský
e-mail: jarsky@fsv.cvut.cz

E. Lysenko et al. (eds.), *Recent Developments in the Field of Non-Destructive Testing, Safety and Materials Science*, Studies in Systems, Decision and Control 433,
https://doi.org/10.1007/978-3-030-99060-2_12

is appropriate to recommend the implementation of plants in the indoor environment for permanently reduction of the VOCs concentration.

1 Introduction

The indoor environment of the buildings starts to be more popular and interesting topic to discuss. Nowadays lots of people are working in big offices with harmful indoor environment. The present article that presents the results of authors research of such kind of buildings from the point of view of bad particles influence, another words called pollutants—volatile organic compounds (VOC) specifically benzene (CAS 71-43-2), trichloroethylene (CAS 79-01-6), toluene (CAS 108-88-3) and octane (CAS 111-65-9).

There are several ways how these particles can reach or get inside the building, for instance during the building process, interior preparation, normal daily operation, maintenance, cleaning or ventilation from outdoor environment [1–3]. Such compounds (VOC) have a very destructive effect to health of humans that are working or living in such kind of buildings. This fact that VOC has a tight connection with sick buildings syndrome (SBS) is already proven [4–7].

Buildings users that are staying in an indoor environment with high or below the limit concentration of VOC for a long time can see a negative influence on human's health, which can cause different chronic diseases [8–10]. High level of CO_2 and VOC have an essential influence on the building users work performance [11–13]. Benzene and trichlorethylene are classified as proven carcinogens of group 1 [14, 15], toluene and octane can cause nausea, fatigue, and hallucinations at higher concentrations, invade the mucous membranes. Increased concentration of CO_2 causes nausea and fatigue. The concentration of 0.1% of CO_2 decreases productivity in basic activities about 25% [11].

In commonly used administrative spaces, it is not assumed that mentioned pollutants in the indoor environment will be in such concentrations that are able to endanger the user's life. Thanks to the ventilation of the indoor environment, the solved pollutants are mostly maintained in permissible concentrations. The measurements showed that, the problem of VOC can touch new buildings or renovated buildings where the VOC concentration can be very high that makes buildings uninhabitable during the first half year [8]. Permissible concentrations of pollutants are varying according to local regulations at the legal national level. The concentration of pollutants must also be monitored regarding the performance of building users, where it is demonstrated that the work performance of building users increases significantly in an environment that meets the stricter quality criteria of the indoor environment [11, 12].

To ensure adequate VOC levels, buildings must get some air regularly, e. g. using air-conditioning equipment or by openable windows. It is important to increase the level of ventilation, especially in situations where the emulation of said high VOCs

Table 1 The ability of cultivars to reduce the observed VOC v $\mu g \cdot m^{-3} \cdot m^{-2} \cdot h^{-2}$ [26]

VOC	Cultivar		
	Hemigraphis alternata	*Hedera helix*	*Tradescania pallida*
Benzene	5.5	3.6	3.9
Toluene	9.6	8.3	9.1
Trichlorethylene	11.1	8.1	8.0
Octane	5.6	5.1	2.8

concentration, for example in the situation of inappropriately used construction products, the application of aggressive cleaning agents, etc. This results in an adverse effect on the overall operating costs of the construction, where the increased amount of supply air must be adjusted to the humidity and temperature required by the operation at the same time [16, 17]. Mainly in places with great temperature differences between the outdoor and indoor environment, this problem is very topical, as the potential and minimal operating savings can have a significant impact on the operating costs of the building.

The ventilation issue is also linked to the demand of sustainability, when developers are placing an increasing emphasis on energy efficient buildings while maintaining a high-quality indoor environment [16–21]. For this reason, it is necessary to find new ways to reduce pollutants in the indoor environment of buildings while maintaining the energy efficiency of buildings as much as possible. In recent years, several measurements show that the implementation of green plants in the indoor environment of buildings can make a positive contribution to the reduction of monitored pollutants [8, 22–25]. Selected research shows that selected plants reduce the solved VOC even in the indoor environment of buildings [26–29]. Table 1 shows selected plants and their ability to reduce the monitored VOCs.

Table 1 is representing the implementation of greening in the indoor space environment for reduction of concentration of selected VOC in addition of ventilation. It has a big potential to decrease the requirement for the total supplied air in the event of the current emission of the pollutants. The cultivar *Hemigraphis alternata* has the greatest ability to adopt all studied VOCs. The *Hedera helix* and *Tradescania pallida* cultivars are less efficient than *Hemigraphis alternata*, but they are also considered very efficient as houseplants [27].

2 Research Method

2.1 Methods Description

For the determination of the theoretical impact of the implemented green plants for indoor environment on the entire amount of supplied air from outdoor space, a basic model of an office room was defined. Before starting the simulation, all the internal parameters of the internal environment that are necessary to determine the ability to reduce VOC by the plant are defined. The methodology specifically examines the amount of total supplied air to the office room in order to reduce the concentration of selected VOCs to a predetermined level. Subsequently, based on the simulation, it determines the differences in the total requirement for the amount of supplied air to the indoor environment in the case when plants are not implemented in the office and in the case when plants are implemented in the office. The simulation clearly shows the differences between the individual situations and based on this, evaluations can be made in terms of the effectiveness of the implementation of plants for the reduction of individual VOCs.

2.2 Model Description

An office room with a total space of 74.7 m^3, a total area of 24.1 m^2 and a height of 3.1 m was selected as a model environment for which the development of the concentration of the investigated VOCs was observed. For simulation purposes, it is assumed that this room is forcibly ventilated by a central air conditioning system.

For the purposes of the simulation, the room considers the condition of the occupied office just after the construction, when a high VOC emulation from the built-in building elements is assumed.

In the office, the temperature of the indoor environment is considered to be stable, for the purposes of the model environment it is considered to be in the range of 21–23 °C, the relative humidity of the indoor environment is considered to be 55%. Atmospheric pressure is set at 101.3 kPa. The illumination of the implemented plants is considered to be a PPFD value of 200 $\mu mol \cdot m^2 \, s^{-1}$. The initial VOC concentration is determined as follows: benzene 3.0 $\mu g \cdot m^{-3}$, toluene 192 $\mu g \cdot m^{-3}$, trichlorethylene 55 $\mu g \cdot m^{-3}$, octane 3.0 $\mu g \cdot m^{-3}$. The following increase in VOC pollutants is also considered during the reduction: benzene 0.15 $\mu g \cdot m^{-3} \cdot h^{-1}$, toluene 9.6 $\mu g \cdot m^{-3} \cdot h^{-1}$, trichlorethylene 2.75 $\mu g \cdot m^{-3} \cdot h^{-1}$, octane 0.15 $\mu g \cdot m^{-3} \cdot h^{-1}$. The specific hourly values of VOC production are difficult to measure, and they are affected by the life cycle phase of the building, so they are set for simulation purposes as 5% of the original initial concentration. The supply air is considered to be filtered, considering a negligible amount of the concentration of the monitored VOCs, therefore these increments are neglected. Leaks and wear of the system, which could potentially impact the efficiency of the ventilation system, are also neglected.

The simulation assumes the implementation of green plants in an area of 3 m^2 in the composition *Hedera helix*, *Hemigraphis alternata*, and *Tradescania pallita*. Therefore, the ability of greenery as a whole to reduce selected pollutants was considered as follows: benzene 3.58 $\mu g \cdot m^{-3} \cdot m^{-2} \cdot h^{-1}$, toluene 9.00 $\mu g \cdot m^{-3} \cdot m^{-2} \cdot h^{-1}$, trichlorethylene 9.00 $\mu g \cdot m^{-3} \cdot m^{-2} \cdot h^{-1}$, octane 4.50 $\mu g \cdot m^{-3} \cdot m^{-2} \cdot h^{-1}$. A schematic representation of the model environment can be seen in Fig. 1.

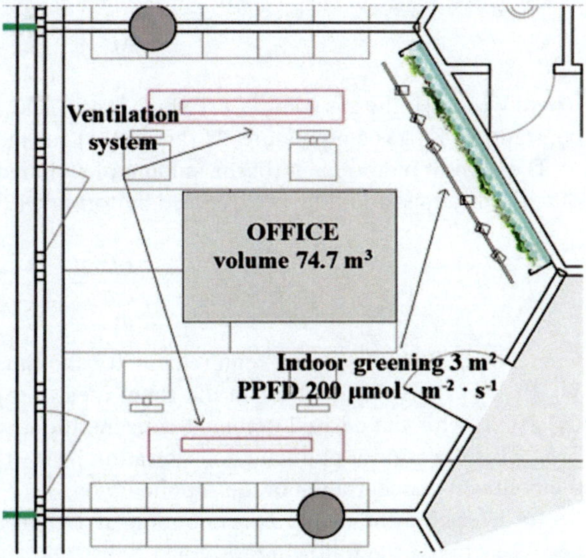

Fig. 1 Schematic representation of the model

It is considered to supply the air to the indoor environment using an air-conditioning distribution system, while for modeling purposes the residual VOC concentrations in the supply air are neglected due to filtration. The inclusion of higher concentrations of VOCs in the supply air could potentially occur in the real environment, especially in cases where the intake of air from the outside environment is close to traffic routes for cars with internal combustion engines, possibly it is located in polluted industrial zones and localities and at the same time in case of neglected maintenance of the filtration equipment in the air-conditioning units in cases where the air-conditioning systems do not have sufficient filtration systems [30, 31].

The simulation researches the amount of ventilated air over 1 h, which must be changed in indoor environment to achieve the required concentration. For the purposes of the simulation, 2 situations of the required concentrations were determined. The first state considers the required concentration as the default at the beginning of the simulation. The second state considers the required concentration significantly lower than the original state, namely the concentration of pollution corresponding to 10% of the original pollution: benzene 0.3 $\mu g \cdot m^{-3}$, toluene 19.2 $\mu g \cdot m^{-3}$, trichlorethylene 5.5 $\mu g \cdot m^{-3}$, octane 0.3 $\mu g \cdot m^{-3}$.

2.3 Computational Relations

To convert the volume production to the monitored pollutants by mass, it is necessary to start from (1):

$$V = \frac{m \cdot R \cdot T}{M \cdot p} \tag{1}$$

where V [dm^3] is the gas volume, m [g·dm^{-3}] gas mass, R [J·K^{-1}·mol^{-1}] universal gas constant, T [K] gas temperature, M [kg·mol^{-1}] gas molar mass, p [Pa] gas pressure.

The current requirement for the amount of supplied air to the space as a function of concentrations and time is expressed by formula (2):

$$Q_{sup} = \frac{V_{in}}{t} \cdot \left(\frac{C_{req} - C_{in}}{C_{out} - C_{in}} \right) \tag{2}$$

where Q_{sup} [m^3·h^{-1}] is the requirement for the amount of supply air for office, V_{in} [m^3] is the volume of air in the room area, t [h] is a defined period of time, C_{req} [g·m^{-3}] is the desired pollutant concentration, where $C_{req} \in (C_{out}, C_{in})$, C_{in} [g·m^{-3}] is the current pollutant concentration inside the room area, C_{out} [g·m^{-3}] is the pollutant concentration of the supplied air.

To determine the actual concentration of C_{in} in the indoor environment, it is necessary to use the following formula (3):

$$C_{in} = m_{ori} + m_{pro} - m_{pla} - m_{vent} \tag{3}$$

where C_{in} [g·m^{-3}] is the current pollutant concentration inside the room area, m_{ori} [g·m^{-3}] is the initial indoor pollutant mass, m_{pro} [g·m^{-3}] is the indoor pollutant mass excess due to emitation by built-in materials and operation, m_{pla} [g·m^{-3}] is the indoor pollutant mass loss due to plant reduction, m_{vent} [g·m^{-3}] is the indoor pollutant mass loss due to air ventilation.

3 Results and Considerations

3.1 Simulation Results

Based on the theoretical modeling, it was shown that plants can make a positive contribution to the reduction of pollutants in the indoor environment. The significant reducing effect of plants can be traced especially when they are planted in a polluted environment for a long time. While with the help of air-conditionig it is possible to perform a flush-out and immediately achieve high indoor air quality at the cost of increased operating costs, with the implementation of plants, a sudden increase in indoor air quality is difficult, given the green space that must be added to the indoor environment. Table 2 shows the results of the simulation showing the first condition under investigation, in which the requirement to maintain the VOC concentration at the level of the initial concentration was simulated. A graphical representation of the simulation results is given in Fig. 1. Simulation results that are listed in Table 3 show

Table 2 Requirement for rate of ventilation air flow for office—case 1 $C_{req} = C_{ori}$

VOC	$C_{req} = C_{ori}$ [$\mu g \cdot m^3$]	Rate of ventilation [$m^3 \cdot h^{-1}$]		Difference [%]
		Without indoor greening	With indoor greening	
Benzene	3.0	3.56	0.15	96
Octane	3.0	3.56	0.00	100
Trichlorethylene	55.0	0.20	0.00	100
Toluene	192.0	0.06	0.00	100

Table 3 Requirement for rate of ventilation air flow for office—case 2 $C_{req} = 10\% C_{ori}$

VOC	$C_{req} = 10\% C_{ori}$ [$\mu g \cdot m^3$]	Rate of ventilation [$m^3 \cdot h^{-1}$]		Difference [%]
		Without indoor greening	With indoor greening	
Benzene	0.3	67.59	67.25	0.50
Octane	0.3	67.59	67.15	0.65
Trichlorethylene	5.5	67.24	67.22	0.03
Toluene	19.2	67.25	67.20	0.07

the investigated second state, where the requirement to reduce the VOC concentration to a level equal to 10% of the original state was simulated.

From Table 2 and Fig. 2 can be seen that in the case where only the requirement to maintain the VOC concentration is set, indoor greenery is very effective. In the case of planting 3m^2 of greenery into the indoor environment, it can be assumed according to the model case to reduce the amount of supplied air by 96% to maintain the required initial benzene concentration. To maintain the required initial concentration of octane, trichlorethylene and toluene, the implemented greenery is fully sufficient, and it is not necessary to additionally ventilate the space from the point of view of VOC reduction.

From Table 3 and Fig. 3 one can see that in the case of an immediate reduction in VOC concentrations in the indoor environment, it is necessary to ventilate the entire area, while the implemented greenery has negligible efficiency in this short period of time, in the order of tenths of a percent. When implementing 3 m^2 of greenery into the indoor environment, if a 90% reduction in the concentration of pollutants is required in 1 h, the plants can influence the amount of supplied air by 0.50% in the case of benzene and by 0.65% for octane. Due to the reduction of toluene concentration, the amount of supplied air will be affected only by 0.07% and trichlorethylene by 0.03%. The results of the simulation confirmed the basic assumptions of the research that the implementation of plants in the indoor environment in order to reduce selected VOCs is important especially from a long-term perspective. With proper implementation, the ability of plants to maintain a stable state can be expected.

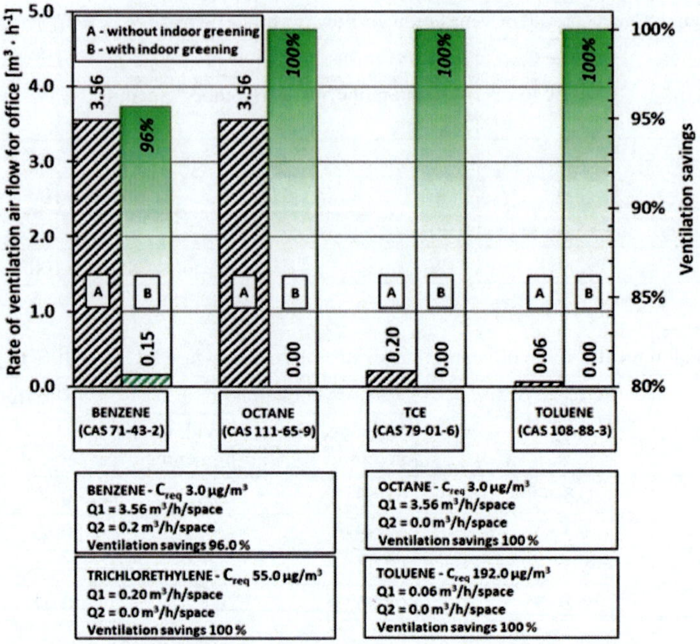

Fig. 2 Graphical representation of savings potential to maintain the required initial concentration of the internal environment of the model office—case 1 $C_{req} = C_{ori}$

Their ability to assimilate VOCs in time can compensate continuously emitted pollutants from the products that are built into the building in time and from the operation of the building itself.

3.2 *Shock Ventilation*

The results of the simulation confirmed the assumption that in the case of shock ventilation, the implemented greening has a negligible effect on the amount of ventilated air. In the case of a requirement for immediate quality assurance of the indoor environment, it is therefore necessary to ensure a so-called flush-out, which will ensure an immediate reduction of the concentrations of the examined VOCs in the indoor environment. However, after such ventilation, the implementation of plants can additionally contribute from the point of view of assimilation of newly emitted VOCs. This fact is shown in Fig. 3.

Figure 4 shows a situation where the space is ventilated during a 1 h period and further monitored for 7 h to maintain the desired concentration. It is clear from the simulation results that the implemented greenery significantly contributes to maintaining the required maximum permissible concentration in the indoor environment.

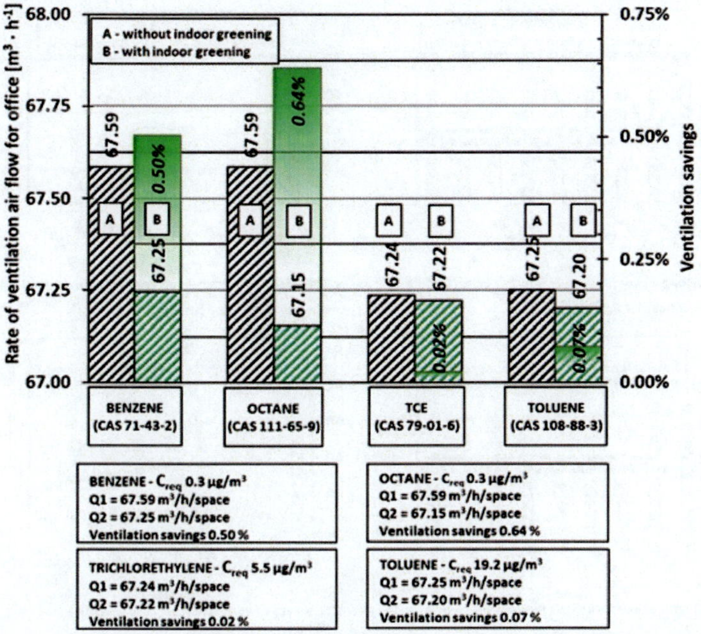

Fig. 3 Graphical representation of the savings potential for maintaining the required concentration of the internal environment of the model office—case 2 $C_{req} = 10\% \ C_{ori}$

To keep the benzene concentration at a maximum of 0.3 $\mu g{\cdot}m^{-3}$, it is necessary to ventilate at a rate of 24.9 $m^3{\cdot}h^{-1}$ for additional hours after additional ventilation, or in the case of implementing a 3 m^2 green wall, it is sufficient to ventilate at 1.5 $m^3{\cdot}h^{-1}$. To ensure a maximum octane concentration of 0.3 $\mu g{\cdot}m^{-3}$, it is necessary to further ventilate at a volume of 24.9 $m^3{\cdot}h^{-1}$ after shock ventilation, for a maximum toluene concentration of 19.2 $\mu g{\cdot}m^{-3}$ it is necessary to further ventilate at 0.6 $m^3{\cdot}h^{-1}$, for a maximum trichlorethylene concentration of 5.5 $\mu g{\cdot}m^{-3}$ continue to ventilate in the amount of 2.0 $m^3{\cdot}h^{-1}$. In the case of the implementation of plants with an area of 3 m^2, it is no longer necessary to ventilate the space after shock ventilation, as the plants assimilate a larger number of pollutants than the built-in material emits.

Figure 3 also shows an interesting trend, where thanks to the implementation of plants, a reduction in the overall requirement for the amount of ventilated air during working hours is achieved and at the same time the quality of the indoor environment is increased in the long run. This is due to the fact that the plant area is sufficient in the model and has the potential to assimilate more VOCs than assumed by imitation.

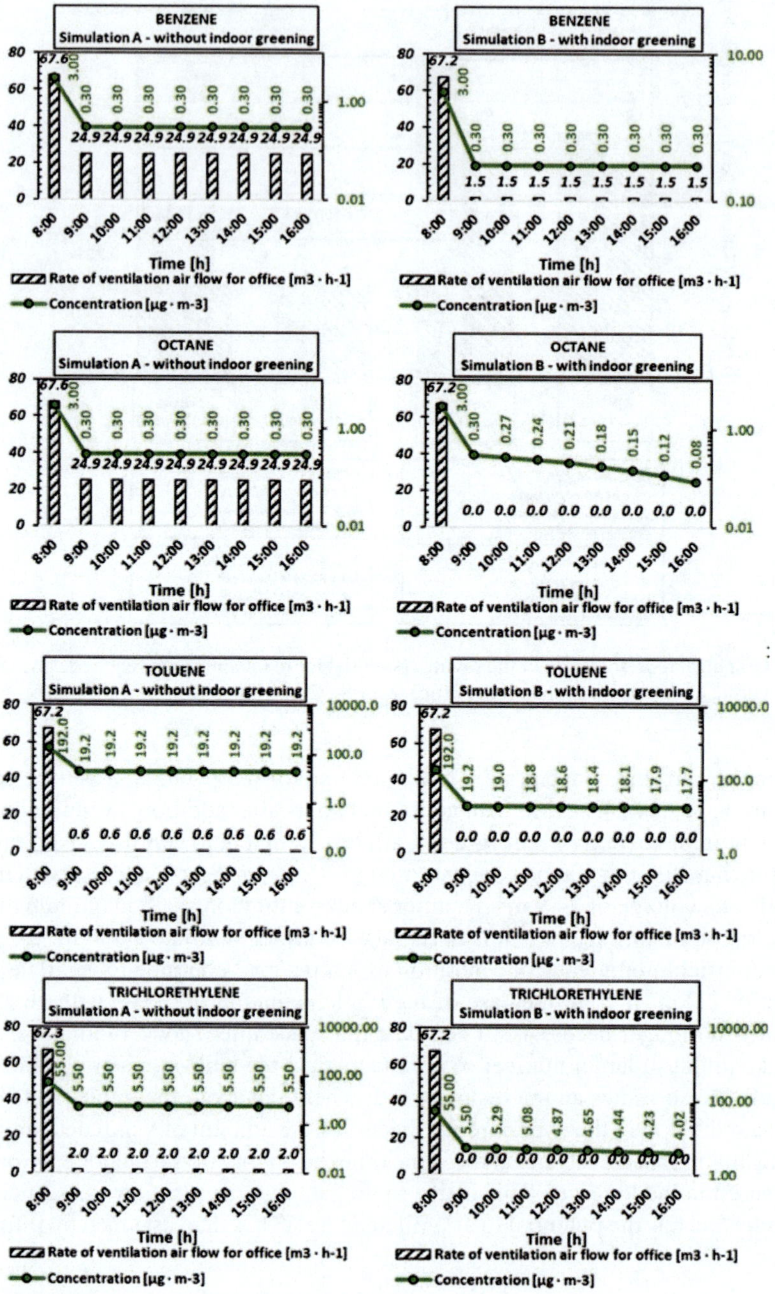

Fig. 4 Influence of implemented greenery on the amount of ventilated air after shock ventilation for the required concentration of max. 10% of the original pollution

4 Discussion

Based on the performed simulations, it was investigated that the implementation of greenery can significantly contribute to reducing VOC concentrations in the indoor environment of buildings. In the first part of the simulation, it was computationally found that the implemented greenery is able to maintain VOC concentrations in the indoor environment at a sustainable level. The environment without implemented greenery was compared with the environment with implemented greenery, while to maintain the VOC concentration in the environment without greenery it was necessary to ventilate air in the amount of 3.56 $m^3 \cdot h^{-1}$ for benzene and octane, 0.20 $m^3 \cdot h^{-1}$ for trichlorethylene and 0.06 $m^3 \cdot h^{-1}$ for toluene. To maintain the VOC concentration in the environment, it was necessary to ventilate the model space with implemented greenery in an area of 3 m^2 only in the case of benzene, in the amount of 0.15 $m^3 \cdot h^{-1}$. The remaining investigated VOCs, ie octane, trichlorethylene and toluene, were able to completely assimilate the implemented plants and it is therefore not necessary to additionally ventilate the space due to these VOCs. The implementation of greenery is sufficient to keep the concentration of monitored VOCs at the same level as the initial concentration.

In the second part of the simulation, the effect of the implemented greenery on the immediate provision of a reduction in the VOC concentration to the level of 10% of the original value within 1 h was investigated. In this case, the plants did not significantly contribute to reducing the concentration of VOCs studied. The environment without implemented greenery had to be ventilated in the amount of 67.59 $m^3 \cdot h^{-1}$ for benzene and octane, 67.24 $m^3 \cdot h^{-1}$ for trichlorethylene, 67.25 $m^3 \cdot h^{-1}$ for toluene. The environment with implemented greenery had to be ventilated in the amount of 67.25 $m^3 \cdot h^{-1}$ for benzene, 67.15 $m^3 \cdot h^{-1}$ for octane, 67.22 $m^3 \cdot h^{-1}$ for trichlorethylene and 67.20 $m^3 \cdot h^{-1}$ for toluene. These values theoretically confirm that in the short term, plants in the model case have a minimal ability to contribute to the reduction of pollutants, the potential to reduce the amount of air supplied for the investigated VOC is only 0.07–0.64%.

In this second part of the simulation, however, the assumption of long-term effect of plants was confirmed, when after burst ventilation during the first simulated hour, the plants significantly contributed in the following hours to ensure maintaining the state corresponding to 10% of the original state concentration of the monitored VOCs. While the environment without implemented greenery had to be ventilated at 24.9 $m^3 \cdot h^{-1}$ due to benzene and octane, 2.0 $m^3 \cdot h^{-1}$ due to trichlorethylene and 0.6 $m^3 \cdot h^{-1}$ due to toluene, the environment with implemented greenery had to be ventilated only at 1.5 $m^3 \cdot h^{-1}$ due to benzene. In the case of octane, trichlorethylene and toluene, after green ventilation, the implemented greenery was sufficient to maintain the required concentration. At the same time, due to the sufficient area of the implemented greenery, a gradual decreasing trend of octane concentrations was found, where during the simulation after impact ventilation the concentration reached over the next 7 h from 0.3 $\mu g \cdot m^{-3}$ to 0.08 $\mu g \cdot m^{-3}$, for toluene from 19.2 $\mu g \cdot m^{-3}$ to 17.7 $\mu g \cdot m^{-3}$ and for trichlorethylene from 5.5 $\mu g \cdot m^{-3}$ to 4.02 $\mu g \cdot m^{-3}$. In the environment without implemented greenery, the condition was maintained at the required concentration for another 7 h due to constant ventilation, namely in the case of octane 0.3 $\mu g \cdot m^{-3}$, toluene 19.2 $\mu g \cdot m^{-3}$, trichlorethylene 5.5 $\mu g \cdot m^{-3}$. Benzene was in the case with implemented greenery and in the case of environment without greenery in the next 7 h at a concentration of 0.3 $\mu g \cdot m^{-3}$.

From the performed calculations it is evident that the dominant effect of plants is observable mainly in the longer terms, not during the impact requirement for high indoor air quality and offers the possibility to combine this method with the method of impact ventilation. From the point of view of the implementation of this new method, it is also necessary to consider the degree of ventilation and the current occupancy of the building. It has long been known that the indoor environment of buildings, in which users stay, must be ventilated primarily due to the production of CO_2 by people. The concentration of CO_2 in the indoor environment of buildings occupied by users must be addressed in the first place, because even at concentrations above 1000 ppm a decrease in human cognitive functions can be observed. While due to VOC in normal operation, to maintain the current concentration, it is sufficient to change the amount of air in the order of m^3 per hour, in terms of CO_2 concentration, the amount of air must be changed in the order of tens of m^3 per hour. It follows that if the environment is forcibly ventilated by air conditioning due to CO_2, sustainable VOC concentrations can also be expected in the indoor environment.

5 Conclusion

The method of implementing green plants in the indoor environment of buildings offers a very interesting view of the possibilities of building sustainability. Based on the simulation, it was found that especially from the point of view of stable VOC concentration in the indoor environment, plants can be very useful in the long term. In the case of environments where forced ventilation is not provided for a long time or in unoccupied environments that are ventilated by a CO_2 sensor, significantly increased VOC concentrations may occur in the long term. If it is required to ensure a quality indoor environment in such areas, it is desirable that greenery be implemented that can ensure sustainable VOC concentrations.

On the contrary, based on the obtained results, plants cannot be recommended as effective VOC reducers in a purely short-term perspective. These are cases where it is required to immediately achieve a high-quality indoor environment at a high initial VOC concentration. Although it can be assumed that due to long-term exposure of plants in the contaminated environment, the rate of VOC emulation into the indoor air from embedded products would equal the rate of VOC assimilation by plants over time, such a fact can be expected areas implemented in green.

On the contrary, based on the results obtained, plants cannot be recommended as effective VOC reducers in a purely short-term perspective. These are cases where it is required to immediately achieve a high-quality indoor environment at a high initial VOC concentration. Although it can be assumed that due to long-term exposure of plants in the contaminated environment, the rate of VOC emulation into the indoor air from embedded products would equal the rate of VOC assimilation by plants over time, such a fact can be expected areas implemented in green. At the same time, the simulation shows that there is significant potential for the combination of shock ventilation with the implementation of greenery, especially in the period after the construction of the building, when increased VOC emissions from built-in products to the indoor environment can be expected. This combination of 2 methods, i.e. shock ventilation, which is often used as a flush-out, together with placing a sufficient area of greenery in the indoor environment can result in an immediate reduction of VOC concentration in the indoor environment while minimizing space ventilation costs, as the implemented greenery is already able to effectively maintain the concentrations of monitored VOCs at the required values.

Acknowledgements Present research was supported by the grant SGS21/007/OHK1/1T/11, Czech Technical University in Prague and by the research program Tomsk Polytechnic University.

References

1. Chen, S., Lipo, T.A.: Bearing currents and shaft voltages of an induction motor under hard- and soft-switching inverter excitation. IEEE Trans. Ind. Appl. **34**(5), 1042–1048 (1998)
2. Campagnolo, D., Dikaia, E.S., Andrea, C., et al.: VOCs and aldehydes source identification in European office buildings—the OFFICAIR study. Build. Environ. **115**, 18–24 (2017). https://doi.org/10.1016/j.buildenv.2017.01.009
3. Lee, C.S., Haghighat, F., Ghaly, W.S.: A study on VOC source and sink behavior in porous building materials—analytical model development and assessment. Indoor Air **15**(3), 183–196 (2005). https://doi.org/10.1111/j.1600-0668.2005.00335.x
4. Jia, C., Batterman, S., Godwin, C., Charles, S., Chin, J.Y.: Sources and migration of volatile organic compounds in mixed-use buildings. Indoor Air **20**(5), 357–369 (2010). https://doi.org/10.1111/j.1600-0668.2010.00643.x
5. Sun, Y., Hou, J., Cheng, R., Sheng, Y., Zhang, X., Sundell, J.: Indoor air quality, ventilation and their associations with sick building syndrome in Chinese homes. Energy Buil. **197**, 112–119 (2019). https://doi.org/10.1016/j.enbuild.2019.05.046
6. Sarkhosh, M., Najafpoor, A.A., Alidadi, H., Shamsara, J., Amiri, H., Andrea, T., Kariminejad, F.: Indoor air quality associations with sick building syndrome: an application of decision tree technology. Build. Environ. 107446 (2021). https://doi.org/10.1016/j.buildenv.2020.107446
7. Tsai, D., Lin, J., C., Chan: Office Workers' sick building syndrome and indoor carbon dioxide concentrations. J. Occup. Environ. Hyg. **9**(5), 345–351 (2012). https://doi.org/10.1080/15459624.2012.675291
8. Yu, C., Jeong, Tai K.: Building pathology, investigation of sick buildings–VOC emissions. Indoor Built Environ. **19**(1), 30–39 (2010). https://doi.org/10.1177/1420326X09358799
9. Pinto, M., lanzinha, J., Viegas, J., Infante, A., Freire, T.: Kindergartens and nurseries in central portugal. assessments of indoor environment quality. Int. J. Sustain. Build. Technol. Urban Dev. **12**(1), 27–43 (2021). https://doi.org/10.22712/susb.20210004
10. Seppanen, O., Fisk, W., Mendell, M.: Association of ventilation rates and CO_2 concentrations with health and other responses in commercial and institutional buildings. Indoor Air **9**(4), 226–252 (1999). https://doi.org/10.1111/j.1600-0668.1999.00003.x
11. Allen, J.G., Macnaughton, P., Satish, U., Santanam, S., Vallarino, J., Spengler, J.: Associations of cognitive function scores with carbon dioxide, ventilation, and volatile organic compound

exposures in office workers: a controlled exposure study of green and conventional office environments. Environ. Health Perspect. **124**(6), 805–812 (2016). https://doi.org/10.1289/ehp.1510037
12. Satish, U., Mendell, M.J., Shekhar, K., Hotchi, T., Sullivan, D., Streufert, S., Fisk, W.J.: Is CO_2 an indoor pollutant? Direct effects of low-to-moderate CO_2 concentrations on human decision-making performance. Environ. Health Perspect. **120**(12), 1671–1677 (2012). https://doi.org/10.1289/ehp.1104789
13. Kowalska, J., Szewczyńska, M., Pośniak, M.: Measurements of chlorinated volatile organic compounds emitted from office printers and photocopiers. Environ. Sci. Pollut. Res. **22**(7), 5241–5252 (2015). https://doi.org/10.1007/s11356-014-3672-3
14. Agents Classified by the IARC Monographs. Volumes 1–123. International Agency for Research on Cancer, Lyon (2018)
15. ARC Monographs—Trichloroethylene, tetrachloroethylene, and some other chlorinated agents. Volume 106. International Agency for Research on Cancer, Lyon (2014)
16. Ürge-vorsatz, D., Cabeza, L.F., Serrano, S., Barreneche, C., Petrichenko, K.: Heating and cooling energy trends and drivers in buildings. Renew. Sustain. Energy Rev. **41**, 85–98 (2015). https://doi.org/10.1016/j.rser.2014.08.039
17. Serrano, S., Ürge-vorsatz, D., Barreneche, C., Palacios, A., Cabeza, L.F.: Heating and cooling energy trends and drivers in Europe. Energy **119**, 425–434 (2017). https://doi.org/10.1016/j.energy.2016.12.080
18. Franek, O., Jarský, Č: On reducing CO_2 concentration in buildings by using plants. Acta Polytechnica **61**(5), 617–623 (2021). https://doi.org/10.14311/AP.2021.61.0617
19. Liu, J., Heidarinejad, M., Nikkho, S., Mattise, N.W., Srebric, J.: Quantifying impacts of urban microclimate on a building energy consumption—a case study. Sustainability **11**(18), (2019). https://doi.org/10.3390/su11184921
20. Tuháček, M., Franek, O., Svoboda, P.: Application of FMEA methodology for checking of construction's project documentation and determination of the most risk areas. Acta Polytech. **60**(5), 448–454 (2020). https://doi.org/10.14311/AP.2020.60.0448
21. Majewski, G., Orman, ŁJ., Telejko, M., Radek, N., Pietraszek, J., Dudek, A.: Assessment of thermal comfort in the intelligent buildings in view of providing high quality indoor environment. Energies **13**(8), 1973 (2020). https://doi.org/10.3390/en13081973
22. Irga, P.J., T., Pettit, Irga, R.F., Paull, N.J., Douglas, A.N.J., Torpy, F.R.: Does plant species selection in functional active green walls influence VOC phytoremediation efficiency? Environ. Sci. Pollut. Res. **26**(13), 12851–12858 (2019). https://doi.org/10.1007/s11356-019-04719-9
23. Franek, O., Jarský. Č.: On implementation of plants in the indoor environment in intelligent buildings. Environ. Res. Eng. Manag. **77**(3), 66–73. (2021). https://doi.org/10.5755/j01.erem.77.3.28928
24. Obuhov S., Plotnikov I., Surkov M., Sumarokova L., Savkin K.: Development of the structural and functional design of the laboratory bench for experimental research diesel generator sets on variable speed. MATEC Web Conf. **91**, 01036 (2017). https://doi.org/10.1051/matecconf/20179101036
25. Shao, Y., Li, J., Zhou, Z., Hu, Z., Zhang, F., Cui, Y., Chen, H.: The effects of vertical farming on indoor carbon dioxide concentration and fresh air energy consumption in office buildings. Build. Environ. **195**, 107766. (2021). https://doi.org/10.1016/j.buildenv.2021.107766
26. Cruz, M.D., Christensen, J.H., Thomsen, J.D., Müller, R.: Can ornamental potted plants remove volatile organic compounds from indoor air?—A review. Environ. Sci. Pollut. Res. **21**(24), 13909–13928 (2014). https://doi.org/10.1007/s11356-014-3240-x
27. Yang, D.S., Pennisi, S.V., Son, K., Kays, S.J.: Screening indoor plants for volatile organic pollutant removal efficiency. HortScience **44**(5), 1377–1381 (2009). https://doi.org/10.21273/HORTSCI.44.5.1377
28. Mikkonen, A., Li, T., Vesala, M., et al.: Biofiltration of airborne VOCs with green wall systems-microbial and chemical dynamics. Indoor Air **28**(5), 697–707 (2018). https://doi.org/10.1111/ina.12473

29. Brilli, F., Fares, S., Ghirardo, A., et al.: Plants for sustainable improvement of indoor air quality. Trends Plant Sci. **23**(6), 507–512 (2018). https://doi.org/10.1016/j.tplants.2018.03.004
30. Polidori, A., Fine, P.M., White, V., Kwon, P.S.: Pilot study of high-performance air filtration for classroom applications. Indoor Air **23**(3), 185–195 (2013). https://doi.org/10.1111/ina.12013
31. Chan, W.R., Xiwang, L., Singer, B.C., Pistochini, T., Vernon, D., Outcault S., Sanguinetti A., Modera M.: Ventilation rates in California classrooms: why many recent HVAC retrofits are not delivering sufficient ventilation. Build. Environ. **167**, 106426. (2020). https://doi.org/10.1016/j.buildenv.2019.106426

Analysis of Factors Affecting the Performance of the Business Process Based on Statistical Analysis

Ludmila Redko, Aleksandra Bykova, Inna Plotnikova, Elena Sheveleva, and Marina Yanushevskaya

Abstract The study is relevant due to the need to improve the process performance. The study object is data on the process of the cycle of well construction in oil production. The process duration is analyzed as a performance parameter. The aim of the study was to develop proposals to improve the performance of the well construction cycle process based on statistical analysis. The study employed such statistical methods as descriptive statistics, analysis of variance, correlation analysis, Shewhart charts, and analysis of process capabilities. In the study, the process stability was analyzed, factors affecting the process parameters were identified, and the analysis of the data variability with regard to the identified factors was performed. As a result of the study, factors affecting the process duration were identified, and recommendations for improving the process were developed.

1 Introduction

The world is moving towards green technologies, however, extraction of hydrocarbons is an important sector of the economy since it supplies raw materials for chemical products.

L. Redko (✉) · A. Bykova · I. Plotnikova · E. Sheveleva
Division for Testing and Diagnostics, National Tomsk Polytechnic University, 30, Lenin Avenue, Tomsk 634050, Russia
e-mail: laredko@tpu.ru

I. Plotnikova
e-mail: inna@tpu.ru

E. Sheveleva
e-mail: vasendina@tpu.ru

M. Yanushevskaya
Faculty of Innovation Technologies, Tomsk State University of Control Systems and Radioelectronics, 74, Vershinina Avenue, Tomsk 634058, Russia
e-mail: vela2007@bk.ru

E. Lysenko et al. (eds.), *Recent Developments in the Field of Non-Destructive Testing, Safety and Materials Science*, Studies in Systems, Decision and Control 433,
https://doi.org/10.1007/978-3-030-99060-2_13

Polymers produced from hydrocarbons are the main material used for production of consumer goods and packaging. Consumers are interested in the lowest possible cost of fuel, electronics products, packaging and other goods, which contain petroleum products as an essential component. Similar to any other economy sector, hydrocarbon production companies struggle to be efficient. These companies study the processes and seek for opportunities to eliminate losses and increase process performance. To attain the goal, statistical methods of analysis are employed [1].

Statistical methods of the analysis of process parameters help, firstly, study the process and obtain data on the process "behavior" in various conditions, secondly, visualize data to observe patterns that are poorly comprehensible from tables of numbers, and thirdly, structure analysis and make decisions based on facts rather than intuition [2, 3].

The paper deals with the process of well construction in oil production. The well construction cycle is a complex multi-stage process that requires techniques to identify losses and reduce the total drilling time. The well construction process, also referred to as the well construction cycle, includes the following operations:

1. Preparatory work for construction—preparation of the area for drilling, preparation of the site, equipping with communications, delivery of equipment.
2. Derrick installation—construction or dragging of a tower, installation of drilling equipment on foundations, and installation of a circulation system, equipment for the wellhead of the future well.
3. Preparatory work for well drilling—testing of the installed equipment, delivery of tools and materials, preparation of drilling tools for work, preparation or delivery of drilling mud.
4. Well drilling—hole-making works.
5. Well lining and formation segregation—preparation of the wellbore, casing and equipment for lowering of casing string, lowering and cementing of casing string, quality control of cementing and tightness of casing string.
6. Sampling from reservoirs and testing the well for oil and gas inflow.
7. Teardown of drilling equipment, derricks and near-derrick structures, and their preparation for transportation to a new site.
8. Transportation of the dismantled equipment and property to a new site.
9. Cleanup of the territory and carrying out environmental protection measures.

The study object is data on the well construction cycle. The study subject is time characteristics of the process (for example, well drilling rate).

The study analyzed the amount of data on the well construction cycle for the period from 2015 to 2020. The data array contained a large amount of information.

(1) Basic information on the well

This information includes its purpose. There are injection and oil wells. The injection well is used for pumping water, gas, heat carriers, etc. into reservoirs.

Oil well is constructed for oil production.

Nine deposits and reservoirs were considered: Jurassic and Paleozoic. By type, wells are divided into horizontal wells (HW), a well with an angle of deviation from the wellbore vertical of 80–90° and directional wells (DW) with greater horizontal deviations, an angle of deviation from the vertical of more than 5°. Well characteristics also include well pad and well number, and drilling year and season.

(2) Data on depth and rate of drilling:

- drilled depth, actual and planned, m;
- vertical depth, m;
- total interval and effective drilling, m;
- deviation from the vertical, m;
- overall rate (meters per month);
- drilling rate, day/1000 m.

(3) Data on equipment and contractors:

- type of drilling rigs with a serial number;
- bottom-hole motors, pumps, cleaning system;
- drilling bits and their contractors;
- supervising contractor;
- drilling contractor;
- directional drilling contractor;
- mortar contractor;
- contractor for geological and technological research;
- cementing contractor;
- contractor for geological and physical works.

(4) Cycle duration.

Cycle duration (total and planned) include production time (PT) and non-production time (NPT). The production time of the well construction cycle includes the following operations: drilling; derrick installation and commissioning works; geophysical work; well lining; buildup; preparation/completion works; waiting-on-cement time; sealing with blowout preventers; round trips/bottom hole assembly; circulation; reaming.

The information provided for the analysis included 253 wells with only 43 horizontal wells. When segmenting the data on the field and reservoir, it was revealed that 58 existing wells belong to the Paleozoic, and 198 belong to the Jurassic. In total, 6 types of installations were used during the analyzed period, of which type number 1 is most commonly used (187 wells). The analysis of the names of equipment and tools was complicated due to the variety per well; these were drilling bits, bottom-hole motors, pumps, cleaning systems.

2 Research Method

For statistical analysis, the following tasks should be performed:

- determination of measures of the central trend, statistical characteristics, average time of work by type, by quarters, by fields;
- determination of the drilling time for wells of different types by years;
- study of the dependence of the work execution time on the field by era (Jurassic and Paleozoic periods), on season and crew;
- determination of changes in the operation variability;
- testing the hypothesis about the existence of the "mud" season, which may affect the total time of losses.

Methods to solve the assigned tasks:

1. Descriptive statistics—summarizes and organizes data.
2. Analysis of variance—identifies significant factors affecting the variables.
3. Correlation analysis—reveals the relationship between variables.
4. Control charts—evaluate process stability.

1. Descriptive statistics is a tool to present quantitative data in a manner that helps to characterize data distribution. This method generalizes the primary results obtained by observation or experiment. All calculations of descriptive statistics imply grouping data according to their values, plotting the distribution of their frequencies, identifying central distribution tendencies, and, finally, assessing data spread in relation to the identified central tendency. The method is used for general review and data description. It is typically the starting point for analyzing quantitative data and using other statistical procedures. The characteristics of the sampled data can be the basis for drawing conclusions about characteristics of the entire array of data with given levels of confidence and error [4].
2. Correlation analysis is a method of statistical data processing, which implies studying the correlation coefficients between variables. The analysis involves comparison of the correlation coefficients between one pair or multiple pairs of features to establish their statistical relationships. Correlation is a statistical relationship between two variables (quantitative or ordinal), which shows that a larger value of one quantity in some cases corresponds to a greater (in the case of positive direct correlation) or a smaller (in the case of negative inverse correlation) value of another quantity [5].
3. Analysis of variance (ANOVA) is used to investigate the effect of one or more qualitative variables on one quantitative dependent variable. The analysis is based on the assumption that some variables can be considered as causes (factors, independent variables), and others are referred to as consequences (dependent variables). Independent variables are sometimes called adjustable factors since in the experiment the researcher has the ability to vary them and analyze the result. The main purpose of ANOVA is determination of the statistical significance of the difference between means (for groups or variables). To do this, the sum of squares is divided into components, i.e. the total variance is divided into parts, one of which is due to a random error, i.e. intragroup variability, and the second one is associated with the difference in mean values. The last variance component is then used to analyze the statistical significance of the difference between the mean values.

4. Shewhart charts are used to identify unnatural changes in data from repetitive processes and provide criteria for detecting lack of statistical control. A process is in a statistically controlled state if the variability is due to random causes [6, 7]. During determination of this acceptable level of variability, any deviation from it is considered to be the result of specific causes that should be identified, eliminated or mitigated. The main criterion for checking for special causes is the point that goes beyond the control limits. Any structure of points that can indicate the manifestation of specific (non-random) causes should be considered. The study employs the charts of individual values considered along with the moving range charts. A moving range is understood as the difference between the first and second dimensions, etc.

3 Description of the Installation

3.1 Statistical Analysis of the Well Construction Process by the Parameter "production Time"

First, an individual/moving-range chart should be constructed for the entire analyzed period (2015–2020) in order to see the variability of the process and assess its stability, i.e. predictability for the analyzed parameter—production time (Fig. 1). Eleven points go beyond the control limits, and this indicates process instability; it is affected by a large number of random factors to be studied. When considering data on wells corresponding to these points, only four of them show the production time less than 80% of the total construction time.

The factors that can affect this process include deposit, reservoir, season, quarter, well type, and crew. Analyze the effect of these factors.

Consider changing production times by quarter and season. The first quarter includes January, February and March. The second quarter includes April, May and June. The third quarter includes July, August and September. The fourth quarter includes October, November and December. Seasons are divided into three types: summer (June–September), winter (November-March) and the "mud" season (April, May, October).

The charts Fig. 2a show no significant differences by quarters. The means are located at approximately the same level and are not observed for more than 40 days, which indicates that wells with this and longer time were evenly distributed throughout the quarters. The seasonal chart shows a similar pattern Fig. 2b. It can also be noted that the interval for the "mud" season is less than that for other seasons, as well as the second quarter of months, which partially overlap.

The analyzed data contains two types of wells: horizontal (HW) and directional wells (DW). Horizontal wells exhibit higher values; this is due to the complexity of the drilling technology at an angle close to 90°. Directional wells show less variability in the production time Fig. 2c.

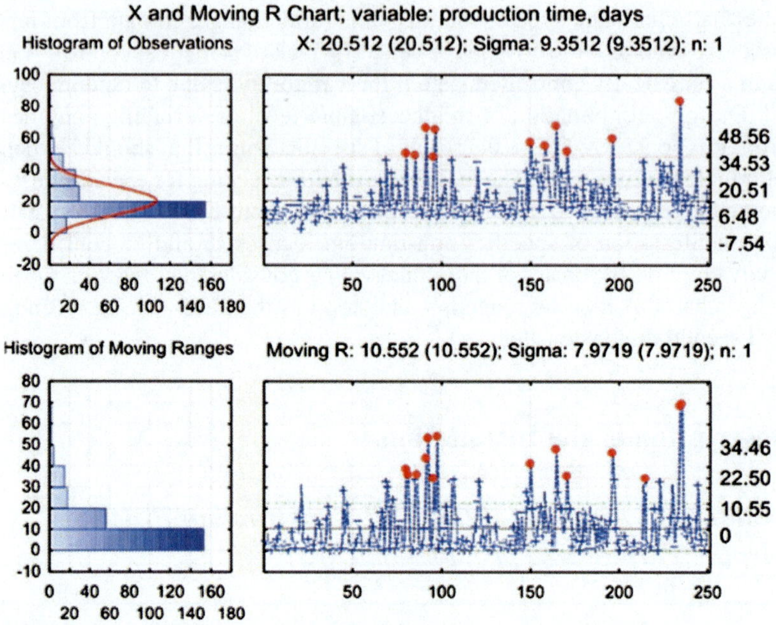

Fig. 1 Shewhart chart for the process parameter "production time"

The maximum production time falls on field 8, well construction on this site shows the minimum time of 55.6 days with a total penetration of 4000 m. Most of the wells fall on fields 1 and 5, the mean values of production time at which are 15 and 19 days, respectively. The most insignificant variation in the parameter is observed for fields 3 and 4; on these fields, the production time has the minimum variance and mean values Fig. 2d.

The distribution of the variable across the reservoir is similar to the distribution of samples by well type Fig. 2e. This can be due the fact that the Paleozoic reservoir accounts for most of the horizontal wells, for which, as shown earlier, the average production time is longer. Throughout the entire period (2015–2020), there have been four drilling contractors Fig. 2f. The Fig. 2 shows the means of samples of the analyzed parameter for these companies. The largest interval of the parameter can be seen for company 3. This company completed 23 wells available in the data array. Less variable crews correspond to smaller means and medians among the samples, and, within the minimums, they show the smallest values of the population.

Statistical significance of the factors was determined by one-way ANOVA Table 1. ANOVA showed that factors such as "quarter" and "season" do not affect the analyzed parameter; the p value exceeds the confidence level (0.05). The "well type", "field", "reservoir" and "drilling contractor" factors affect the duration of the well construction process, namely, the production time.

ANOVA revealed the interaction of the factors "well type" and "reservoir". Thus, the production time value changes unevenly when these factors change.

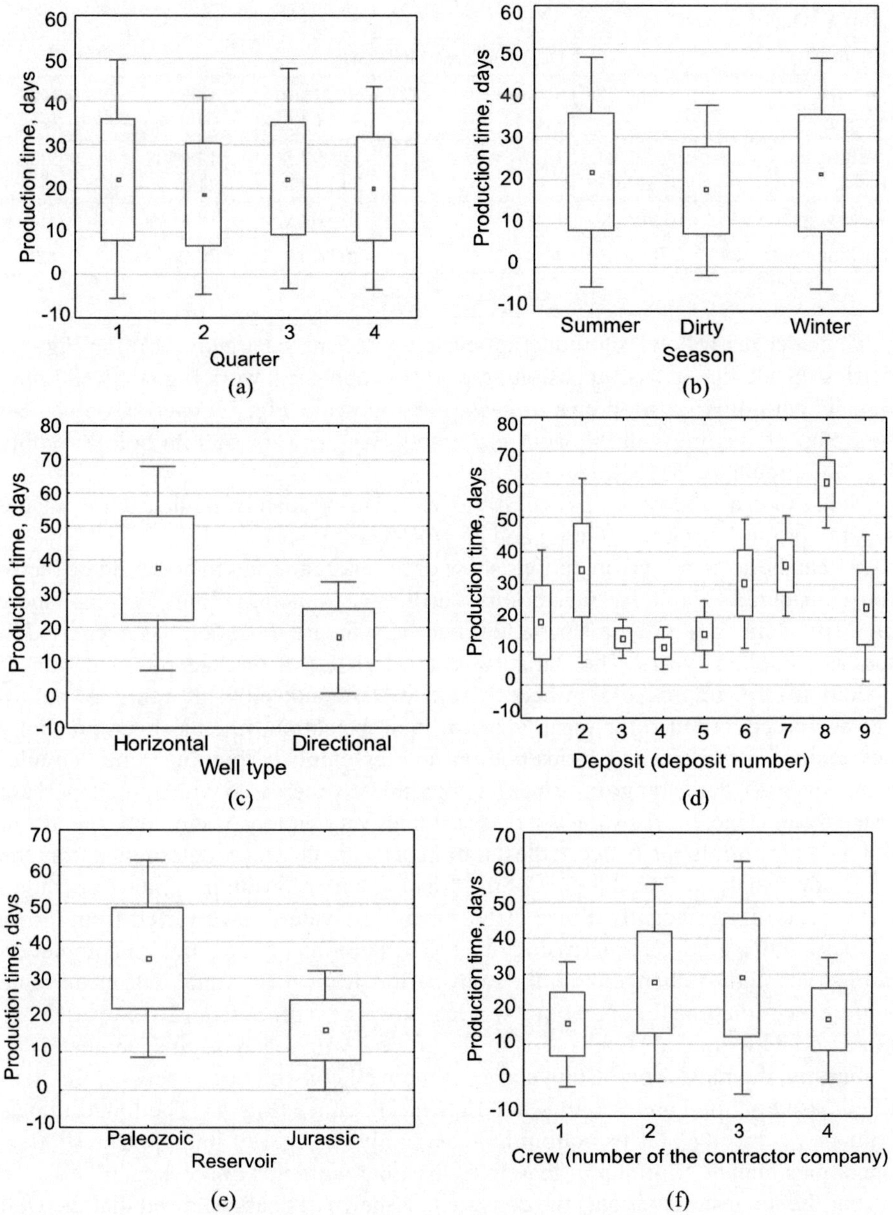

Fig. 2 Box-and-whisker charts to assess the variation due to the effect of factors: **a** Quarter, **b** Season (summer, winter, mud), **c** Well type (horizontal, directional), **d** Deposit (deposit number), **e** Reservoir (Jurassic, Paleozoic), **f** Crew (number of the contractor company)

Table 1 One-way ANOVA

Effect	SS	Degree of freedom	MS	F	P
Quarter	656.72	3	218.9	1.3701	0.2524
Season		410.1	2.5880	0.077203	820.2
Well type	14,992.8	1	14,992.8	148.588	0.00
Field	20,391.13	8	246.195	31.2695	0.00
Reservoir	16,908.2	1	16,908.2	181.399	0.00
Drilling contractor	7866.06	3	2622.02	20.0810	0.0000

Consider the following production cycle operations separately: drilling Fig. 3a; derrick installation and commissioning works; geophysical work Fig. 3b; well lining Fig. 3c; build-up Fig. 3d; preparation/completion works Fig. 3e; waiting-on-cement time Fig. 3f; sealing with blowout preventers Fig. 3g; trips/bottom hole assembly Fig. 3h; circulation Fig. 3j; reaming Fig. 3k.

Study the variability of operations by years. Unfortunately, no data are available separately for derrick installation and commissioning works.

As can be seen, all the operations show different trends and they should be therefore considered separately. "Operations" "drilling", "well lining" and "buildup" show no distinct changes in the mean values, but the standard deviation has decreased in the last couple of years. The statistical characteristics of the "geophysical works" operation have improved significantly. Since 2016, not only the range of values has decreased, but also the mean value of their execution time has been gradually decreasing. For operations "trips/bottom hole assembly", "buildup" and "circulation", since 2015, the range of values has significantly increased with a small increase in the mean value, i.e. the values in these samples vary greatly. It should be noted that a significant correlation between the duration of these stages was determined from the correlation analysis. The duration of the "sealing with blowout preventer" operation has increased significantly. Since 2017, mean time values have varied from 0.6 to 0.8 days with a gradually increasing range. The duration of "preparation/completion works" has slightly changed over the study period by the mean value. The mean value of the execution time for operation "waiting-on-cement time" from 2015 to 2017 has increased with gradual stabilization of the process with regard to this parameter.

Further, the most long-lasting stages of the well construction process were identified based on the Pareto chart plotted for mean values (Fig. 4). The longest operations are "drilling", "trips/bottom hole assembly" and "well lining". The shortest operations include "buildup", "reaming", "sealing with blowout preventer".

For the longest operations, the constructed Shewhart charts showed that the well construction process for the indicated operations is not stable in terms of the time parameter Fig. 5.

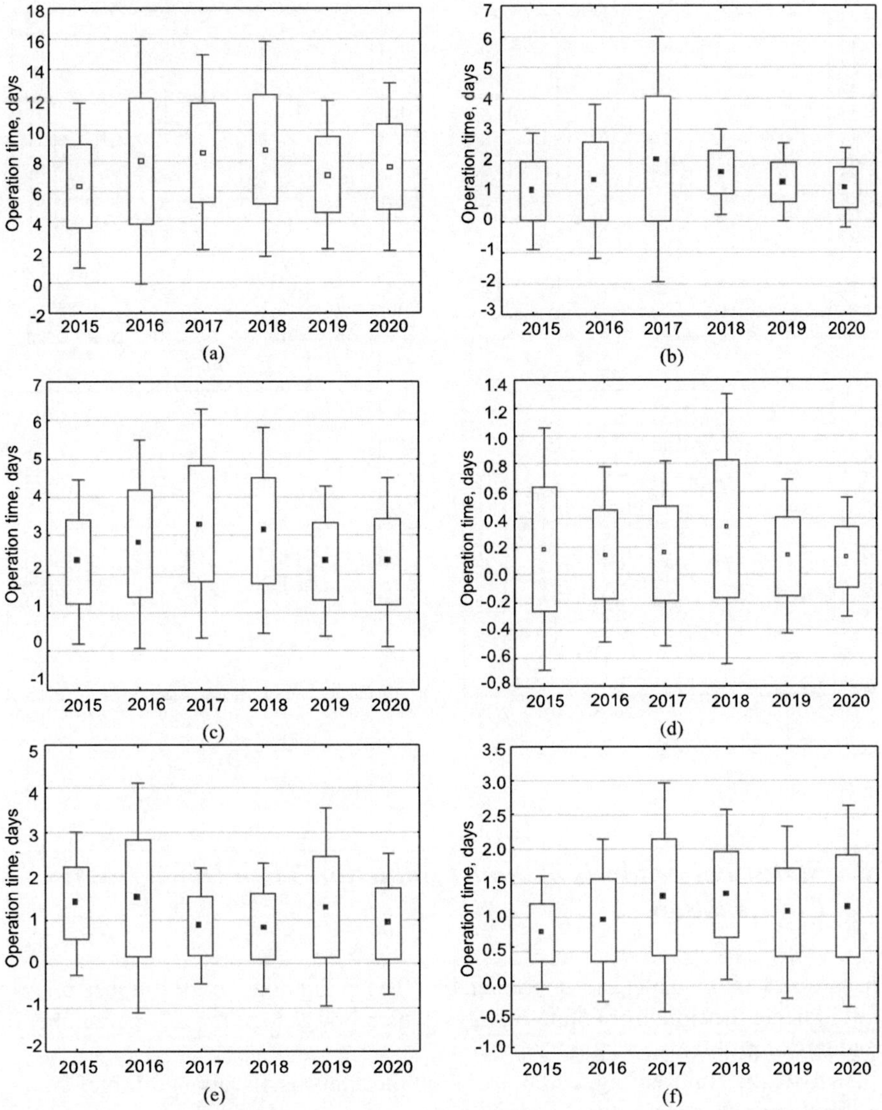

Fig. 3 Production cycle operations: **a** drilling, **b** geophysical work, **c** well lining, **d** buildup, **e** preparation/completion works, **f** waiting-on-cement time, **g** sealing with blowout preventer, **h** trips/bottom hole assembly, **j** circulation, **k** reaming

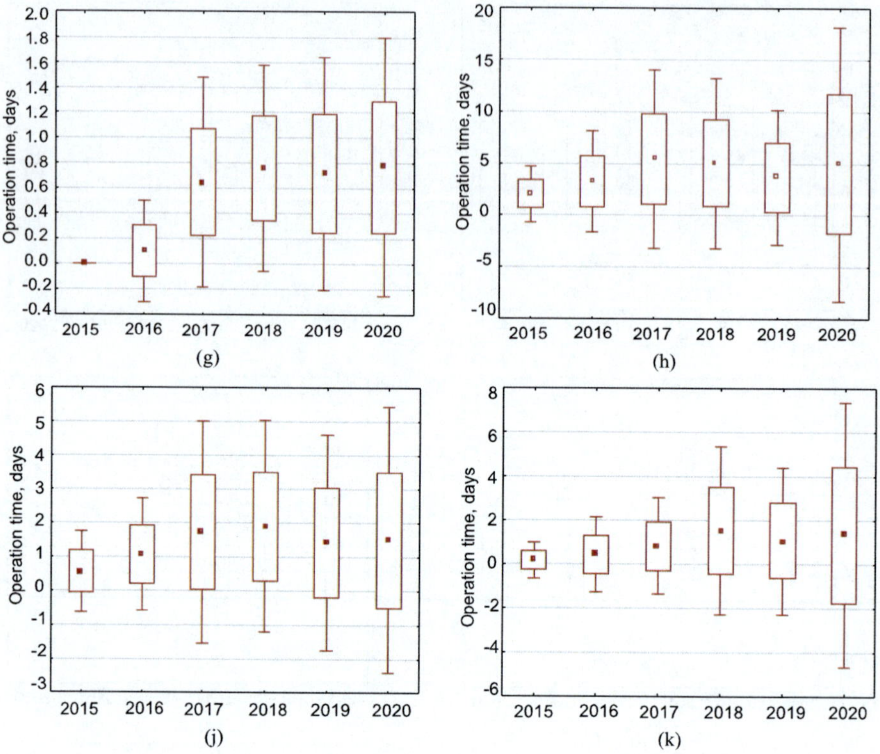

Fig. 3 (continued)

3.2 *Statistical Analysis of Non-Production Time During Well Construction*

The process is unstable and unpredictable. The points beyond the upper control limit indicate the time from 30 to 50 days, which brings together a large number of accidents or problems.

Analyze the contributing factors based on the analysis of variance Table 2.

The factor "season" does not exhibit statistical evidence for non-production time. Other factors affect non-production time: "field", "well type" and "drilling contractor".

According to the results of the correlation analysis, there is no relationship between the duration of the non-production time components. A moderate correlation is observed between the time due to accidents and the repair.

To determine the longest periods in non-production time, the Pareto chart was constructed Fig. 6. The chart shows that repairs, accidents and downtime contribute significantly to the total non-production time.

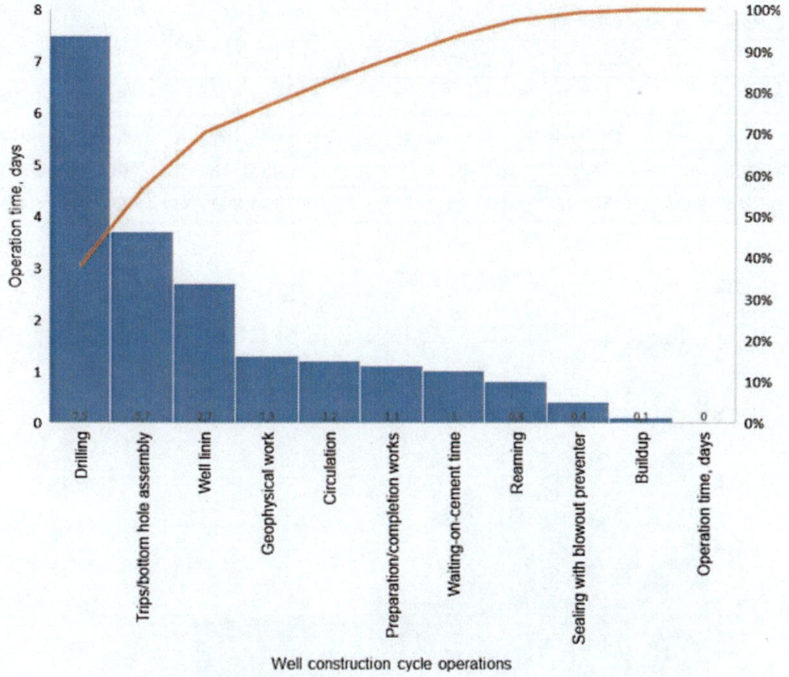

Fig. 4 Pareto charts plotted by operation duration

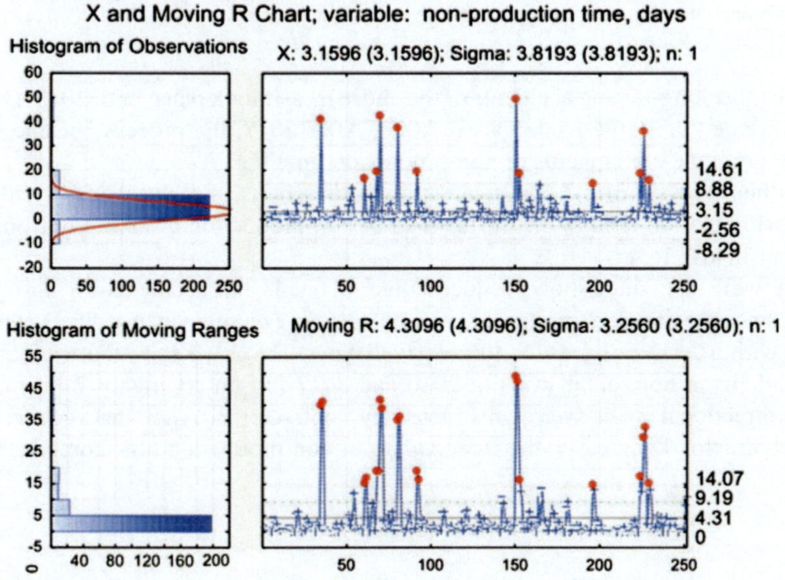

Fig. 5 Shewhart charts for non-production time

Table 2 One-way ANOVA

Effect	SS	Degree of freedom	MS	F	P
Season	127.30	2	63.652	1.46454	0.2331
Field	1969.564	8	246.195	6.66706	0.0000
Well type	483.32	1	483.318	11.54665	0.0079
Drilling contractor	587.67	3	195.891	4.68928	0.0033

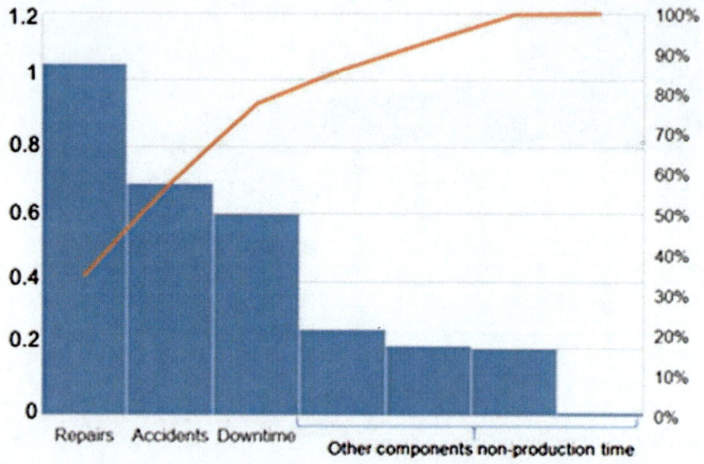

Fig. 6 Pareto chart for non-production time by arithmetic means

The most long-lasting accidents of the entire massif took place until 2018. Downtime revealed the largest intervals in 2020. After 2017, the process became more stable for all the components of non-production time Fig. 7.

Further analysis of the components using the charts of mean values yields the conclusion that on field 8, the mean value of non-production time differs from that on other fields Fig. 8.

The well type chart shows reduced time of repair and accidents for horizontal wells, since repairs are a constant non-production time component Fig. 8b. Downtime varies within the same limits for different well types. The charts show that the majority of accidents at horizontal wells in 2016 and 2017 fall on contractor 2 Fig. 8c. In 2017, directional wells were constructed by contractor 3. After that, cooperation with contractor 4 decreased the mean values of non-production time components.

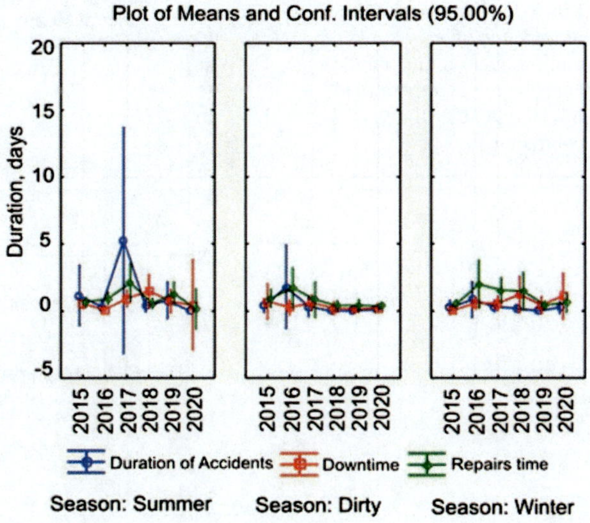

Fig. 7 Mean value of the non-production time components by years and seasons

3.3 *Statistical Analysis of the Drilling Rate During Well Construction*

Consider the drilling rate variation using the Shewhart individual/moving-range chart Fig. 9.

The chart shows that eight points go beyond the upper control limit, therefore, the process is unstable and uncontrollable, and the value of the analyzed parameter cannot be predicted.

Consider the variability of the drilling rate mean value by years Fig. 10. In 2016, the largest range observed for the variable is followed by a decline in the mean value. In the last year, the interval increased due to one well with a relatively high drilling rate (20.7 d/1000 m).

ANOVA Table 3 shows that the change in the drilling rate is affected by a number of factors such as "drilling contractor", "field", and "well type". The "season" factor does not affect the drilling rate.

The variability of the rate depending on the crew is similar to the drilling operation as they are directly related. For fields, the discrepancies are represented by the increased interval and accordingly the variance. The variability depending on the well type corresponds to the production time, since the rate is calculated from the depth and time of drilling; a horizontal well exhibits a higher rate.

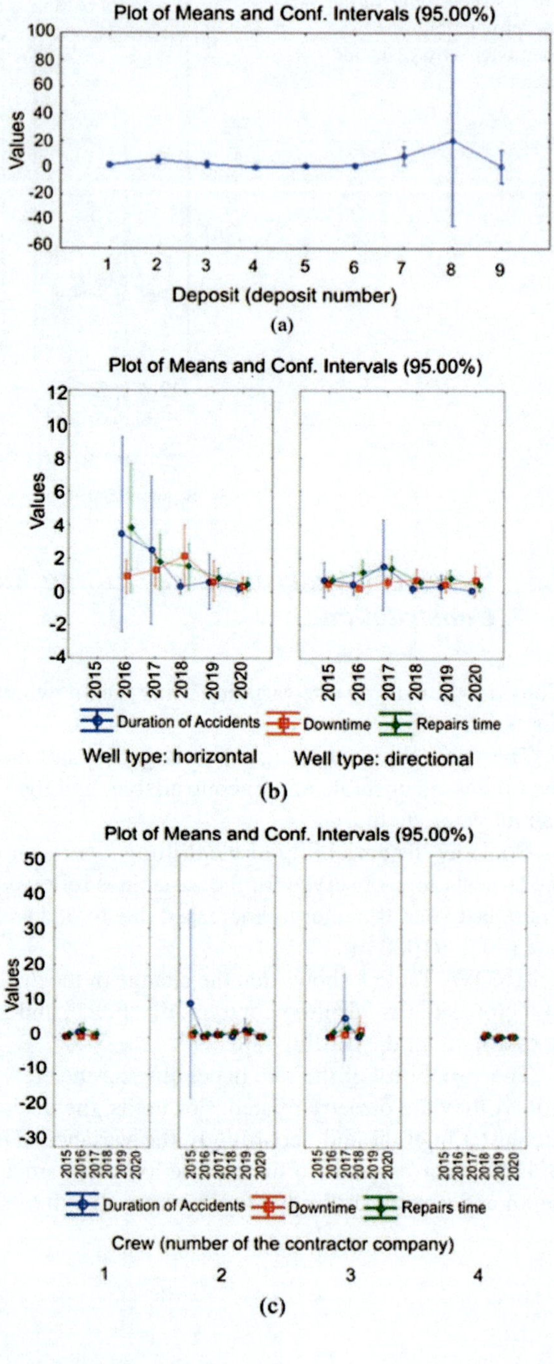

Fig. 8 Dependence of the mean value of non-production time on the contributing factors: **a** field, **b** well type, **c** drilling contractor

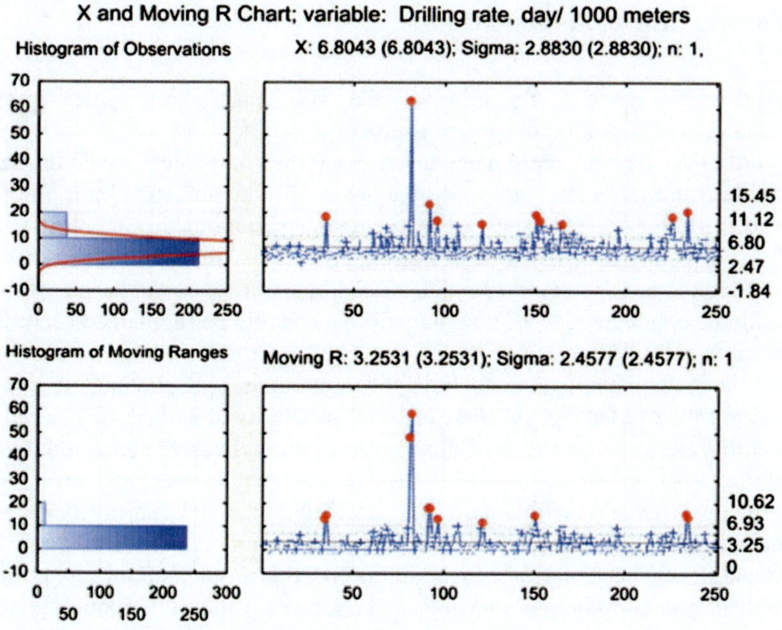

Fig. 9 Shewhart charts for the drilling rate

Fig. 10 Mean and confidence intervals for the drilling rate by years

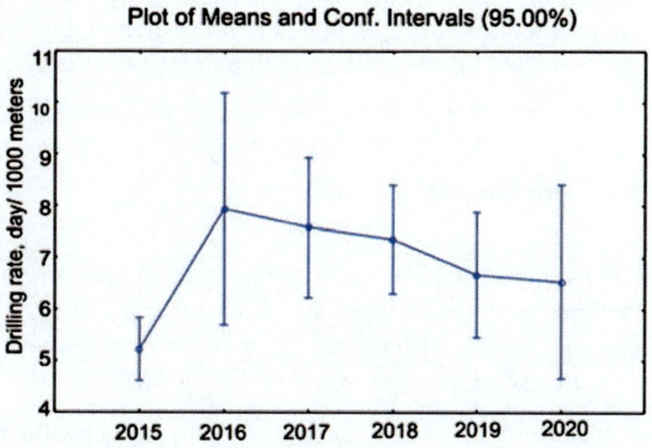

Table 3 One-way ANOVA

Effect	SS	Degree of freedom	MS	F	P
Drilling contractor	963.602	3	321.201	14.6995	0.0000
Field	2118.174	8	246.195	15.1025	0.00
Well type	483.32	1	1433.51	72.4420	0.0000
Season	6.51	2	3.26	0.1271	0.8807

4 Results and Considerations

Improved performance of the process—the well construction cycle—implies a rational decrease in production and non-production time.

According to the construction and analysis of Shewhart charts, well construction is unstable in terms of the analyzed parameters: production time, non-production time and drilling rate. This shows that parameter variations are unpredictable. To find the causes of variability, a series of experiments with changing parameters is typically carried out. However, the cycle is a complex process, and such experiments can hardly be conducted [8, 9]. For analysis, data should be segmented according to the contributing factors.

Control charts should also include the components of production and non-production time to identify specific causes of variability and trace the general trend of variability between operations. This requires comprehensive and qualitative data on wells.

The analysis revealed erroneous or incorrect values in the data array. For example, field or season are indicated erroneously, or there are empty cells. Electronic document management can contribute to the solution of this problem. Forms for collecting and analyzing data on inconsistencies and requests for their elimination should be implemented.

In terms of the analyzed process, data collection forms enable specification of the non-production time and provide not only the duration and number of accidents and problems. Currently, there are no sufficient data on causes for occurrence of the components of non-production time, place of origin, etc.

5 Conclusion

The well construction cycle has been analyzed in terms of timing and drilling rate. Methods of statistical analysis (descriptive statistics, ANOVA, charts and correlation analysis) used in specialized software were employed for analysis [10, 11].

Study of the process revealed the following:

- instability of these parameters and uncontrollability (low process capability indicators);
- the most time-consuming operations such as drilling, tripping, bottom configuration of the drilling rig, well lining, and fast performing operations such as buildup, reaming and sealing the blowout preventer;
- production time is affected by well type, drilling contractor, field and reservoir;
- no general trend of variability for all operation by years;
- season is of no statistical significance with respect to variables, etc.

Based on this analysis, proposals have been elaborated to address improvement in the electronic document management, quality messages, search for causes of process variability, and parameters for correction factors to the efficiency formula. Currently, efficiency is calculated as the ratio of production time to well depth that yields extremely small values. Correction factors to the factors affecting production time can be used to improve this formula. ANOVA revealed the parameters that can be used as parameters for the factors: well type, field and reservoir. The factor “drilling contractor” was of statistical significance, but the formula is designed to evaluate and compare factors, and this factor should be neglected.

References

1. Wheeler, D., Chambers, D.: Statistical Process Management. Business optimization using Shuhart control cards. Alpina Publisher, Moscow (2016)
2. Montgomery, D.C.: Introduction to Statistical Quality Control. Wiley, NY (2000)
3. Deming, U.E.: The Way Out of the Crisis: A New Paradigm of Managing People, Systems and Processes. Bizneskom, Moscow (2009)
4. Taguchi, G., Chowdhury, S., Wu, Y.: Taguchis Quality Engineering Handbook. Publ. by John Wiley and Sons, NY (2005)
5. Viorel-Mihai Nani.: Statistical control of the technological process stability to manufacturing cylindrical parts into high series. MMS Eng. J. **7**https://doi.org/10.13140/RG.2.2.33528.65284 (2017)
6. Ishikawa, K., Loftus J.H.: Introduction to Quality Control, 3rd edn. 3A Corp, Tokyo (1990
7. Ryan, T.P.: Statistical Methods for Quality Impovement. Wiley, NY (2000)
8. Plotnikova, I.V., Redko, L.A., Chicherina, N.V.: The analysis of the process of complaints consideration using statistical methods of control. IOP Conf. Ser. Mater. Sci. Eng. **516**(1), 101–105 (2019). https://doi.org/10.1088/1757-899X/516/1/012013
9. Huang, X., Xu, J., Zhou, Q.: Multi-scale diagnosis of spatial point interaction via decomposition of the k function-based t2 statistic. J. Qual. Technol. **49**(3), 213–227 (2017). https://doi.org/10.1080/00224065.2017.11917991
10. Woodall, W.H., Montgomery, D.C.: Some current directions in the theory and application of statistical process monitoring. J. Qual. Technol. **46**(1), 78–94 (2014). https://doi.org/10.1080/00224065.2014.11917955
11. Plotnikova, I.V., Redko, L.A. et al., Study of the technological process parameters basis on a mathematical model. IOP Conf. Ser.: Mater. Sci. Eng. **457**(1), 77–85. (2019). https://doi.org/10.1088/1757-899X/457/1/012021

Numerical Solution of the Crown Forest Fires Spread Taking into Account Fire Barriers and Breaks

Valeriy Perminov

Abstract Using the method of mathematical modeling, the process of the emergence and spread of crown forest fires, as well as their termination in the presence of fire breaks and barriers, consisting of deciduous species, is studied. The Reynolds equations for turbulent flow were solved numerically, taking into account chemical reactions. The research was based on numerical solution of two dimensional Reynolds equations. The boundary-value problem is solved numerically using the method of splitting according to physical processes. A discrete analogue for the system of equations was obtained by means of the control volume method. The distributions of the fields of velocity, temperature, oxygen concentrations, volatile products of pyrolysis, combustion, and volume fractions of the condensed phase are obtained with time. The model made it possible to obtain in dynamics the contours of the spread of crown forest fires, which depend on the bulk and type of forest combustible materials, moisture content, wind speed and direction, etc. It was also possible to determine the dependence of the size of fire breaks and barriers on the above parameters, at which the crown fire stops spreading. Based on this data, we can estimate the minimum sizes of fire breaks and barriers, or their combined application, to stop the propagation of the flame front.

1 Introduction

Forest fires are an important and integral part of the dynamics of forest vegetation. Intense forest fires lead to the replacement of forest reserves of different ages and compositions. Frequent low-intensity fires maintain the optimal specific amount of forest fuels, which in the long term reduces the risk of catastrophic fires. There are also some types of conifers (for example, sequoias), whose cones open much more actively in extreme temperatures [1]. However, the current state of forests and

V. Perminov (✉)
Department of Control and Diagnostics, Tomsk Polytechnic University, Lenin Avenue, 30, Tomsk 634050, Russia
e-mail: perminov@tpu.ru

E. Lysenko et al. (eds.), *Recent Developments in the Field of Non-Destructive Testing, Safety and Materials Science*, Studies in Systems, Decision and Control 433,
https://doi.org/10.1007/978-3-030-99060-2_14

the anthropogenic nature of forest fires suggests that there are much more negative aspects than positive ones. As an example of such negative aspects, one can single out the established relationship between global warming and massive CO_2 emissions from forest fires [2–6], smoke in cities, and a possible transition of a forest fire to buildings located near the forest border. In particular, cases of burnout of entire settlements are known [7]. In addition, with the spread of large fires, it is possible to cross the borders of states, both of the combustion process, and the transboundary transfer of combustion products, which brings this problem to the international level. In this regard, the development of preventive fire-fighting measures is an urgent problem. The purpose of this work is to develop a mathematical model for the spread of raised forest fires, taking into account fire breaks and barriers made of deciduous trees.

2 Physical and Mathematical Formulation of the Problem

Let there be an area of elevated temperature of finite dimensions in a forest with a height h, and the origin of the XYZ coordinate system is associated with the center of this focus. The OZ axis is directed vertically upward, and the OX and OY axes are parallel to the earth's surface. The OX axis coincides with the direction of the wind. Under the influence of the wind, the seat of the crown fire begins to spread over the forest. Figure 1 schematically shows the location of a forest area and fire-prevention facilities in it.

In this case, the gap and the barrier are adjacent and have finite dimensions, but this is not required. In the course of calculations, various cases of relative position or absence of one of these objects were considered. The mathematical formulation of the problem is based on the physical one and is an amended and modified model given in [5]. The main assumptions made in the derivation of the system of equations of initial and boundary conditions: the wind is directed along the x-axis, the flow has a developed turbulent character, and molecular transfer is neglected in comparison with turbulent; the forest canopy is assumed to be a non-deformable porous-dispersed

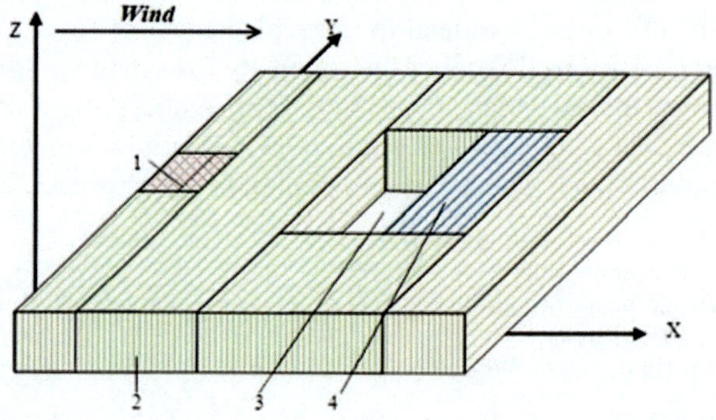

Fig. 1 Coordinate system with an ignition source in the center, where 1 is an ignition source, 2 is a coniferous forest, 3 is a fire gap, 4 is a fire barrier made of deciduous trees

medium; the environment in the forest canopy is considered to be two-temperature, because the temperatures of the gas and condensed phases differ; the speed of air movement over the forest canopy in a stationary state is known; the so-called blown forest area is considered, when the volume fraction of the condensed phase of forest combustible materials (FCM), consisting of dry organic matter, water in a liquid-droplet state and ash, can be neglected in comparison with the volume fraction of the gas phase, which includes air and gaseous components pyrolysis and combustion products; the density of the gas phase does not depend on pressure, because the speed of the current is small compared to the speed of sound; energy transfer is carried out by turbulent convection and radiation; to describe the process of energy transfer by radiation, the diffusion approximation is used; to describe the convective transfer caused by the action of gravity, the Reynolds equations are used.

Thus, the problem under consideration is reduced to solving a three-dimensional system of equations proposed by Grishin [5] in a forest with a height h located above the roughness level z_0 and below the level of the upper boundary of the forest canopy. The equations express conservation laws for a multiphase multicomponent continuous reacting medium. Proceeding from the fact that the vertical dimensions of the forest are incomparably smaller than its horizontal dimensions, the general system of differential equations of the heat and mass transfer process can be integrated along the vertical coordinate. For this, the original system of equations was reduced to a divergent form, then integrated over the height over the height of the forest canopy [5].

$$\int_0^h \phi dz = \overline{\phi} h, \tag{1}$$

where $\overline{\phi}$ is the average value of ϕ. The integrated system of equations looks like this:

$$\frac{\partial \rho}{\partial t} + \frac{\partial}{\partial x}(\rho v) + \frac{\partial}{\partial y}(\rho w) = \dot{m} - (\dot{c}^- - \dot{c}^+)/h, \tag{2}$$

$$\rho\frac{dv}{dt} = -\frac{\partial p}{\partial x} + \frac{\partial}{\partial x}\left(-\rho\overline{v'^2}\right) + \frac{\partial}{\partial y}(-\rho\overline{v'w'}) - \rho s c_d v|\vec{v}| - \dot{m}v + (\tau_x^- - \tau_x^+)/h, \tag{3}$$

$$\rho\frac{dw}{dt} = -\frac{\partial p}{\partial y} + \frac{\partial}{\partial x}(-\rho\overline{v'w'}) + \frac{\partial}{\partial y}\left(-\rho\overline{w'^2}\right) - \rho s c_d w|\vec{v}| - \dot{m}v + (\tau_y^- - \tau_y^+)/h, \tag{4}$$

$$\rho c_p\frac{dT}{dt} = \frac{\partial}{\partial x}(-\rho c_p\overline{v'T'}) + \frac{\partial}{\partial y}(-\rho c_p\overline{w'T'}) + q_5R_5 - \alpha_V(T - T_s) + \\ (q_T^- - q_T^+)/h + k_g(cU_R - 4\sigma T^4), \tag{5}$$

$$\rho\frac{\mathrm{d}c_{\alpha}}{\mathrm{d}t}=\frac{\partial}{\partial x}(-\rho\overline{v\prime c\prime_{\alpha}})+\frac{\partial}{\partial y}(-\rho\overline{w\prime c\prime_{\alpha}})+\mathrm{R}_{5\alpha}-\dot{m}c_{\alpha}+\left(J_{\alpha}^{-}-J_{\alpha}^{+}\right)/\mathrm{h};\ \alpha=\overline{1,3},\tag{6}$$

$$\frac{\partial}{\partial x}\left(\frac{c}{3k}\frac{\partial U_R}{\partial x}\right)+\frac{\partial}{\partial y}\left(\frac{c}{3k}\frac{\partial U_R}{\partial y}\right)-kcU_R+4k_g\sigma T^4$$
$$+4k_s\sigma T_s^{\,4}+\frac{(q_R^- - q_R^+)}{h}=0,\ k=k_g+k_s,\tag{7}$$

$$\sum_{\mathrm{i}=1}^{4}\rho_i\varphi_i c_{pi}\frac{\partial T_s}{\partial t}=q_3R_3-q_2R_2+k_s\left(cU_R-4\sigma T_s^{\,4}\right)+\alpha_V(T-T_s)\tag{8}$$

$$\rho_1\frac{\partial\varphi_1}{\partial t}=-R_1,\ \rho_2\frac{\partial\varphi_2}{\partial t}=-R_2,\rho_3\frac{\partial\varphi_3}{\partial t}=\alpha_cR_1-\frac{M_c}{M_1}R_3,\ \rho_4\frac{\partial\varphi_4}{\partial t}=0,\tag{9}$$

$$\sum_{\alpha=1}^{3}c_{\alpha}=1,\ p_e=\rho RT\sum_{\alpha=1}^{3}\frac{c_{\alpha}}{M_{\alpha}},\ \dot{m}=(1-\alpha_c)R_1+R_2+\frac{M_c}{M_1}R_3,\tag{10}$$

$$R_{51}=-R_5-\frac{M_1}{2M_2};\ R_{53}=v_c(1-\alpha_c)R_1-R_5;\ R_{53}=0.\tag{11}$$

To describe the rates of pyrolysis reactions, moisture evaporation, combustion of condensed and volatile pyrolysis products, the following formulas were used:

$$R_1=k_1\rho_1\varphi_1\exp\left(-\frac{E_1}{RT_s}\right),\ R_2=k_2T^{-0,5}\rho_2\varphi_2\exp\left(-\tfrac{E_2}{RT_s}\right),$$
$$R_3=k_3\rho s_{\sigma}\varphi_3c_1\exp\left(-\frac{E_3}{RT_s}\right),\ R_5=k_5M_2\left(\frac{c_1M}{M_1}\right)^{0,25}\left(\frac{c_2M}{M_2}\right)T^{-2,25}\exp\left(-\frac{E_5}{RT}\right).\tag{12}$$

When solving the presented system of equations, the following initial and boundary conditions should be taken into account:

$$t=0:v=0;\ w=0;\ T=T_e;\ c_{\alpha}=c_{\alpha e};\ T_s=T_e;\ ;\ \varphi_i=\varphi_{ie}\tag{13}$$

$$x=0:v=V_e;\ w=0;\ T=T_e;\ c_{\alpha}=c_{\alpha e};\ -\frac{c}{3k}\frac{\partial U_R}{\partial x}+\frac{cU_R}{2}=0,\tag{14}$$

$$x=x_e:\frac{\partial v}{\partial x}=0;\ \frac{\partial w}{\partial x}=0;\ \frac{\partial c_{\alpha}}{\partial x}=0;\ \frac{\partial T}{\partial x}=0;\ \frac{c}{3k}\frac{\partial U_R}{\partial x}+\frac{cU_R}{2}=0,\tag{15}$$

$$\mathrm{y}=0:\frac{\partial v}{\partial y}=0;\ \frac{\partial w}{\partial y}=0;\ \frac{\partial c_{\alpha}}{\partial y}=0;\ \frac{\partial T}{\partial y}=0;\ -\frac{c}{3k}\frac{\partial U_R}{\partial y}+\frac{cU_R}{2}=0,\tag{16}$$

$$y = y_e : \frac{\partial v}{\partial y} = 0; \frac{\partial w}{\partial y} = 0; \frac{\partial c_\alpha}{\partial y} = 0; \frac{\partial T}{\partial y} = 0; \frac{c}{3k}\frac{\partial U_R}{\partial y} + \frac{cU_R}{2} = 0 \quad (17)$$

The temperature value in the ignition source was set in the following form:

$$T = T_s = \begin{cases} T_e + \frac{t}{t_0}(T_0 + T_e), t \le t_0 \\ T_e + \left(T_0 - T_e \exp\left[-k\left(\frac{t}{t_0} - 1\right)\right]\right), t > t_0 \end{cases} \quad (18)$$

Here t_0 is the time of the maximum heat release from the source. Considering that heat and mass transfer in the fire front with the surface layer of the atmosphere and the lower tier of the forest occurs according to Newton's law, then the corresponding terms of the equations can be written in the following form:

$$\frac{J_\alpha^- - J_\alpha^+}{h} = -\frac{\alpha}{c_{ps}h}(c_\alpha - c_{\alpha e}), \frac{q_T^- - q_T^+}{h} = -\frac{\alpha}{h}(T - T_e), \quad (19)$$

Here and above $\frac{d}{dt}$ is the symbol of the total (substantial) derivative; α_v is the coefficient of phase exchange; ρ—density of gas—dispersed phase, t is time; v_i—the velocity components; T, T_S,—temperatures of gas and solid phases, U_R—density of radiation energy, k—coefficient of radiation attenuation, P—pressure; c_p—constant pressure specific heat of the gas phase, c_{pi}, ρ_i, φ_i—specific heat, density and volume of fraction of condensed phase (1—dry organic substance, 2—moisture, 3—condensed pyrolysis products, 4—mineral part of forest fuel), R_i—the mass rates of chemical reactions, q_i—thermal effects of chemical reactions; k_g, k_S—radiation absorption coefficients for gas and condensed phases; T_e—the ambient temperature; c_α—mass concentrations of α—component of gas—dispersed medium, index $\alpha = 1$, 2, 3 where 1 corresponds to the density of oxygen, 2—to carbon monoxide CO, 3—to carbon dioxide and inert components of air; R—universal gas constant; M_α, M_C, and M molecular mass of α—components of the gas phase, carbon and air mixture; g is the gravity acceleration; c_d is an empirical coefficient of the resistance of the vegetation, s is the specific surface of the forest fuel in the given forest stratum. To close the system (1)–(11), the components of the tensor of turbulent stresses, and the turbulent heat and mass fluxes are determined using the local-equilibrium model of turbulence [5]. The thermodynamic, thermophysical and structural characteristics correspond to the forest fuels in the canopy of a different (for example pine [5]) type of forest.

3 The Results of Calculation and Their Analysis

For the numerical integration of the original system of equations, the control volume method is used [7, 8]. We divide the computational domain into a certain number of non-intersecting control volumes, and each control volume must correspond to a nodal point, relative to which the equations of state will be considered. The obtained values will be valid for the entire area of the control volume. Then we integrate the original system of equations over each control volume. The resulting nonlinear system of algebraic equations was solved iteratively by the SIP method [7, 8]. Test checks of the calculation program were carried out by comparison with analytical solutions, as well as with known numerical solutions of problems in combustion theory. In addition, to assess the accuracy of the difference schemes used and to check the correctness of the program as a whole, the method of a priori given analytical solutions was used, which consists in the fact that analytical expressions of the sought functions were substituted into the equations used, the discrepancy of the equations was calculated, which was then interpreted as a fictitious source in each equation. Then the values of the functions were restored. The accuracy of restoring these functions was at least 0.5%. The stability and accuracy of the obtained solutions was also checked by decreasing the steps in time and space. In serial calculations, an algorithm for automatic time step selection was used.

On the basis of the above model, numerical calculations were carried out and the distributions of the desired functions were obtained at different points in time in the considered area. To obtain reliable calculations, it is important that the fire front, approaching the barrier or rupture, has a constant propagation speed, that is, it reaches a stationary mode. Therefore, the initial boundary of the areas of ruptures and barriers is set at a distance of 40 m from the ignition source. As a result of calculations, isolines of temperature, oxygen concentration and concentration of pyrolysis products were obtained. Table 1 shows the results of calculating the size of the fire barrier depending on the moisture content of the deciduous trees. Figure 2 illustrates this relationship.

Here x_1, x_2, y_1, y_2 are the coordinates of the start and end of the screen along the x and y axes, respectively. X and Y are the dimensions of the screen along the corresponding axes. S is the area of the screen. The calculations were carried out with the following parameters: the bulk and moisture content of FCM in the coniferous forest of 0.2 kg/m^3 and 0.66, respectively, the wind speed in an undisturbed environment is 5 m/s. The moisture content of the deciduous trees can vary in the

Table 1 Dependence of the size of the fire barrier on the moisture content of deciduous trees

	Moisture content of deciduous forest, W							
	0.8	0.9	1	1.1	1.2	1.3	1.4	1.5
x_1, m	40	40	40	40	40	40	40	40
x_2, m	48	48	45	44	43	42	41	41
y_1, m	1	15.5	20	21	22.5	23	23.5	23.5
y_2, m	49	34.5	30	26	26.5	27	26.5	26.5
X, m	8	8	5	4	3	2	1	1
Y, m	48	19	10	5	4	4	3	3
S, m^2	768	304	100	40	24	16	6	6

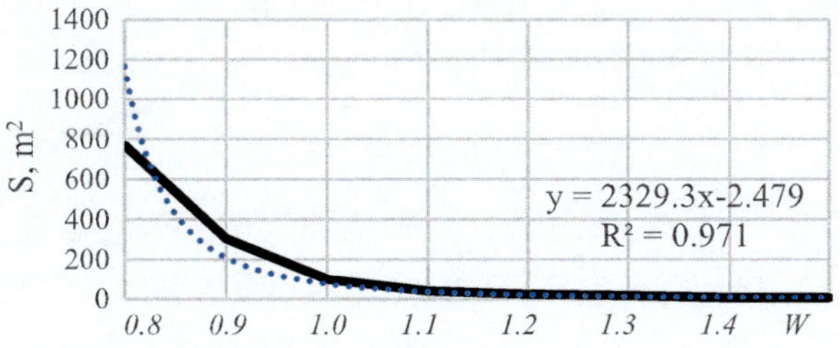

Fig. 2 Dependence of the size of the area of the fire barrier on its moisture content

range from 0.8 to 1.5. Outside of these values, the fire either continues to spread regardless of the size of the barrier, or stops burning at the minimum size of the barrier (less than 0.5 m). Figure 2 shows that the dependence of the dimensions of the area of the fire barrier has the form of a power function.

With an increase in the moisture content of the deciduous trees in the range of 0.7–1.0, the size of the barrier area sharply decreases. In the future, the changes are not so significant. Also, according to the data from Table 1, an additional dependence is traced—with an increase in moisture content, the linear dimensions along the *X* and *Y* axes change according to similar laws. A decrease in the size of the barrier naturally follows from the fact that wetter FCM ignite worse, because the size of the barrier through which the fire does not pass logically decreases. Let us consider the dependence of the size of the fire barrier on the wind speed with the remaining parameters unchanged: the bulk of FCM fuels in the coniferous trees is 0.3 kg/m^3, the moisture content of fuels in the coniferous forest is 0.66, the moisture content in the deciduous trees is 1. The data are presented in Table 2 and Fig. 3.

x_1, x_2, y_1, y_2—coordinates of the beginning and end of the screen along the X and Y axes, respectively. X and Y are the dimensions of the screen along the corresponding axes. *S* is the area of the screen. As can be seen from the graph, with an increase in the wind speed, the area of the minimum required barrier decreases nonlinearly, and the changes mainly affect the dimensions along the Y axis.

This is due to the fact that with a small wind the fire front actively expands, and with an increase in the wind speed, expansion in the perpendicular direction

Table 2 Dependence of the size of the fire barrier on the wind speed

	Wind velocity, m/s				
	4	5	6	7	8
x_1, m	40	40	40	40	40
x_2, m	47	47.5	49	50	52
y_1, m	3	9.5	14	16	18.5
y_2, m	47	40.5	36	34	31.5
X, m	7	7.5	9	10	12
Y, m	44	31	22	18	13
S, m^2	616	465	396	360	312

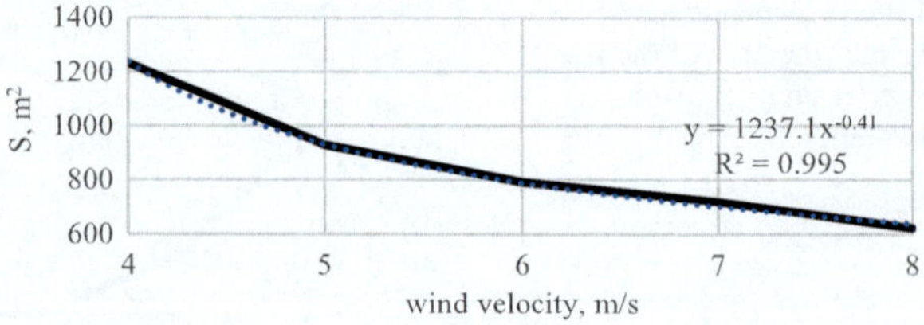

Fig. 3 Dependence of the area of the fire barrier on the wind speed

decreases. This can be seen if we compare the isotherms in the fire front in front of the barrier (Figs. 4 and 5). The numbers 1, 2, 3, 4, 5, 6 denote dimensionless isotherms 1.5, 2, 2.5, 3, 4, 5, defined as $\overline{T} = T/T_0$, where $T_0 = 300$ K. Isotherms corresponding to a wind speed of 4 m/s form wider areas, especially for a temperature of 1500 K.

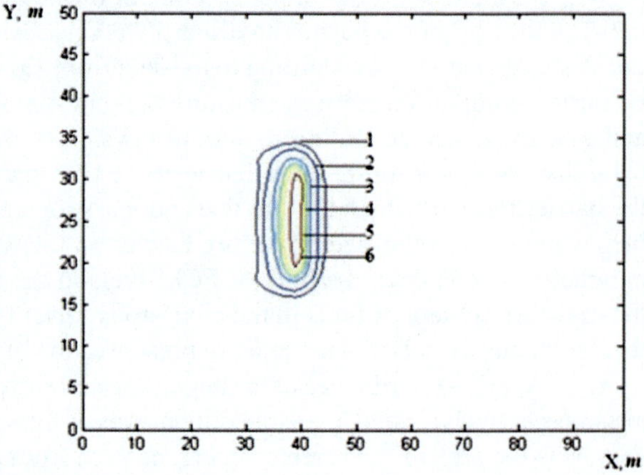

Fig. 4 Distribution of isotherms in the fire front at $V_e = 4$ m/s

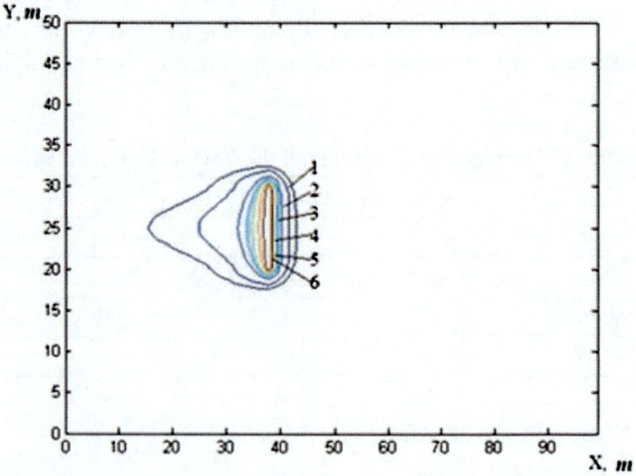

Fig. 5 Distribution of isotherms in the fire front at $V_e = 8$ m/s

Let us compare the effectiveness of fire breaks, barriers and their combination in limiting the spread of fire. Figures 6, 7 and 8 show the results of calculations with break, barrier, break + barrier and barrier + break.

When analyzing the data, it can be seen that the fire barrier has the lowest efficiency, and the fire break—the highest. When these methods were combined, the gap + barrier combination turned out to be more effective. This is because the decrease in the energy emitted in the fire front occurs according to power laws, in other words, the bulk of the energy is spent in the area where there is no FCM and there can be no ignition. The energy that reaches the area of the deciduous trees is not enough to ignite it, especially if its moisture is high.

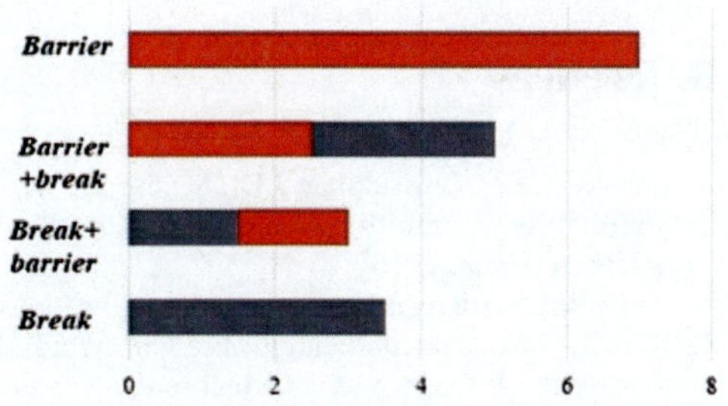

Fig. 6 Comparison of the dimensions of the fire gap and the barrier, as well as their combinations for the parameters: FCM bulk—0.5 kg/m^3, moisture content of coniferous fuel—0.66, wind speed 7 m/s, moisture content of deciduous trees

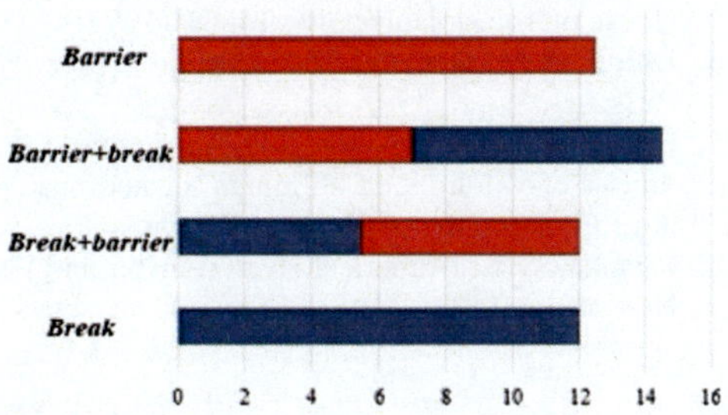

Fig. 7 Comparison of the dimensions of the fire gap and the barrier, as well as their combinations for the parameters: FCM bulk—0.66 kg/m^3, moisture content of coniferous fuel—0.2, wind speed 7 m/s, moisture content of deciduous trees 1

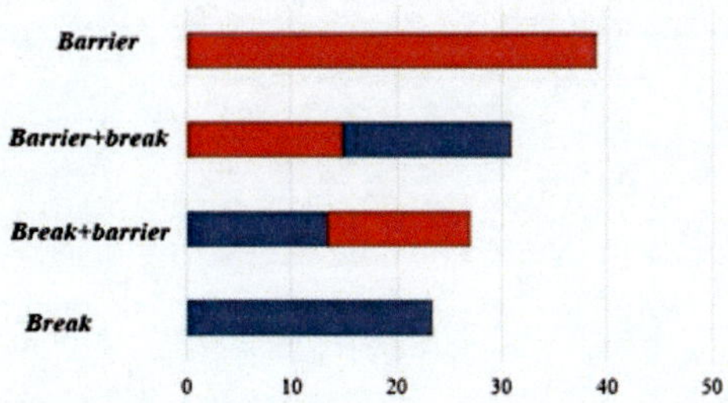

Fig. 8 Comparison of the dimensions of the fire gap and the barrier, as well as their combinations for the parameters: FCM bulk—0.2 kg/m^3, moisture content of coniferous fuel—0.2, wind speed 7 m/s, moisture content of deciduous trees 1

4 Conclusion

A multiphase mathematical model of wind-aided crown forest fires propagating through heterogeneous fuel beds has been performed. It allows to study the spread of crown forest fires in the presence of fire breaks and barriers under the influence of various external conditions. under the influence of various external conditions: (a) meteorology conditions (air temperature, wind velocity etc.), (b) type (various kinds of forest combustible materials) and their state (load, moisture etc.). Based on these calculations, the dependences of the critical sizes of fire breaks and barriers were determined, at which the crown fire stops spreading.

References

1. Harold, S.J.Z., Christopher, J.D.: Severe fire weather and intensive forest management increase fire severity in a multi-ownership landscape. Ecol. Appl. **28**(4), 1068–1080 (2018). https://doi.org/10.1002/eap.1710
2. Fons, W.L.: Analysis of fire spread in light forest fuels. J. Agric. Resour. **7**(3), 93–121 (1946)
3. Byram, G.M.: The modeling of fire whirlwinds. For. Sci. **16**(4), 586–398 (1970)
4. Rothermel, R.C.: A mathematical model for fire spread predictions in wildland fuels. USDA For. Serv. Res. Pap. INT-115, Ogden (1972)
5. Grishin, A.M.: Mathematical Modeling Forest Fire and New Methods Fighting Them. Publishing House of Tomsk University, Tomsk (1997)
6. Dorrer, G.A.: Dynamics of forest fires. Siberian Department of RAS, Novosibirsk (2008) (in Russian)
7. Perminov, V.A., Fraynova, R.O., Lukianov, A.K.: Numerical modeling of the process of thermal impact of wildfires on buildings located near forests. Mater. Sci. Forum **970**, 82–87 (2019). https://doi.org/10.4028/www.scientific.net/MSF.970.82
8. Patankar, S.V.: Numerical Heat Transfer and Fluid Flow. Hemisphere Publishing Corporation, New York (1981)

Innovations in Technical Diagnostics and Materials Science

Modeling of Microcomposition Structure of Crystals

Yury Borodin

Abstract The study was carried out within the framework of the development of physicochemical methods for the study of nanostructured materials. The development of the science of materials with a low-dimensional and aperiodic structure has again highlighted the problem of modeling the nucleation of growth and deformation of lattice structures. When considering structural changes in modified materials, one has to face the contradictions between the infinity of the ideal and the finiteness of the corresponding real lattices. The problem of the dimensional ordering of the structure in the modified layers is closely related to the internal geometry of the crystal structure, since the structural changes occurring under any forms of exposure are related to the internal crystallographic model. In the subsurface modified layers, which have nonequilibrium properties and exchange mass and energy with the environment, various types of self-organization and ordered behavior of structural elements are observed, which today cannot be explained within the framework of the accepted crystal-chemical models. The use of new crystallographic models allows us to consider in a new way the organization of the filling of the crystal space by atoms, the dynamic modeling of polyhedral rearrangements, and the effect of interimpurity interaction on the formation of modulated structures in modified layers.

1 Introduction

The development of the science of materials with low-dimensional and aperiodic structures again highlighted the problem of the existing inconsistency of the patterns of electron density distribution with the crystallographic description of infinitely broadcast lattices [1–3]. There remains some well-known convention between model distortions and lattice defects on the one hand and electron microscopic images on the other. Scanning tunneling microscopy contributes to understanding the structure

Y. Borodin (✉)
Division for Control and Diagnostics, School of Non-Destructive Testing, National Research Tomsk Polytechnic University, 30 Lenin Avenue, 634050 Tomsk, Russia
e-mail: uryborodin@tpu.ru

E. Lysenko et al. (eds.), *Recent Developments in the Field of Non-Destructive Testing, Safety and Materials Science*, Studies in Systems, Decision and Control 433,
https://doi.org/10.1007/978-3-030-99060-2_15

of low-dimensional structures, but the explanation remains within the framework of existing crystallographic provisions [4, 5]. Nanostructured materials [6], polytypic formations [7], and other manifestations of the aperiodic arrangement of atoms in the lattice are poorly described. Additional methods for modeling lattice structures are needed, one of which may be microcompositional (MC), which provides for a spiral lattice organization [8]. The MC model more adequately describes the properties of crystals and especially the quantum collective phenomena of transfer and plasticity [3, 9]. The spiral organization of the lattice is based on the specific behavior of atoms, taking into account the exchange interactions of their electrostatic fields, extending over several periods of the lattice and a thin outer shell [2, 8, 10, 11]. Such a shell ensures rotation and implementation of the principle of self-similarity of atoms with the same chemical nature, as well as the possibility of their closest packing. In the MC model, for different schemes of distribution of centers of self-similar atomic particles (C-particles), the same external faceting with the corresponding symmetry for a given crystal will always be obtained [10]. It seems possible with the help of computer simulation to consider the regularities of the organization of MC structures from one and many crystallization centers.

2 Crystal Growth from a Single Crystallization Center

Modeling of the microcomposition (MC) of crystalline substances was carried out taking into account the fields of exchange interactions of atomic particles, the principles of closest packing, and self-similarity [8, 10]. A point is considered as the center of an atom, a point system is studied without taking into account the chemical nature of atoms, and crystal particles are interpreted as balls with a movable thin outer shell and forces of attraction and repulsion between themselves. A plane lattice of Euclidean space is associated with a lattice on a Clifford surface (K-surface) in an elliptic Riemann space [10]. The organization of the lattice is carried out in stages using C-particles at a minimum distance from the zero point in accordance with the chosen Fedorov symmetry group. C-particles are connected by arcs of outer shells and are spaced from each other by several lattice periods.

The centers of C-particles are located on the (111) face perpendicular to the third-order symmetry axis. When constructing a system of C-particles, the zero point is chosen arbitrarily within a triangular contour (for example, at the vertex point), and with the help of the K-system, the points of location of C-particles of a given level of self-similarity are found. Moving the K-system parallel to itself in a spiral from the periphery to the center and fixing points from the zero to the nth level within the triangular contour, we obtain a structure corresponding to the Fedorov group with fixed lattice nodes (Fig. 1). Fragments of such a structure of C-particles are identified as minimal scattering volumes in X-ray diffraction and are called structural elements (SE) [11, 12].

The organization of the lattice system before the faceting of the crystal polyhedron consists of successive structure-forming and structure-filling stages. Starting from

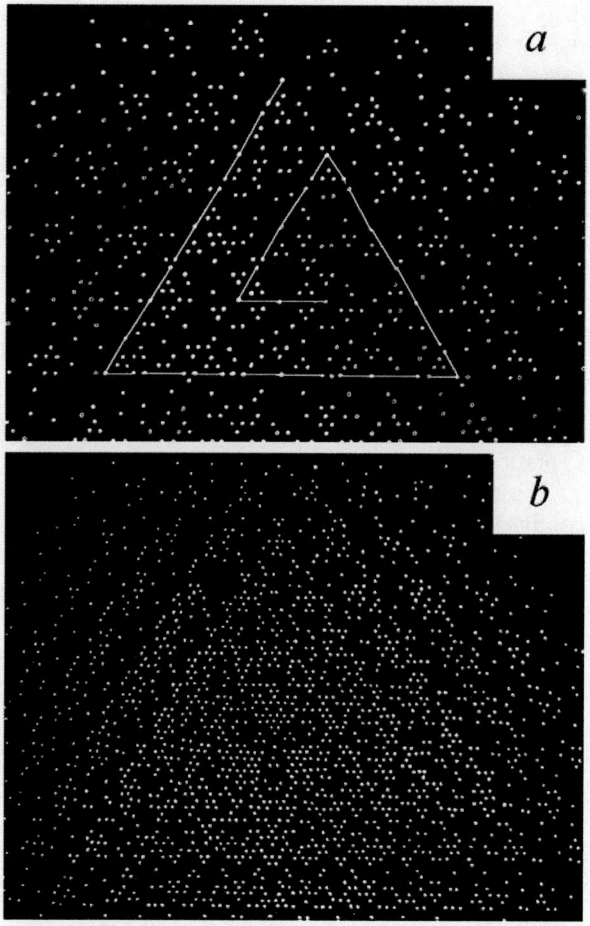

Fig. 1 Organization of a structure of self-similar atomic particles at the initial (**a**) and final (**b**) stages of the growth space captured by the growing crystal

the zero point, the propagation of the lattice in space is carried out by attaching new C-particles to the fixed at the nodes, each of which behaves as a new center of the lattice organization. The bonds are realized along the K-lines with a period significantly exceeding the lattice period. This ensures the propagation of the lattice system in the form of a rapidly contoured closed lattice not yet filled with C-particles of considerable size (Fig. 1a). Subsequently, the filling of the vacant lattice sites proceeds without significant increases in the linear dimensions, and the correct distribution of C-particles within the contoured lattice system is observed. The organization of the lattice is completed by the final filling of the vacant nodes, the correct distribution of the SE throughout the system, and the outline of the external shape.

The distribution of C-particles in the lattice sites in the process of its organization depends on the dynamics and the filling method. Figure 1a shows that the SE distribution pattern for the same sections is different for different methods of arranging the lattice. At the final stage of organizing the lattice distribution system, the SE does not depend on the method of its implementation and is determined only by the selected Fedorov group (Fig. 1b). In lattices organized in spiral and uniformly sequential ways, the final pattern of particle distribution is outwardly the same. In the MC structure, vacancies and cavities are not defects of the crystal lattice and are interconnected by the symmetry elements of the outer shape, the third-order symmetry axis, and the inversion center. Their distribution in the complete lattice of crystalline polyhedra is determined by the SE and the finite symmetry of this polyhedron. In the MC structure, for each class of finite symmetry, there is a specific distribution of vacancies, voids, and cavities, and only the general motive of the ideal SE distribution should appear in the crystal.

The organization of the lattice of a crystalline polyhedron can take place both by growing a monolayer and by an active layer equal to the size of the SE. Spreading along the K-lines in crystalline space, by the force of long-range action, they capture all space and are closed. To organize the lattice in the crystal space between the C-particles, therefore, only repulsive and fixing forces are sufficient.

The forces of interaction between infected C-particles of the same name in a confined space bring them to diametrically opposite points of the K-line. They fall into mutually fixing positions, and, for each of them, the repulsive forces from different sides are balanced. In the case of more than two C-particles, the resultant vector directed along the K-line divides each diametrically opposite point of the K-line with subsequent fixation. The repulsive forces between the infected C-particles of the same name play, therefore, an important structure-forming role. If the crystal lattice consists of several elements, then it is considered to consist of SE of the same-named infected particles. The forces of attraction between oppositely charged C-particles play a subordinate role, providing interaction between like-charged SE.

3 Growth of Crystals from Many Centers of Crystallization

With the simultaneous growth of several self-similar SEs, it is necessary to take into account their fields of exchange interactions and the organizing influence of this process on the formation of the entire structure. Modeling the lattice of a polycentric crystal proceeds by analogy with growing from one crystallization center, observing the principles of closest packing, and hierarchical subordination of self-similar micro blocks. In Fig. 2a rounded triangles mark SEs, from which the subsequent organization of the lattice structure proceeds; dots—the location of the fixed C-particles. Although the exchange fields of SEs are superimposed on the C-particles of the nearest SEs, this has no effect. Further growth of a polycentric crystal is achieved by the interaction of the SE exchange fields and the formation of a denser structure (Fig. 2b). The field of each FE and the aggregate field acquire the correct form; the general structure begins to show certain motives of the zonal-sectorial structure, center, and faceting. At a given level of self-similarity, as a result of the structure-filling process, the structure of a polycentric crystal is achieved (Fig. 2c).

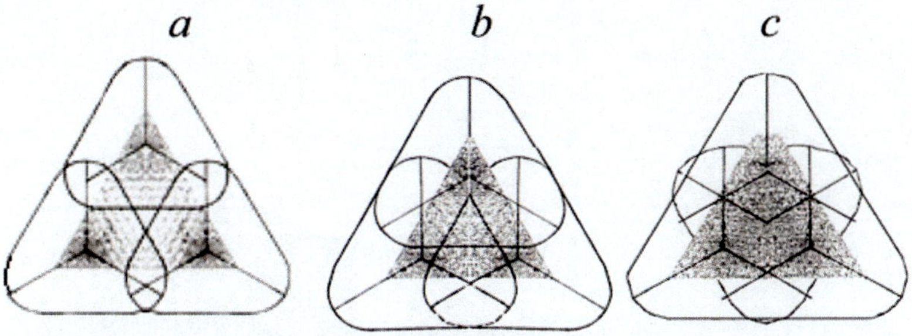

Fig. 2 Organization of the structure with the minimum (**a**), average (**b**), and maximum (**c**) superposition of exchange electrostatic fields of self-similar structural elements

From Fig. 2a, it can be seen that when the general structure is organized, a shadow crystal appears. Here, the superposition of SE fields is minimal, the densest packing is weakly expressed, and the growth of individual microcrystals proceeds largely independently. The aggregate structure is characterized by the presence of microvoids that can be filled with other material. With an increase in the packing density of SE, the polycentric crystal acquires a more saturated and regular structure and a distinctly expressed microblock and their increasing mobility. Such a crystal can be hollow and have high strength, and the shape of the exchange fields of the hollow has the same symmetry as the crystal polyhedron.

With the closest packing, the resulting polycentric crystal has a high perfection and maximum filling of cavities and vacancies (Fig. 2c). The fields of exchange interactions between SE and microblocks do not lose their isolation in the total field of the crystal. The magnitude of the shear rotations of microblocks with the closest packing has a maximum value and is a multiple of the length of the segments between the initial and final positions of the corresponding crystallization centers. The structural features of a polycentric crystal organized from six crystallization centers are in many respects similar to those obtained from one crystallization center (Fig. 3). Energetically, such a polycentric crystal is more advantageous, since a denser structure, occupying a smaller volume, arises for the same number of C-particles. In this case, there is a common center of convergence of microblocks, which subsequently becomes the center of the crystal.

4 Details of MK Structure

Already at the initial stages of growth from one crystallization center, the crystal acquires certain motifs of structural zoning, which are expressed in the division of the space of the growing crystal according to the degree of ordering, distribution density and the correctness of the lattice sites. From Fig. 3 it can be seen that the apical, near-costal and central sectors are distinguished. At the peaks, the maximum disorder and the minimum density of the arrangement of C-particles are observed.

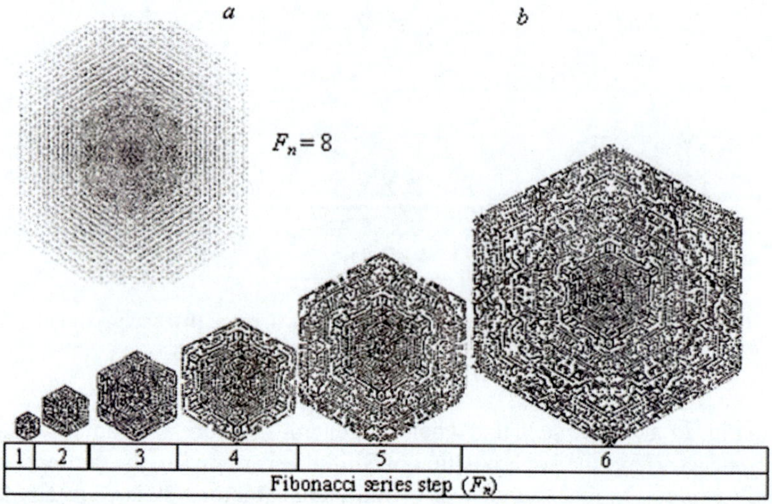

Fig. 3 General view (**a**) and detailed patterns of self-similar microblocks (**b**) according to the Fibonacci series of the MC structure of a crystal with a hexagonal system growing from one crystallization center

In the central sector of the inner zone, the maximum density and the highest degree of perfection of the lattice are recorded. It can also be seen from Fig. 3 that a high packing density of C-particles is observed along the boundaries of the central sector. The central sector has the symmetry of a crystalline polyhedron and can be identified as SE. Selecting fragments of the lattice with the closest packing of C-particles along their boundaries; one can form a set of microblocks with edge sizes obeying the Fibonacci sequence. This indicates the self-similar nature of each of the microblocks and their hierarchical subordination, including SE.

When organizing a polycentric crystal at the initial stages of crystallization, zoning does not appear and, as it grows, is fixed in the peripheral zones, as when growing from one crystallization center. Approaching the closest packing with the limiting overlap of exchange fields from different crystallization centers leads to zoning and microblocking of the internal structure and is the external state of the MC structure. In a hollow crystal, the exchange fields of crystallization centers intersect with small overlaps, and reverse zoning is observed (Fig. 2b). Structural parameters of the SE fields, when they are superimposed, provide the crystal with the same parameters and, depending on the location of the crystallization centers, one can have its variants from strictly symmetric to irregular. The MC model does not exclude the zoning of crystals when changing the physicochemical growth conditions. If the crystal grows with an attached SE, then the structure has a microblock character. If this condition is not met, the blockiness is regulated by the preservation of the electroneutrality of the growing crystal. With a change in the electrical environment in the growth cut, the electrical stability of the edge changes and habitus zoning appears. The direction of cleavage is determined by the anisotropy of mechanical properties caused by improper placement of the SE.

5 Model Description of Block Structures

The zonal and sectorial structure of crystals is confirmed for most natural minerals. Ideally, a single crystal grown from one crystallization center can be represented by one mycopolyhedron. The structural features of polycentric crystals grown from many centers of crystallization are determined by the structural characteristics of micropolyhedra and the way they are combined into a common structure. The structure of the surrounding regions of imperfect structure (interstitial spaces) [11] depends on the mutual orientation of micropolyhedra and compliance with the principle of closest packing. Interimpurity interaction leads to additional deformation of the surrounding regions of micropolyhedra and, when the ultimate strength of the material is exceeded, to the growth of dislocations. The different distribution of the interstitial space regions in depth is accompanied by a more fragile organization of micropolyhedra with complex boundaries that do not correspond to crystallographic faceting.

Three types of states of the interstitial space are distinguished in [10, 11], depending on the degree of conformity to the principle of the closest packing. In Fig. 2 shows a state in the form of a microlacuna between individual micropolyhedrons with the formation of reverse zoning. Decompression occurs in the arrangement of particles between micropolyhedrons without the formation of microlacunas with and without preservation of reverse zoning [10, 11]. Disorder zones with reverse zoning create a crystallographically correct system in accordance with the symmetry of the general structure, and oriented micropolyhedra are located with varying degrees of compliance with the closest packing law. When the micropolyhedra are misoriented at small angles, the principle of closest packing is not observed and the interstitial space is characterized by a chaotic change of zones of condensation and decompaction in the arrangement of particles (Fig. 2).

The unit cell often does not satisfy the principle of electroneutrality and, therefore, cannot act as a micropolyhedron. So, in GaAs, it consists of 18 Ga and As atoms and meets all the requirements, except for the principle of electroneutrality, although in many publications it is recognized as the minimum organization of atoms in a structure that reflects the physical properties of a crystal. The next micropolyhedron can be formed from 2 Ga and As atoms, where electroneutrality is preserved, but the principle of external faceting is violated. All the requirements put forward are met for micropolyhedra of 94 Ga and As atoms [10]. The outer faceting of such a micropolyhedron consists of the cube (100) and tetrahedron (111) faces, and the dimensions along the main axes are $a = b = c = 1.33$ nm. The size distribution of micropolyhedra often obeys the Fibonacci law [1, 12].

The model description of block structures, in contrast to uniform-atomic ones, studied on the basis of X-ray structural analysis, can be carried out by quantization levels from large to small-block ones, by introducing a normalization of the golden ratio and related Fibonacci numbers. A feature of these model concepts is dynamism, since crystals can be subjected to tensile, compression, and torsion deformations, which makes it possible to analyze the block structure at one stage or another of its growth or the impact of energy factors. It becomes possible to move from the old two-stage scheme crystal—atomic structure to the new three-stage crystal—block structure—atomic structure.

Based on the model of block structures of crystalline and glassy materials, not only the data of IR and Raman spectroscopy, but also other methods of analysis are interpreted in different ways. For example, IR and Raman spectra of crystalline and glassy Na_2SiO_4, LaB_3O_6, VPO_4, and other oxide systems are similar and have the form of an envelope of the crystal spectrum and indicate that the structures consist of the same structural units or blocks [13, 14]. At the same time, a layered structure characteristic of α VPO_4 is observed in the glass, but there is a sufficient number of VO_6 structural elements with parallel bonds. The concentration of boron rings in glasses B_2O_3, B_2O_3–GeO_2, B_2O_3–SiO_2, which depends on the cycle period and heating temperature, is estimated by the change in the intensity of the cattle band in the region of 808 cm^{-1}. In crystalline structural units based on Voronov polyhedra, the volume of space is taken, where all points are located closer to a given center than to others [15–17]. In three-dimensional space, they are convex and formed by planes perpendicular to the vectors connecting the centers and passing through the midpoints of these vectors. On the basis of the concept of crystal modules, it turned out to be convenient to describe twin crystals $LiNbO_3$, $LiTaO_3$, and their arrangement is determined and described by the crystal group, and not by the class of the crystal [18].

6 Conclusion

The creation of new geometric means for studying crystals allows us to consider in a new way the organization of filling the crystal space with atoms. The theoretical crystal structure in the Riemannian interpretation of the Fedorov group has a micropolyhedral structure, which contributes to the separation of crystalline structural units, the internal structure of which, despite its small size, must satisfy all the physical properties of a real crystal. Dynamic modeling of micropolyhedral rearrangements, allowing the simultaneous transformation of all micro- and macropolyhedra without violating the integrity of the structure.

Simulation of MC structures of crystalline substances is characterized by the spiral growth of crystals from C-particles. During the growth of crystals from one crystallization center, self-similar structural elements and self-similar microblocks from them with sizes along the Fibonacci series were isolated. The highest packing density of C-particles along the boundaries of structural elements correlates with the observed bands of maximum electron density. In an MC structure with different schemes of distribution of C-particles, the same outer faceting with the corresponding symmetry for a given crystal is always obtained.

Acknowledgements This research was supported by the TPU Development Program.

References

1. Vereshchagin, V.I., Sergeev, M.A., Semukhin, B.S., Borodin, Y.V.: Boron nitride with packets of nanotubes for microcomposite ceramics. Refract. Ind. Ceram. **41**(11–12), 440–444 (2000). https://doi.org/10.1023/A:1011322504412
2. Rothen, F., Pieranski, P.: Mechanical equilibration of conformal crystals. Phys. Rev. E **53**(3), 2828–2842 (1996). https://doi.org/10.1103/PhysRevE.53.2828
3. Visscher, M.I., Baner, C.E.W.: Mesosopic charge-density wave junctions. Phys. Rev. B **54**(4), 2798–2805 (1996). https://doi.org/10.1103/PhysRevB.54.2798
4. Kulakov, M.A., Henn, G., Bullemer, B.: SiC (0001) 3×3-Si surface reconstruction—a new insight with STM. Surf. Sci. **346**(1–3), 49–54 (1996). https://doi.org/10.1016/0039-6028(95)00919-1
5. Edelman, V.S.: STM observation of twin microlayers on cleaved bismuth surfaces. Phys. Lett. A **210**(1–2), 105–109 (1996). https://doi.org/10.1016/0375-9601(95)00864-0
6. Gleiter, H.: Nanocrystalline solids. J. Appl. Crystallorgr. **24**(2), 79–90 (1991). https://doi.org/10.1107/S0021889890011013
7. Michalski, E., Kasczmarek, S., Demianiuk, M.: The diffraction of X-rays by close-packed polytypic crystals containing singe stacking faults. II. Theory for hexagonal and rhombohedral structures. Acta Crystallogr. Sect. A Found. Crystallogr. **44**(5), 650–657 (1988). https://doi.org/10.1107/S0108767388003241.
8. Santra, S.B., Bose, I.: Scaling behaviour of cluster hulls in spiral site percolation. J. Phys. A: Math. Gen. **26**(16), 3963–3971 (1993). https://doi.org/10.1088/0305-4470/26/16/013
9. Borodin, Y.V.: Low-temperature nanodoping of protonated $LiNbO_3$ crystals by univalent ions. Tech. Phys. **60**(1), 107–111 (2015). https://doi.org/10.1134/S1063784215010065
10. Rudnev, S.V.: Application of elliptic Riemannian geometry to problems crystallography. Comput. Math. Appl. **16**(5–8), 597–616 (1988). https://doi.org/10.1016/0898-1221(88)90249-0
11. Sadoc, I.- F., Charvolin, I.: Crystal structures built from highly symmetrical units. J. Phys. Sec. I **2**(6), 845–859 (1992). https://doi.org/10.1051/jp1:1992183
12. Wang, C., Barrio, R.A.: Theory of the Raman response in Fibonacci superlattices. Phys. Rev. Lett. **61**, 191 (1988). https://doi.org/10.1103/PHYSREVLETT.61.191
13. Malinovsky, V.R., Novikov, V.N., Sokolov, A.P., Dodonov, V.G.: Low-frequency Raman scattering on surface vibrational modes of microcrystals. Solid State Comm. **67**(7), 725–729 (1988). https://doi.org/10.1016/0038-1098(88)91015-0
14. Condrate, R.A., Sr.: Vibrational spectra structural units in glass. J. Non-Cryst. Solid **84**(1–3), 26–33 (1986). https://doi.org/10.1016/0022-3093(86)90759-3
15. Medvedev, N.N.: The algorithm for three-dimensional Voronoi polyhedra. J. Comput. Phys. **67**(1), 223–229 (1986). https://doi.org/10.1016/0021-9991(86)90123-3
16. Surzhikov, A.P., Pritulov, A.M., Lysenko, E.N., Sokolovskiy, A.N., Vlasov, V.A., Vasendina, E.A.: Calorimetric investigation of radiation-thermal synthesized lithium pentaferrite. J. Therm. Anal. Calorim. **101**(1), 11–13 (2010). https://doi.org/10.1007/S10973-010-0788-7

17. Surzhikov, A.P., Lysenko, E.N., Malyshev, A.V., Pritulov, A.M., Kazakovskaya, O.G.: Influence of mechanical activation of initial reagents on synthesis of lithium ferrite. Russ. Phys. J. **55**(6), 672–677 (2012). https://doi.org/10.1007/s11182-012-9865-7
18. Lemeshko, V.V., Obukhovsky, V.V., Rez, I.S.: Twinning of crystals and crystal moduli conception. Cryst. Res. Technol. **23**(7), 855–862 (1988). https://doi.org/10.1002/CRAT.2170230704

Study of Fire-Retardance Effectiveness of Wood Treated with Fire Retardants

Natalia Verner, Konstantin Alekseev, Olga Nazarenko, and Yulia Amelkovich

Abstract The article is devoted to the study of the flammability of wood impregnated with fire retardants: melamine polyphosphate, melamine poly(magnesium phosphate), aluminum hydroxide, and boric acid. A serious drawback of wood, limiting its use, is its fire hazard. One of the methods for increasing the fire protection of wood is the application of fire retardants to the surface of wooden structures. To determine the effect of wood treatment with fire retardants on the fire performance properties of wood, the method for determining the ash content and the method of the quality control of fire-retardant treatment were used in this work. The study showed that the fire retardant efficiency of wood with a double application of the solution provides only melamine polyphosphate coating of wood. The treatment of wood with aqueous solutions of boric acid, aluminum hydroxide, and melamine poly(magnesium phosphate) requires multiple coating of these fire retardants. the ash content of the wood samples coated with fire retardants is higher than that of a sample of untreated wood.

1 Introduction

Wood is a widespread and affordable building material, convenient to use, practical, and environmentally friendly [1, 2]. A serious drawback limiting the use of wood in the building industry is its increased fire hazard. One of the factors that determine the fire hazard of wood is its ability to ignite and spread combustion when heated in the air. The combustion process of wood is well studied [2–5]. In general, there

N. Verner · O. Nazarenko (✉) · Y. Amelkovich
Division for Testing and Diagnostics, National Research Tomsk Polytechnic University, 30 Lenin Ave, Tomsk 634050, Russia
e-mail: olganaz@tpu.ru

Y. Amelkovich
e-mail: amely@tpu.ru

K. Alekseev
FSBI Forensic Institution of the Federal Fire Service "Test Fire Laboratory", 7a Vershinin str, Tomsk 634050, Russia

E. Lysenko et al. (eds.), *Recent Developments in the Field of Non-Destructive Testing, Safety and Materials Science*, Studies in Systems, Decision and Control 433,
https://doi.org/10.1007/978-3-030-99060-2_16

are several stages of the thermal decomposition of wood. The first stage is characterized by the removal of free and bound moisture, with a maximum of around 100 °C. At temperatures in the range of 180–350 °C, hemicelluloses decompose, and at 225–350 °C, cellulose decomposes. The most thermostable component is lignin, its decomposition occurs gradually, up to 500 °C [2, 6, 7]. The high thermal stability of lignin is explained by its highly cross-linked structure and high molecular weight [8]. Oxidative destruction of wood includes two successive stages: thermal decomposition of components and oxidation of the resulting coke. In both cases, the process is accompanied by the release of a significant amount of heat. In this case, depending on the rate of the oxidation reaction, the ratio of the rates of heat release and heat loss, decomposition of wood can proceed without the occurrence of combustion, when the intensity of the thermo-oxidative process is determined only by the value of the external heat flux. When the rate of heat release exceeds the rate of heat loss from the reaction zone, the oxidation process goes into a glow or flame combustion mode.

Fire protection of wooden structures is an important stage in the implementation of a set of measures to prevent fires in houses and buildings where walls, roofs, ceilings, and interior covering of premises are made of wood. An effective way of fire protection of wood is its impregnation with an aqueous solution of a fire retardant. Fire retardants are the chemicals that suppress the combustion reaction at the stage of ignition, being combustion inhibitors. In the case of wood treatment with fire retardant, the substance penetrates deep into the wood, filling the pores and fibers [2, 3]. The effect of fire retardants on wood combustion is to delay the time to ignition of wood, to reduce the rate of heat release during combustion, and to reduce the spread of the flame over the surface [2, 9].

As fire retardants, substances are usually used, the molecules of which contain elements such as halogens, phosphorus, nitrogen, boron, and metals [10, 11]. With the help of fire retardants, it is possible to slow down or suppress certain stages of the combustion of wood material. Effective fire retardants change the pyrolysis mechanism, reduce the yield of combustible products, or inhibit flame combustion, as a result of which the amount of heat during the exothermic oxidation process in the gas phase decreases.

The most effective from the point of view of providing fire protection in the solid phase and during smoldering are phosphorus-containing compounds that can decompose when heated to form phosphoric acid. The presence of phosphoric acid changes the CO/CO_2 ratio in the direction of inhibiting the direct oxidation of carbon to CO_2, significantly reducing the exothermic effect of the process [12, 13]. When using compositions based on phosphorus acids and nitrogen-containing compounds, the amount of heat required for deep decomposition of the material increases and at the same time, the amount of heat released decreases. In the gas phase, the most effective are inorganic and organic nitrogen-containing compounds, which decompose to form nitrogen and ammonia. When nitrogen- and phosphorus-containing compounds are used together, a synergistic effect is observed, that is, the effect of such compositions is higher than the effectiveness of compositions in which only phosphorus or nitrogen-containing compounds are used. Inorganic fire retardants (aluminum hydroxide, ammonium polyphosphate, silicon compounds, etc.), when heated, decompose endothermically and release water vapor, which cools the material to a temperature lower than that required to maintain combustion processes, and form an oxygen-deficient zone [14, 15]. They also catalyze the processes of char formation and contribute to the formation of a protective layer on the burning surface. Boron-containing compounds increase the amount of char during combustion; the decomposition of boron compounds occurs in a wide temperature range in several stages [16–19]. Also, with the same boron content, hydrogenated compounds, for example, boric acid, are more effective than anhydrous ones, such as boric anhydride.

The aim of this work is to study the effect of wood treatment with such fire retardants as aluminum hydroxide, boric acid, melamine polyphosphate, melamine poly(magnesium phosphate) using the methods for quality control and determination of ash content.

2 Research Method

Materials. In this work, aluminum hydroxide, boric acid, melamine polyphosphate, melamine poly(magnesium phosphate) were used as fire retardants. It is known that each substance under consideration is used both independently and in the number of fire retardants.

Aluminum hydroxide is an inorganic compound that is widely used as a flame retardant. At the heating, aluminum hydroxide decomposes in the endothermic reaction with the release of water, absorbs heat energy, cooling the combustion zone, and forms a non-combustible aluminum oxide layer on the surface of the material.

Boric acid is also a substance with fire retardant properties due to endothermic decomposition upon heating and the formation of a glassy coating.

Melamine polyphosphate and melamine poly(magnesium phosphate) are the phosphorus–nitrogen systems and act by several mechanisms. The released phosphoric acid promotes char formation on the combustion surface. The release of nitrogen dilutes the combustible volatile products of the wood.

Quality control of fire-retardant treatment. To control the fire retardant treatment in practice, the quality control of fire-retardant treatment is widely used. This method is regulated by the national standard of Russian Federation GOST R 53292–2009 "Fire retardant compositions and substances for wood. General requirements. Test methods". The standard also specifies requirements for sampling for testing.

To control the quality of fire retardant processing, wood samples treated with fire retardants are exposed to fire, while the following indicators are recorded:

- discoloration, shrinkage, warping, smoldering, etc.;
- the appearance of signs of ignition (flame combustion outside the zone of exposure to the flame of a gas burner);
- self-combustion after turning off the gas burner;
- burn-through to hole formation;
- full-depth charring in the zone of exposure to the flame of a gas burner;

– complete or incomplete charring of the side of the specimen treated with fire retardant within the area enclosed by the clamping frame.

The samples were made from straight-layered air-dry pinewood with a moisture content of 8–15% and a density of 400–550 g/m^3. The wood had the form of rectangular bars with a cross-section of 30 × 60 mm and a length of 150 mm along the fibers, the deviation from dimensions was not exceeded 1 mm.

To carry out the test to the quality control of fire-retardant treatment, a wooden block was used, which was covered on four sides with aqueous solutions of fire retardants through painting. Preparation of the fire retardants solution was made by dissolving them in water in a weight ratio of 7:3. Fire retardant solutions were applied to the surface in two layers. After applying the fire retardant solutions, the bar was left to dry completely. Then, the samples for research were prepared according to the recommendations of GOST R 53292–2009, which represented a surface layer of fire-resistant wood (wood chips) with a length of at least 50–60 mm, a width of at least 25–35 mm, and a thickness of at least 1.5–2.5 mm. For testing, 4 samples for each substance were prepared (Fig. 1).

Before testing, the samples were kept indoors at room temperature, since it is unacceptable to test the sample as wet wood chips. The installation for the quality control of the fire-retardant treatment of wood structures (Fig. 2) was prepared for work before the test: the gas burner was ignited, the flame height was adjusted.

Each sample was kept under the burner flame for 40 s, after which the flame was removed, observations were recorded.

Ash content determination. When conducting fire-technical expertise, the question often arises whether the wooden structures of the building before the fire had fire protection and, in particular, whether any object (wall, beam, packing shavings, etc.) was impregnated with fire retardants. This question can be answered by examining the charred remains of these objects. The method for determining ash content allows defining the presence of a fire retardant content in wood and its char residues. An indicator of the presence of a fire retardant is an increased ash content. A sample treated

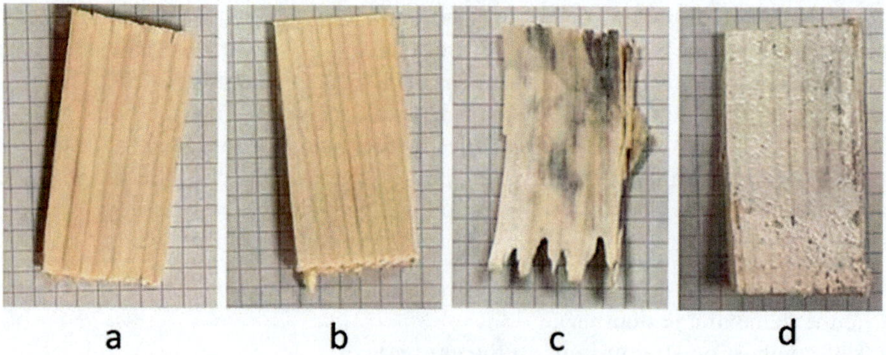

Fig. 1 Wood samples coated with aluminum hydroxide (**a**), boric acid (**b**), melamine poly(magnesium phosphate) (**c**), melamine polyphosphate (**d**)

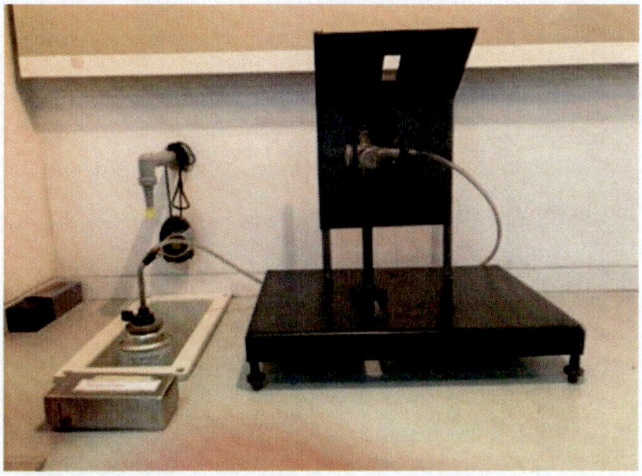

Fig. 2 Installation for testing according to the method of quality control of fire-retardant treatment

with a fire retardant has higher ash content than the original (untreated) wood. This method can be used both to establish the fact of fire protection (qualitative analysis) and to determine the content of fire retardant in wood (quantitative analysis).

To carry out the method for determining ash content, the samples were prepared as follows: the samples were crushed into small chips (Fig. 3), and then weighed.

The samples were placed in a muffle furnace for 2 h at a temperature of (800 ± 10) °C. After calcination, the samples were taken out and cooled for 10 min. After cooling, the samples were weighed. The ash content (A) of the sample was calculated by the formula:

$$A = [(m_3 - m_1)/(m_2 - m_1)] \cdot 100\%,$$

where

m_1 is the empty crucible mass, g;
m_2 is the mass of the crucible with the initial weighed portion of the sample, g;
m_3 is the mass of the crucible with ash, g.

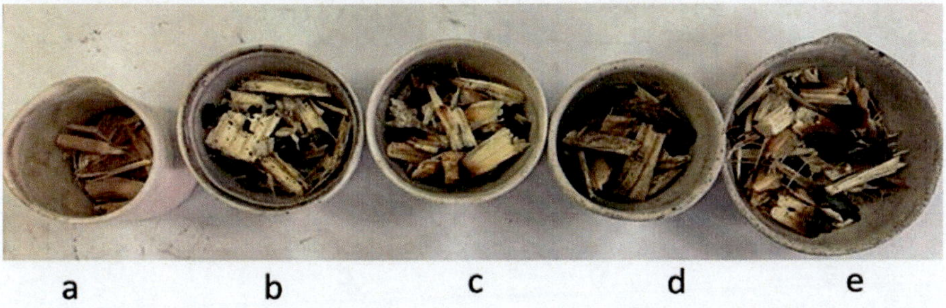

Fig. 3 Shredded wood samples: untreated wood (**a**); wood coated with melamine polyphosphate (**b**); melamine poly(magnesium phosphate) (**c**); boric acid (**d**); aluminum hydroxide (**e**)

3 Results and Considerations

The quality control of fire-retardant treatment. The appearance of the wood samples after the quality control tests is shown in Fig. 4.

The test results were evaluated according to the following criteria:

1. Self-combustion after turning off the gas burner: it is allowed to have local combustion in the zone of influence of the gas burner for no more than 5 s after turning it off);
2. Burn-through to the formation of a hole;
3. Charring of the sample side treated with fire retardants over the entire area bounded by the frame of the clamping device;
4. Charring to the full depth in the zone of exposure to the flame of a gas burner in the presence of signs of ignition: flame combustion outside the zone of influence of the flame of a gas burner.

The results obtained by the method of the quality control of fire-retardant treatment are presented in Table 1. Table 1 shows the compliance of the samples with the specified criteria—match or not.

The results obtained indicate that only melamine polyphosphate provides the fire-retardant efficiency of wood with a double application of the substance since it has been tested according to all criteria. For the wood samples treated with aluminum hydroxide, boric acid, and melamine poly(magnesium phosphate) the results were negative since these samples have not passed the tests according to the criteria. Charring to the full depth in the zone of exposure to the flame of a gas burner in the presence of signs of ignition was observed on some wood samples treated with these fire retardants.

Ash content determination. Figure 5 shows the remains of wood samples after calcination.

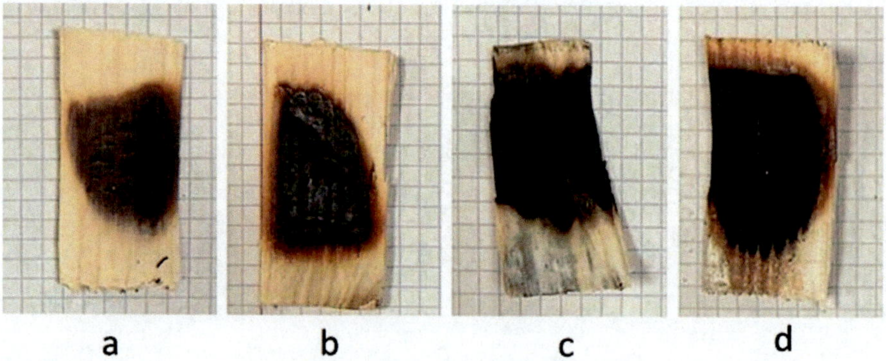

Fig. 4 Wood samples after the quality control of fire-retardant treatment testing: coated with aluminum hydroxide (**a**), boric acid (**b**), melamine poly(magnesium phosphate) (**c**), melamine polyphosphate (**d**)

Table 1 Quality control of fire-retardant treatment test results

Sample	Sample no	Self-burning time, (s)	Burn-through to the formation of a hole	Charring of the sample side treated with fire retardants over the entire area bounded by the frame of the clamping device	Charring to the full depth in the zone of exposure to the flame of a gas burner in the presence of signs of ignition	Test result
Aluminum hydroxide	1	–	–	–	–	Match
	2	2	–	–	–	Match
	3	7	–	–	+	Does not match
	4	3	–	–	–	Match
Boric acid	1	8	–	–	+	Does not match
	2	2	–	–	–	Match
	3	6	–	–	+	Does not match
	4	4	–	–	–	Match
Melamine poly(magnesium phosphate)	1	3	–	–	–	Match
	2	8	–	–	+	Does not match
	3	2	–	–	–	Match
	4	8	–	–	+	Does not match
Melamine polyphosphate	1	1	–	–	–	Match
	2	–	–	–	–	Match
	3	3	–	–	–	Match
	4	2	–	–	–	Match

The results of weighing wood samples and calculating ash content are presented in Table 2.

It can be seen from Table 2 that the table shows that all wood samples treated with fire retardants are characterized by a higher ash content in comparison with the untreated wood sample. The highest ash content was obtained for a wood sample

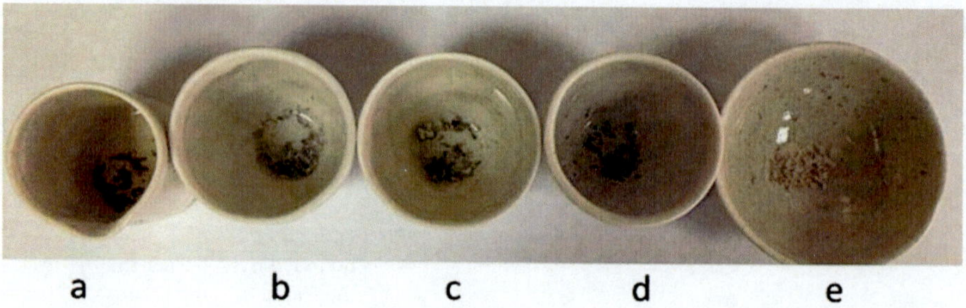

Fig. 5 Wood samples after calcination: untreated wood (**a**); wood coated with melamine polyphosphate (**b**); melamine poly(magnesium phosphate) (**c**); boric acid (**d**); aluminum hydroxide (**e**)

Table 2 Ash content test results

Sample	m_1, (g)	m_2, (g)	$m_2 - m_1$, (g)	$m_3 - m_1$, (g)	A, (%)
Untreated wood	22.58	23.99	1.41	0.0033	0.23404
Wood coated with melamine polyphosphate	27.16	29.75	2.59	0.0212	0.81853
Wood coated with melamine poly(magnesium phosphate)	25.84	28.43	2.59	0.0194	0.74903
Wood coated with boric acid	29.53	31.59	2.06	0.0142	0.68932
Wood coated with aluminum hydroxide	23.87	26.49	2.62	0.0483	1.84351

treated with aluminum hydroxide. Determination of ash content made it possible to establish the fact of wood processing with fire retardants by comparison of the ash content of the treated and untreated surface. According to the results of this study, it can be concluded that the ash content of the wood samples coated with fire retardants is higher than that of a sample of untreated wood.

4 Conclusion

In this work, a study was carried out of the wood treated with the following fire retardants: aluminum hydroxide, boric acid, melamine polyphosphate, melamine poly(magnesium phosphate). For the study, two methods were used: a method of the quality control of fire-retardant treatment and a method for determining ash content.

The results of the study by the method of the quality control of fire-retardant treatment showed that double impregnation of wood with aluminum hydroxide, boric acid, and melamine poly(magnesium phosphate) does not provide fire protection. To achieve reliable fire protection, repeated application of these fire retardants to wood and additional research is required.

The analysis of the results obtained during the study by the method of determining the ash content showed that by this method it is possible to establish, in most cases, only the fact of carrying out fire retardant treatment, by comparing the analytical indicator of the ash content of the treated and untreated wood surface.

Acknowledgements This research was supported by TPU development program.

References

1. White, R.H., Dietenberger, M.A.: Fire safety of wood construction. Wood Handbook. Department of Agriculture, Forest Service, Madison, WI, U.S. (1999)
2. Lowden, L.A., Hull, T.R.: Flammability behaviour of wood and a review of the methods for its reduction. Fire Sci. Rev. **2**, 1–19 (2013)
3. Dietenberger, M.: Update for combustion properties of wood components. Fire Mater. **26**, 255–267 (2002)
4. Babrauskas, V.: Ignition of wood: a review of the state of the art. J. Fire. Prot. Eng. **12**(3), 163–189 (2002)
5. Friquin, K.L.: Material properties and external factors influencing the charring rate of solid wood and glue-laminated timber. Fire Mater. **35**, 303–327 (2010)
6. Kim, H.S., Kim, S., Kim, H.J., Yang, H.S.: Thermal properties of bio-flour-filled polyolefin composites with different compatibilizing agent type and content. Thermochim. Acta **451**, 181–188 (2006)
7. Bryś, A., Bryś, J., Ostrowska-Ligęza, E., Kaleta, A., Gorniski, K., Glowaski, S., Koszon, P.: Wood biomass characterization by DSC or FT-IR spectroscopy. J. Therm. Anal. Calorim. **126**, 27–35 (2016)
8. Yang, H., Yan, R., Chen, H., Zheng, C., Lee, D., Liang, D.T.: In-depth investigation of biomass pyrolysis based on three major components: hemicellulose, cellulose and lignin. Energy Fuels **20**, 388–393 (2006)
9. Yao, C.H., Wu, Y.Q., Hu, Y.C.: Flame-retardation characteristics and mechanisms of three inorganic magnesium compounds as fire-retardant for wood. J. Central South Univ. For. Technol. **32**(1), 18–23 (2012)
10. Li, Q., Jiang, P., We, P.: Synthesis, characteristic, and application of new flame retardant containing phosphorus, nitrogen, and silicon. Polym. Eng. Sci. **46**, 344–350 (2006)
11. Fudang, S., Zhiming, D.U., Xiaomin, C., Linshuag, Z., Ye, Y., Linming, L.: Experimental study on fires extinguishing properties of melamine phosphate powders. Procedia Eng. **84**, 535–542 (2014)
12. Van der Veen, I., De Boer, J.: Phosphorus flame retardants: properties, production, environmental occurrence, toxicity and analysis. Chemosphere **88**, 1119–1153 (2012)
13. Kandola, B.K., Horrocks, A.R., Price, D., Coleman, G.V.: Flame retardant treatments of cellulose and their influence on the mechanism of cellulose pyrolysis. J. Macromol. Sci. **36**, 721–794 (1996)
14. Hull, T.R., Witkowski, A., Hollingbery, L.: Fire retardant action of mineral fillers. Polym. Degrad. Stab. **96**, 1462–1469 (2011)
15. LeVan, S.L.: Chemistry of fire retardancy. In: Rowell, R. (Ed.) The Chemistry of Solid Wood, vol. 207, pp. 531–574. American Chemical Society, Washington (1984)
16. Atar, M., Keskin, H.: Impacts of coating with various varnishes after impregnation with boron compounds on the combustion properties of Uludag fir. J. Appl. Polym. Sci. **106**, 4018–4023 (2007)
17. Blasi, C.D., Branca, C., Galgano, A.: Flame retarding of wood by impregnation with boric acid—pyrolysis products and char oxidation rates. Polym. Degrad. Stab. **92**, 752–764 (2007)

18. Pedieu, R., Kouba, A., Riedl, B., Wang, X.-M., Deng, J.: Fire-retardant properties of wood particleboards treated with boric acid. Eur. J. Wood Wood Prod. **70**, 191–197 (2012)
19. Popescu, C.-M., Pfriem, A.: Treatments and modification to improve the reaction to fire of wood and wood based products—an overview. Fire Mater. **44**(1), 100–111 (2020)

Determination of Correlation Between Photoelectric Properties and Structural Features of Nanocomposite Photoelectrode $ZnO:SnO_2$ by Impedance Spectroscopy

Dinara Kambar, Togzhan Seisembekova, Assylbek Zeinidenov, Aitbek Aimukhanov, and Yury Borodin

Abstract The objects of the study were nanocomposite materials and inverted solar cells based on metal oxides $ZnO:SnO_2$. The aim of the work was to study the influence of structural features of the nanocomposite photoelectrode $ZnO:SnO_2$ on the efficiency of charge injection by impedance spectroscopy. Experimental studies were carried out by methods of optical spectroscopy, voltamperometry and impedance measurement. Thin films of zinc oxide and tin were synthesized by the sol–gel method. The optical properties of the obtained nanostructured ZnO and SnO_2 films are investigated. Optimal technological conditions for the production of ZnO nanocomposite films have been $ZnO:SnO_2$ with a developed surface. Studies have been carried out to establish a correlation between the photoelectric properties and structural features of the nanocomposite photoelectrode $ZnO:SnO_2$. From the analysis of the cell impedance spectra, it was found that there is an optimal thickness of ZnO and SnO_2 films, at which a balance is observed between the efficiency of transport and recombination of charge carriers. In this case, the electrons in the photoactive layer have the maximum lifetime of charge carriers and a low probability of recombination. Analysis of the obtained VAC showed that the use of a nanocomposite layer of $ZnO:SnO_2$ leads to an increase in the efficiency of solar energy conversion. Thus, with a thickness of SnO_2 films equal to 54 nm, the cell current density is 5.2 mA/cm^2. With a photoelectrode thickness of SnO_2 equal to 62 nm in the composite structure of $ZnO:SnO_2$ in comparison with SnO_2, there is a decrease in the current density to a value of 5.1 mA/cm^2. At the same time, there is an increase in the no-load voltage by 2.7 times, which leads to an increase in the efficiency of the cell with a composite electrode $ZnO:SnO_2$ by 1.5 times.

D. Kambar · T. Seisembekova · A. Zeinidenov · A. Aimukhanov
Center of Nanotechnology and Functional Nanomaterials, Buketov Karagandy University, University str. 28, Karaganda 100028, Kazakhstan
e-mail: tosh_0809@mail.ru

Y. Borodin (✉)
Division for Testing and Diagnostics, National Research Tomsk Polytechnic University, 30 Lenin Avenue, Tomsk 634050, Russia
e-mail: uryborodin@tpu.ru

E. Lysenko et al. (eds.), *Recent Developments in the Field of Non-Destructive Testing, Safety and Materials Science*, Studies in Systems, Decision and Control 433,
https://doi.org/10.1007/978-3-030-99060-2_17

1 Introduction

Currently, compact layers, mesoporous and nanostructured films of ZnO, TiO_2 and SnO_2 are the most widely used materials in organic solar cells (OSCs) [1]. The energy conversion efficiency obtained in ZnO-based OSE is usually significantly lower compared to TiO_2 or SnO_2-based cells [2, 3]. Most research on ZnO-based solar cells is focused on two main problems: the first is the development of simple methods for the synthesis of one-dimensional ZnO nanostructures that have good conductive properties, and the second is the search for suitable photoactive materials for ZnO [4]. To solve this problem, we propose to use a nanocomposite electrode ZnO:SnO_2. This is due to the fact that the electric transport electrode with a composite structure of ZnO:SnO_2 will have a number of advantages over an electrode based on a single ZnO or SnO_2. Firstly, SnO_2 has a better hole conductivity than ZnO, which makes it possible to transfer photoinduced holes from the active layer to the electrode more quickly, thereby reducing the probability of recombination of charge carriers and increasing the photocurrent of the cell [5]. Thirdly, it is known that the SnO2-based OSE electrodes do not have a sufficiently high photovoltage due to the lower position of the quasi-Fermi electron level (EF) in SnO_2, when as a composite of ZnO:SnO_2 has a higher EF position, and hence a ZnO-based OSE ZnO:SnO_2 will have a greater value of the generated photovoltage [6]. Thus, the use of ZnO composite films ZnO:SnO_2 with a developed surface structure will increase the photovoltaic parameters of polymer solar cells.

In this paper, the ways of increasing the quantum efficiency of photogeneration of charge carriers in the organo-inorganic heterojunction are determined by establishing a correlation between the photoelectric properties and structural features of the nanocomposite photoelectrode ZnO:SnO_2 of a semiconductor film [7]. Based on the data obtained, the optimal technological parameters of the solar cell were determined, providing a high efficiency index.

2 Research Method

The preparation of substrates for photosensitive cells based on FTO was carried out according to the method [8]. The production of ZnO:SnO_2 films was carried out as follows: $Zn_5(OH)_8Cl_2$ (purity 99.9% Sigma Aldrich, weight m = 98.7 mg) was dissolved in a volume (V = 1 ml) of isopropanol (purity 99.9% Sigma Aldrich). Monoethanolamine (Sigma Aldrich) was added to the resulting solution after 20 min in an amount (V = 75 μl). After that, the solution was stirred at a temperature of T = 60 °C for 2 h, and then kept for 24 h at room temperature.

To obtain tin oxide films, the following technique was used: 50 mg of $SnCl_2$ (Sigma Aldrich) was added to 0.5 ml of ethyl alcohol, after which the solution was stirred on a magnetic stirrer at a stirring speed of 300 rpm for three hours at a temperature of 80 °C. After that, the solution was stored at room temperature for 12 h. To obtain films, the resulting solution was applied to the FTO surface by centrifugation (SPIN150i model, Semiconductor Production System) at rotational speeds of 2000, 3000, 4000, 5000, 6000 rpm. After the film was annealed in an air atmosphere at a temperature of 500 °C, annealed for one hour.

3 Description of the Installation

The impedance spectra were measured using a P-45X potentiostat (Elins) with an optional FRA-24 M frequency analyzer module. This module is designed to measure impedance characteristics at alternating current [9]. The frequency analyzer consists of two high-speed analog-to-digital converters (ADCs). The first ADC module is used to register an alternating current signal, the second is used to measure the potential value. In addition, the frequency analyzer module has a highly stable sinusoidal signal generator. The processing of the measured values of current and applied voltage is carried out by the main microprocessor. By analyzing the phase shift and the ratio of the amplitudes of these values, the values of the resistances of the imaginary and real parts of the impedance spectrum of the sample under study are calculated.

4 Results and Considerations

Figure 1 shows SEM images of transverse cleavage of films (a) and energy dispersion analysis of $ZnO:SnO_2$ composite films (b). The thickness of the deposited polymer ZnO was 86 nm, and the thickness of SnO_2 was 93 nm. To determine the composition of substances on the surface of the substrates, an energy dispersion analysis was carried out. Elemental analysis of the samples was carried out using the energy dispersion analyzer INCAPentaFET–x3 (Oxford Instruments, England). As

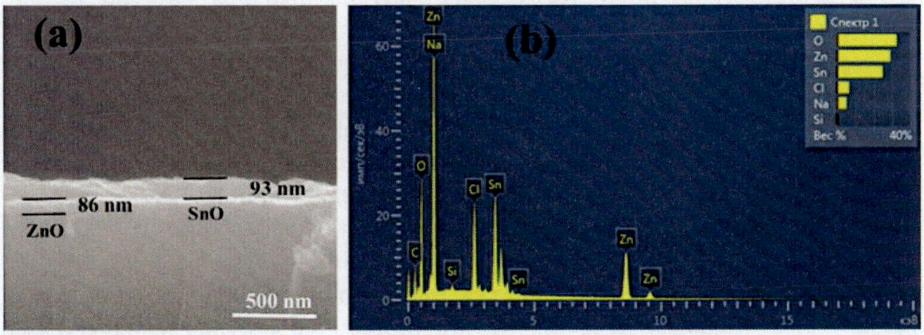

Fig. 1 Energy dispersion analysis of the surface of $ZnO:SnO_2$ composite films

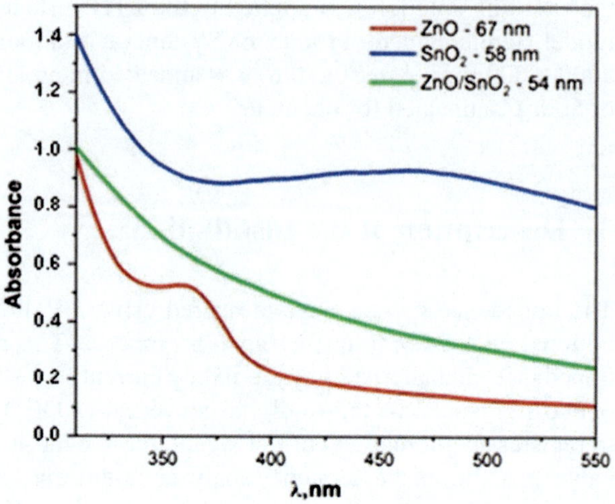

Fig. 2 Absorption spectra of ZnO, SnO_2 films and their ZnO:SnO_2 composites

the data of energy dispersion analysis showed, Zn, Na, Sn, O, Si, C are present in the composition of the film.

Figure 2 shows the absorption spectra of ZnO, SnO_2 films and their ZnO:SnO_2 composites. The edge of the absorption band falls for all films at 380–400 nm, which corresponds to the optical transition of the band gap of semiconductors. Figure 2 shows that the shape of the absorption spectra of composite films does not differ much from the absorption spectra of individual semiconductor oxides.

The impedance spectra were measured using a P-45X potentiostat (Elins) with an optional FRA-24M frequency analyzer module. This module is designed to measure impedance characteristics at alternating current. The frequency analyzer consists of two high-speed analog-to-digital converters (ADCs). The first ADC module is used to register an alternating current signal, the second is used to measure the potential value. In addition, the frequency analyzer module has a highly stable sinusoidal signal generator. The processing of the measured values of current and applied voltage is carried out by the main microprocessor. By analyzing the phase shift and the ratio of the amplitudes of these values, the resistances of the imaginary and real parts of the impedance spectrum of the sample under study are calculated [10].

To interpret the impedance spectra, an equivalent electrical circuit of a photovoltaic cell was used, where R1 (R_w) is the equivalent resistance of a multilayer film ($R_{ITO} + R_{Al}$ + resistance of a photoactive layer), R2 (R_{rec}) is the resistance characterizing the recombination of localized electrons with holes.

Figure 3 shows the impedance hodographs of inverted nanocomposite OSC based on ZnO:SnO_2 films. Using the ZnO composite ZnO:SnO_2 will contribute to the rapid transport of electrons. And this, in turn, will have an impact on the electrophysical characteristics of solar cells. Electric transport properties of polymer solar cells based on ZnO films ZnO:SnO_2 was analyzed using the impedance spectroscopy method.

The table shows that there is an optimal thickness of the SnO_2 film on the ZnO surface, at which the ratio R_{rec}/R_w increases and the ratio k_{eff}/τ_{eff} decreases (Table 1).

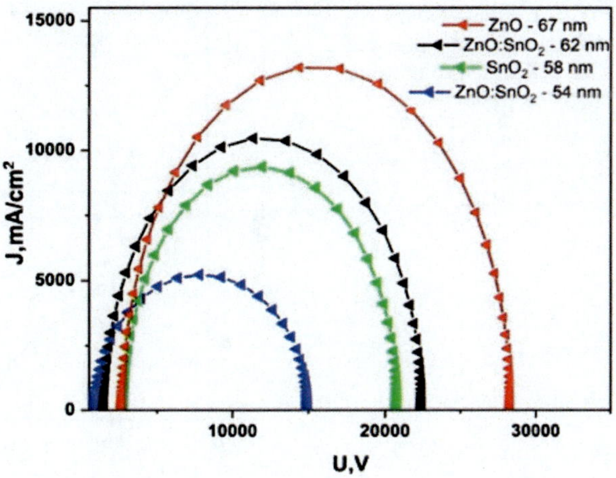

Fig. 3 Impedance hodographs of inverted nanocomposite OSC based on $ZnO:SnO_2$ films

Table 1 Value of electrophysical parameters of ZnO/SnO_2 films

Type of film	R_w, (Ohm)	R_{rec}, (Ohm)	R_{rec}/R_w	τ_{eff}, (mc)	k_{eff}, (c^{-1})
ZnO—67 nm	1142	15,393	13.4	1.3	754
SnO_2—58 nm	1501	20,952	14.1	1.1	910
ZnO/SnO_2—54 nm	1450	27,080	18.6	0.7	1325
ZnO/SnO_2—62 nm	1235	22,671	18.3	0.8	1199

Moreover, the system $ZnO:SnO_2$ has a shorter lifetime of charge carriers compared to individual films.

Analysis of the hodograph showed that the composite $ZnO:SnO_2$ has better electronic conductivity than ZnO, which makes it possible to transfer photoinduced holes from the active layer to the electrode more quickly and thereby reduces the likelihood of recombination of charge carriers, thereby increasing the photocurrent of the cell.

To improve the efficiency of the work, studies of the effect of the SnO_2 layer on the performance of the organic cell were carried out. Figure 4 shows the VAC of a polymer cell with an inverted nanocomposite photoelectrode with different thickness SnO_2. The Fig. 4 shows that the SnO_2 layer affects the parameters of the volt-ampere characteristics of the cells. Thus, the short-circuit current density of the cell increased by 4% as much as possible when using the SnO_2 layer, and the efficiency value was 0.9%. Organic solar cells based on the SnO_2 electrode do not have a sufficiently high photovoltage due to the lower position of the quasi-Fermi electron level (EF) in SnO_2 when as a composite of $ZnO:SnO_2$ has a higher EF position, and hence a ZnO-based cell $ZnO:SnO_2$ has a greater value of the generated photovoltage. The parameters of the VAC cells are given in Table 2.

The analysis of the obtained graphs allows us to conclude that the use of the nanocomposite layer $ZnO:SnO_2$ leads to an increase in the efficiency of solar energy conversion. In the case of an inverted structure with a ZnO photoelectrode, the

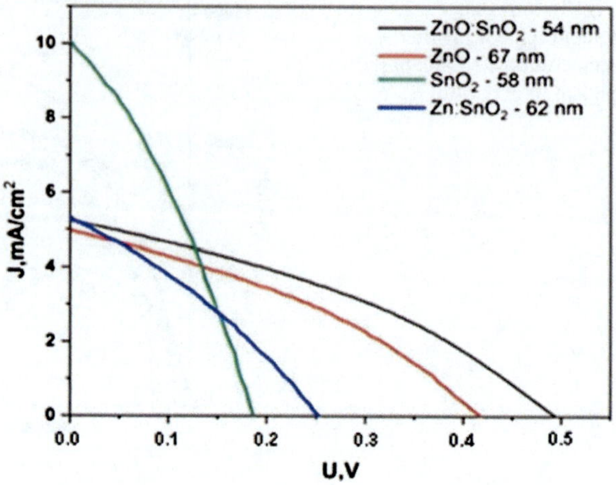

Fig. 4 Current–voltage characteristics of a polymer solar cell with a composite photo electrode

Table 2 Photovoltaic characteristics of organic solar cells

Type of film	J_{sc} mA/cm^2	U_{oc} V	J_{max} mA/cm^2	U_{max} V	*FF* %	*H* %
ZnO	4.9	0.42	2.9	0.26	0.36	0.7
SnO_2	9.9	0.18	5.7	0.11	0.35	0.6
ZnO/SnO_2—54 nm	5.2	0.25	2.9	0.14	0.31	0.4
ZnO/SnO_2—62 nm	5.1	0.5	3.2	0.29	0.36	0.9

maximum current density (JSCs) was 4.9 mA/cm^2 and for a SnO_2 photoelectrode it was 9.9 mA/cm^2. The current density for composite photoelectrodes depends on the thickness of the SnO_2 films, so with a film thickness of SnO_2 equal to 54 nm, the current density is 5.2 mA/cm^2. With a photoelectrode thickness of 62 nm, a decrease in current density to 5.1 mA/cm^2 is observed. The data obtained showed that the current density of the composite photoelectrode is ZnO:SnO_2 is reduced by 2 times compared to the SnO_2 photoelectrode. At the same time, there is an increase in the idle voltage by 2.7 times, which leads to an increase in efficiency by 1.5 times.

5 Conclusion

Methods of synthesis of ZnO:SnO_2 films have been developed by the sol–gel method. Optimal technological conditions for the production of nanocomposite films have been developed ZnO:SnO_2 with a developed surface. It is shown that the edge of the absorption band of composite films falls at 380–400 nm. Studies have been carried out to establish a correlation between the photoelectric properties and structural features of the nanocomposite photoelectrode ZnO:SnO_2. Analysis of the obtained VAC showed that the use of a nanocomposite layer of ZnO:SnO_2 leads to an increase in the efficiency of solar energy conversion. It is established that the current density for composite photoelectrodes depends on the thickness of the SnO_2 films. Thus, with a film thickness of SnO_2 equal to 54 nm, the current density of the cell is 5.2 mA/cm^2. With a photoelectrode thickness of SnO_2 equal to 62 nm in the composite structure of ZnO:SnO_2 in comparison with SnO_2, there is a decrease in the current density to a value of 5.1 mA/cm^2. At the same time, there is an increase in the no-load voltage by 2.7 times, which leads to an increase in the efficiency of the cell with a composite electrode ZnO:SnO_2 by 1.5 times.

References

1. Kim, S.A., Abbas, M.A., Lee, L., Kang, B., Kimc, H., Bang, J.H.: Control of morphology and defect density in zinc oxide for improved dye-sensitized solar cells. Phys. Chem. Chem. Phys. **18**(44), 30475–30483 (2016). https://doi.org/10.1039/C6CP04204J
2. Rogé, V., Didierjean, J., Crêpellière, J., Arl, D., Michel, M., Fechete, I., Dinia, A., Lenoble, D.: Tuneable functionalization of glass fibre membranes with ZnO/SnO_2 heterostructures for photocatalytic water treatment: effect of SnO_2 coverage rate on the photocatalytic degradation of organics. Catalysts **10**(733), 1–18 (2020). https://doi.org/10.3390/catal10070733
3. Surzhikov, A.P., Frangylyan, T.S., Ghyngazov, S.A.: A dilatometric study of the effect of pressing on the kinetics of compression of ultrafine zirconium doxide powders under thermal annealing. Russ. Phys. J. **55**(4), 345–352 (2012)
4. Yan, Y., Yao, B., Cao, B., Ma, W.: Morphology-controlled fabrication of SnO2/ZnO nanocomposites with enhanced photocatalytic. Perform. Photochem. Photobiol. **95**, 1131–1141 (2019). https://doi.org/10.1111/php.13101
5. Sommeling, P.M., Spath, M., Roosmalen, J.A.M., Meyer, T.B., Meyer, A.F., Kohle, O.: Dye-sensitized nanocrystalline TiO_2 solar cells on flexible substrates. In: Proceedings of the 2nd WCE PSEC '98: 2nd World Conference and Exhibition on Photovoltaic Solar Energy Conversion. JRC Publications, Vienna, Austria (1998)
6. Nikiforov, I., Timofeev, V.A., Mashanov, V.I., Azarov, I.A., Loshkarev, I.D., Korolkov, I.V., Gavrilova, T.A., Yesin, M.Y.: Physics of semiconductors and quantum electronics morphology, structure, and optical properties of SnO (x) films. Russ. Phys. J. **63**(2), 276–281 (2020). https://doi.org/10.1007/s11182-020-02032-4
7. Banik, A., Ansari, M.S., Qureshi, M.: Efficient energy harvesting in SnO_2 based dye-sensitized solar cells utilizing nano-amassed mesoporous zinc oxide hollow microspheres as synergy boosters. ACS Omega **3**(10), 14482–14493 (2018). https://doi.org/10.1021/acsomega.8b02520
8. Galil, A.A., Mohamed, R., Balboul, J.: Nano ZnO thin films synthesis by sol–gel spin coating method as a transparent layer for solar cell applications. Sol-Gel Sci. Technol. **82**, 269–279 (2017). https://doi.org/10.1007/s10971-016-4277-8
9. Sergeev, V., Yurchenko, V., Vavilova, G., Belik, M., Bezkorovainy, P.: Detection of damages in electrically dead power lines. Stud. Syst. Decis. Control **351**, 51–57 (2021). https://doi.org/10.1007/978-3-030-68103-6_5
10. Bisquert, J., Mora-Sero, I., Fabregat-Santiago, F.: Diffusion-recombination impedance model for solar cells with disorder and nonlinear recombination. ChemElectroChem **1**(1), 1–9 (2014). https://doi.org/10.1002/celc.201300091

Curie Temperature Control of Magnetic Materials Using Thermogravimetric Measurements in Magnetic Field

Elena Lysenko, Anatoliy Surzhikov, Evgeniy Nikolaev, and Oldřich Starý

Abstract The influence of the heating rate at thermogravimetric measurements in a magnetic field on recorded values of the Curie temperature of magnetic materials was investigated. A reference nickel sample was chosen as the object of research. The thermogravimetric measurements was conducted using a Netzsch STA 449C Jupiter thermal analyzer and two permanent magnets, which were connected to the furnace of the thermal analyzer. The heating rate varied from 1 to 50 K/min. The corundum crucibles of high density were used. The Curie temperature of investigated sample was determined using the temperature position of the derivative thermogravimetric curve in the region of the magnetic-paramagnet transition. It was shown that the heating rate at thermogravimetric analysis affects the measured T_C that ranges from 356.7 to 360.9 °C. Thus, a change in the heating rate leads to fluctuations in the Curie temperature up to ±4 °C for nickel. It was found that by controlling T_C of magnetic materials at a high heating rate, the obtained value is closer to the literature data.

E. Lysenko (✉) · A. Surzhikov · E. Nikolaev
Tomsk Polytechnic University, 30 Lenin av., 634050 Tomsk, Russia
e-mail: lysenkoen@tpu.ru

A. Surzhikov
e-mail: surzhikov@tpu.ru

E. Nikolaev
e-mail: nikolaev0712@tpu.ru

O. Starý
Czech Technical University, Prague, Czech Republic
e-mail: oldstar@zentrum.cz

E. Lysenko et al. (eds.), *Recent Developments in the Field of Non-Destructive Testing, Safety and Materials Science*, Studies in Systems, Decision and Control 433,
https://doi.org/10.1007/978-3-030-99060-2_18

1 Introduction

Thermogravimetric analysis (TG) allows automatic registration of changes in the mass of samples depending on the temperature–time regime, including non-isothermal and isothermal heating, as well as the composition of the gaseous atmosphere. Thermogravimetry is widely used to study the kinetics of physical and chemical processes, as well as to study the nonstoichiometry of materials [1–5].

As known [6–8], during thermogravimetric measurements in a magnetic field, when a ferro- or ferrimagnetic material is heated, a change in the sample mass is observed on the TG curve in the region of the magnetic phase transition, which determines the transition temperature from the magnetic state to the paramagnetic state of the material. Wherein, an increase or decrease in the mass depends on the direction of the magnetic field gradient. Thus, from the derivative of the TG curve (DTG curve), it is possible to determine the Curie temperature of the sample. This method of determining the Curie temperature of magnetic materials using TG in the literature is called the thermomagnetometric (TM) method [9, 10]. A more detailed description of this method is given in [11].

Thermomagnetometric measurements are used in practice, both for determining the Curie temperature in known and unknown materials, and for temperature calibration of thermal analysis devices [12–15]. The TM method was used to analyze magnetic phase transitions in metals [16], ferromagnetic metallic glasses [17], and rare-earth iron garnets [18]. In our previous works [19–21] it was found that using TM measurements it is possible to determine the Curie temperature of materials and, from their values, to estimate the phase composition of complex ferrites.

Today, there are a number of thermal analysis instruments that allow thermogravimetric measurements in a magnetic field. However, along with a fairly simple execution of the measurement technique, there are difficulties with the correct determination of the obtained Curie temperature values. First of all, before experimental measurements, it is necessary to calibrate the device using reference samples. As a rule, such a calibration is carried out using data on the melting points of metals [22]. Secondly, the measurement signal can be influenced by many factors, including the atmosphere in the heating furnace, heating rate, sample mass, etc. [23–25]. There is very limited information in the literature on the influence of these factors on the measured Curie temperature of materials.

In this work, the influence of the heating rate during thermogravimetric measurements in a magnetic field on the recorded values of the Curie temperature of nickel was investigated.

2 Research Method

Thermogravimetric measurements were carried out using a highly sensitive STA 449C Jupiter thermal analyzer (Netzsch-Geratebau GmbH, Germany) with a balance

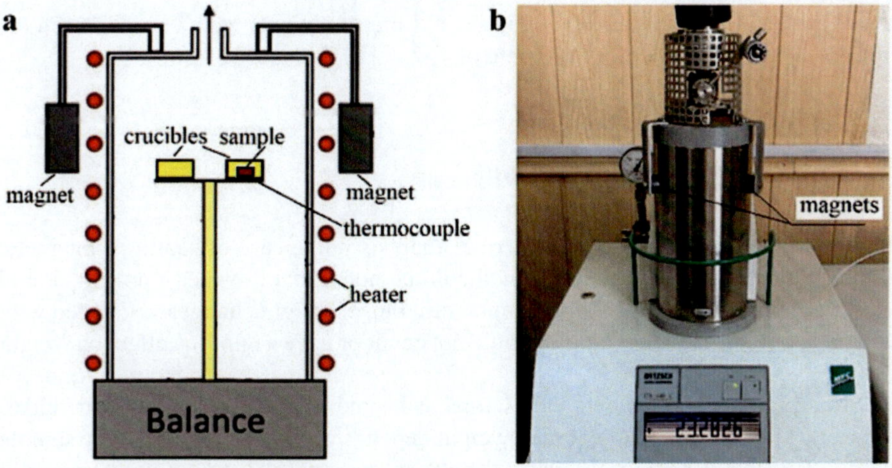

Fig. 1 Scheme of the measuring cell (**a**) and image (**b**) of thermal analyzer

resolution of 0.1 μg. It should be noted that this device allows simultaneous thermogravimetric and calorimetric measurements for the same sample in the temperature range (20–1500) °C and the heating or cooling rate (0.01–50) K min^{-1}. This synchronous analysis greatly improves measurement performance and simplifies the interpretation of results.

Figure 1 shows the weight part of the analyzer (Fig. 1a) and the view of the device (Fig. 1b). In this work, we used a sensor with flat platinum–rhodium thermocouple, on which crucibles for the reference and the sample under study were placed. The measurements were carried out using high density corundum crucibles made of Al_2O_3 with a volume of 0.085 ml, which are widely used for thermogravimetric and calorimetric measurements [26, 27]. The crucible was covered with a lid with a hole, ensuring the interaction of the material with the surrounding atmosphere.

The thermal analysis device was calibrated according to the temperature and melting enthalpy of reference samples based on high-purity metals, specially supplied by the manufacturer of the device.

Nickel was used as the test material, which is included in the list of calibration materials. Before TG measurement of the test sample, the baseline was measured on an empty crucible under identical conditions as the heating of the test sample.

Since the results of thermal studies can depend on the mass of the sample, the experimental samples were prepared with the same initial mass, equal to about 5 mg. The nickel samples were heated in an air atmosphere and nitrogen protective gas with blowing rates of 20 and 10 ml min^{-1}, respectively.

As indicated above, thermomagnetometric analysis is the thermogravimetric measurement of a sample when an external magnetic field is applied. Thus, this analysis makes it possible to control the magnetic state of the samples [28–30]. In this work, an external magnetic field was created by attaching two permanent magnets to the outer part of the furnace. The schematic position of the magnets and the experimental cell is shown in Fig. 1. TM measurements were carried out in the non-isothermal heating mode at a rate of 1, 2, 5, 10, 20, and 50 K min^{-1}.

3 Results and Considerations

Figure 2 shows typical curves of thermal analysis in the case of heating a magnetic sample during TM measurements. It should be noted that this measurement should be carried out when, in the Curie temperature range, weight changes associated with chemical processes in the sample should not occur or have a minimal effect on weight changes.

From Fig. 2, when the magnetic sample is heated in an external magnetic field to the Curie temperature, the recorded weight depends on the interaction of the sample with the magnetic field. In this case, the obtained weight value does not correspond to the true mass of the sample, since the magnetic sample is drawn into the magnetic field. When the Curie temperature is reached, the sample ceases to interact with the magnetic field, as it becomes non-magnetic. In this case, a weight jump occurs on the TG curve and the further recorded weight corresponds to the true mass of the sample.

When using the layout of the magnets and the sample in the crucible shown in Fig. 1a, the observed weight jump on the TG curve corresponds to an increase in weight. Using the derivative thermogravimetric curve (the first derivative of the TG curve), it is easy to determine the temperature of the magnetic-paramagnet transition corresponding to the Curie point.

Figures 3 and 4 show the results of thermomagnetometric analysis of nickel obtained at different heating rates. One can notice that the shape of the TG and DTG curves is similar to the typical behavior of the curves shown in Fig. 2.

It was found that the change in weight at the Curie temperature varies in the range 0.68–0.69% and does not depend on the heating rate. It should be noted that the obtained value depends on the magnetization of the sample. As shown in [31, 32], the higher the magnetization of the sample, the greater the change in weight occurs during TM measurement. In certain cases, when examining multiphase materials, calibration of the thermal analyzer using weight values allows to determine the quantitative content of the magnetic phase in the sample.

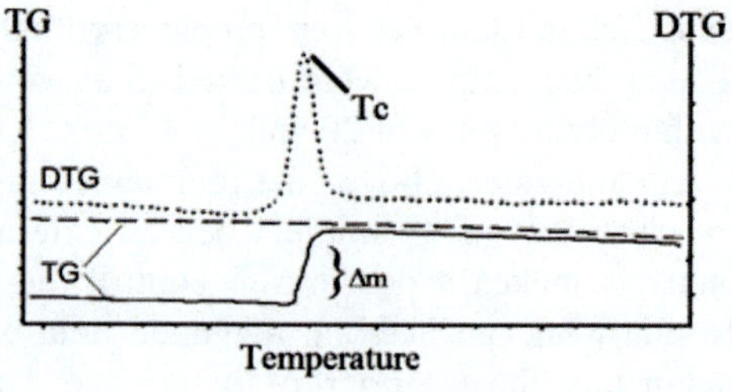

Fig. 2 Behavior of TG and DTG curves for magnetic sample at heating in magnetic field (solid TG curve) and without magnetic field (dashed TG curve)

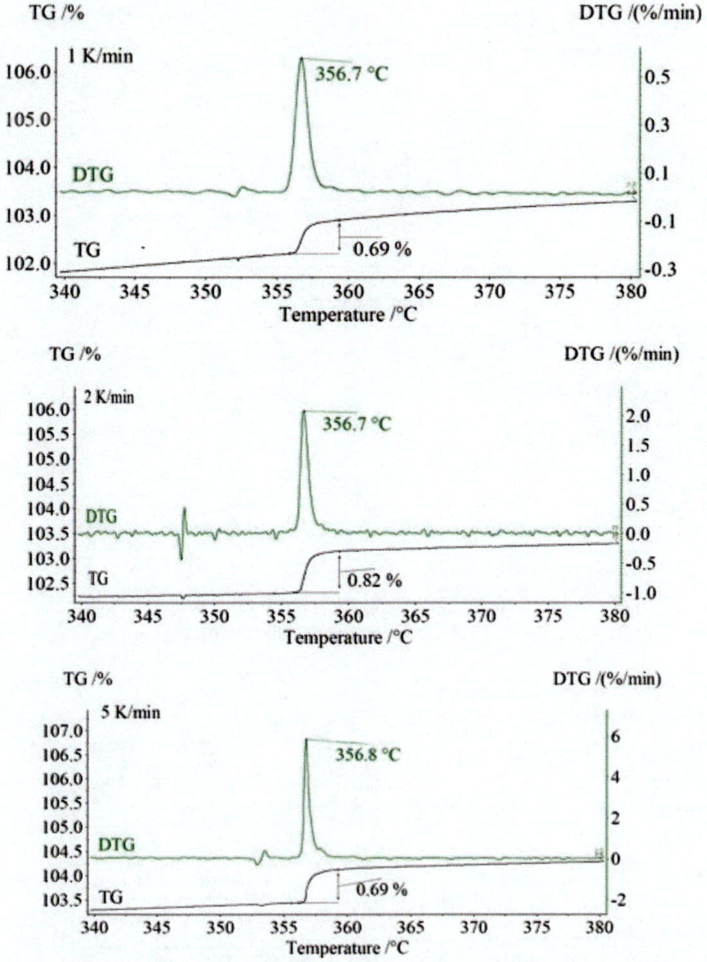

Fig. 3 Thermomagnetometric analysis obtained for Ni at heating rate of 1, 2, 5 K min^{-1}

Figures 3 and 4 also show that the intensity of the peak on the DTG curves increases with increasing heating rate. This fact indicates that the magnetic-paramagnetic transition occurs in a narrower temperature range with an increase in the heating rate. Conversely, the magnetic phase transition spreads over temperature with a decrease in the heating rate.

Using the temperature position of the DTG peak, the Curie temperature of each sample was determined. Figure 5 shows the dependence of the measured Curie temperature value on the heating rate used in the TM measurement. It can be seen that with an increase in the heating rate, the Curie temperature increases linearly from 356.7 to 360.9 °C. Thus, the variation of the obtained value is ± 4 °C for nickel.

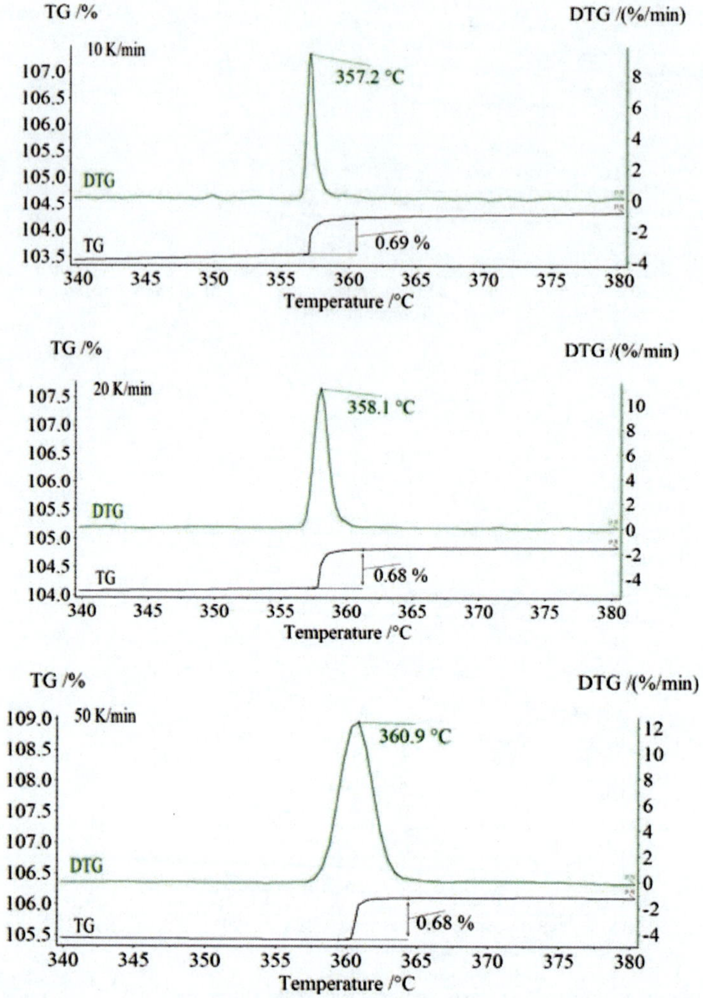

Fig. 4 Thermomagnetometric analysis obtained for Ni at heating rate of 10, 20, 50 K min^{-1}

It should be noted that the obtained values of the Curie temperature are similar to the values of 358–365 °C, which are indicated in the literature [33–35]. However, given the clearer magnetic phase transition as the heating rate increases, we recommend using the high heating rate that the thermal analyzer instrument can provide. In our case, this is 50 K min^{-1}. At the same time, the recorded value of the Curie temperature is the closest to the literary values.

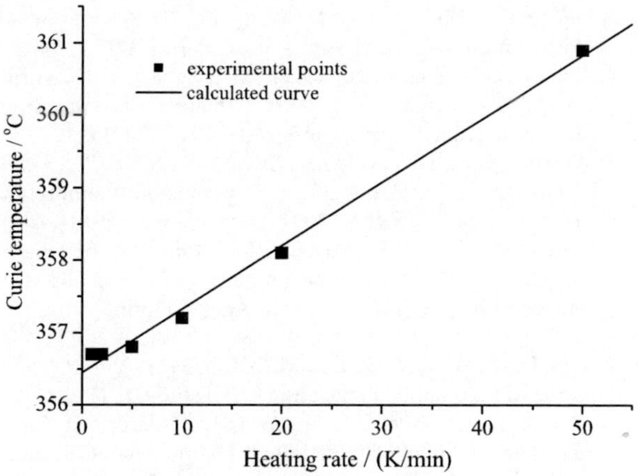

Fig. 5 Dependence of the registered Curie temperature of nickel on the heating rate used in thermomagnetometric measurement

4 Conclusion

In this work, the influence of the heating rate during thermogravimetric measurements in a magnetic field on the recorded values of the Curie temperature of nickel was investigated.

It was shown that the heating rate affects the recorded Curie temperature. Thus, with an increase in the heating rate from 1 to 50 K min^{-1}, the Curie temperature increases from 356.7 to 360.9 °C. However, the obtained values correspond to the literature data of the Curie temperature for nickel.

Based on thermogravimetric data for nickel, obtained in the region of the magnetic-paramagnet magnetic phase transition, a high heating rate was recommended to control the Curie temperature using thermomagnetometric measurements.

It should be noted that the regularities obtained in this work can be observed when measuring the Curie temperature of other magnetic materials.

Acknowledgements The research was supported by the Russian Foundation for Basic Research, project No. 20-07-00662. The work was carried out using the equipment of the Center for Sharing Use "Nanomaterials and Nanotechnologies" of Tomsk Polytechnic University supported by the RF Ministry of Education and Science, project 075-15-2021-710.

References

1. Barbosa, G.F., Machado, F.L.A., Rodrigues, A.R., Azevedo, W.M.: On the influence of the milling time in the Curie temperature of $Fe_{56}Co_7Ni_7Zr_{10}B_{20}$ investigated by magnetic thermogravimetric analysis. J. Therm. Anal. Calorim. **109**, 825–829 (2012)
2. Moukhina, E.: Determination of kinetic mechanisms for reactions measured with thermoanalytical instruments. J. Therm. Anal. Calorim. **109**, 1203–1214 (2012)

3. Wang, G., Min, X., Peng, N., Wang, Zh.: The isothermal kinetics of zinc ferrite reduction with carbon monoxide. J. Therm. Anal. Calorim. **146**, 2253–2260 (2021)
4. Zhou, L.H., Zhang, B., Zhao, S., Asuha, S.: Octyltrimethylammonium bromide-assisted synthesis of maghemite powder by thermal decomposition of ferric nitrate and its properties. J. Therm. Anal. Calorim. **146**, 2403–2410 (2021)
5. Peixoto, E.B., Duque, J.G.S., Silva, L.S., Silva, R.A.G.: The effect of the thermal heating on the magnetic properties of a high permeability amorphous $Fe_{76.5}Cu_1Nb_3Si_{13.5}B_6$ ribbon. J. Thermal Anal. Calorim. (2021). https://doi.org/10.1007/s10973-021-10711-8
6. Oraon, A., Das, B.P., Michael, M., Adhikary, T., Dhar, P., Aich, S., Ghosh, S.: Impact of magnetic field on the thermal properties of chemically synthesized Sm-Co nanoparticles based silicone oil nanofluid. J. Therm. Anal. Calorim. (2021). https://doi.org/10.1007/s10973-021-10572-1
7. Lin, D.M., Wang, H.S., Lin, M.L., Lin, M.H., Wu, Y.C.: TG (M) and DTG (M) techniques and some of their applications on material study. J. Therm. Anal. Calorim. **58**, 347–353 (1999)
8. Suñol, J., Gonzőlez, A., Escoda, L., Vilaró, A.: Curie temperature in Fe(Ni)Nb based mechanically alloyed materials. J. Therm. Anal. Calorim. **80**, 257–261 (2005)
9. Gallagher, P.K.: Thermomagnetometry. J. Therm. Anal. Calorim. **49**, 33–44 (1997)
10. Warne, S.St.J., Gallagher, P.K.: Thermomagnetometry. Thermochimica Acta **110**, 269–279 (1987)
11. Brown, M.E., Gallagher, P.K.: Handbook of Thermal Analysis and Calorimetry. Applications to Inorganic and Miscellaneous Materials. Elsevier, Amsterdam (2003)
12. Sanders, J.P., Gallagher, P.K.: Thermomagnetometric evidence of γ-Fe_2O_3 as an intermediate in the oxidation of magnetite. Thermochim. Acta **406**, 241–243 (2003)
13. Norem, S.D., O'Neill, M.J., Gray, A.P.: The use of magnetic transitions in temperature calibration and performance evaluation of thermogravimetric systems. Thermochim. Acta **1**, 29–38 (1970)
14. Wang, T., Wang, H., Wang, F., Li, J., Zhang, Q., Huang, X.: Certification of reference materials of Alumel, nickel and iron for Curie point. J. Therm. Anal. Calorim. **131**, 1979–1985 (2018)
15. Fischer, H.: Calibration of micro-thermal analysis for the detection of glass transition temperatures and melting points. J. Therm. Anal. Calorim. **92**, 625–630 (2008)
16. Weddle, B.J., Robbins, S.A., Gallagher, P.K.: Further studies on the use of simultaneous TM/DTA to establish magnetic transition temperatures. Pure Appl. Chem. **67**, 1843–1847 (1995)
17. Luciani, G., Costantini, A., Branda, F., Scardi, P., Lanotte, L.: Thermal evolution of ferromagnetic metallic glasses. A study using TG(M) technique. J. Thermal Anal. Calorim. **72**, 105–111 (2003)
18. Opuchovic, O., Niznansky, D., Kareiva, A.: Thermoanalytical (TG/DSC/EVG-GC-MS) characterization of the lanthanide (Ho) iron garnet formation in sol-gel. J. Therm. Anal. Calorim. **130**, 1085–1094 (2017)
19. Astafyev, A.L., Lysenko, E.N., Surzhikov, A.P., Nikolaev, E.V., Vlasov, V.A.: Thermomagnetometric analysis of nickel-zinc ferrites. J. Therm. Anal. Calorim. **142**, 1775–1781 (2020)
20. Lysenko, E.N., Astafyev, A.L., Vlasov, V.A., Surzhikov, A.P.: Analysis of phase composition of LiZn and LiTi ferrites by XRD and thermomagnetometric analysis. J. Magn. Magn. Mater. **465**, 457–461 (2018)
21. Surzhikov, A.P., Lysenko, E.N., Malyshev, A.V., Vlasov, V.A., Vasendina, E.A.: Analysis of the phase composition and homogeneity of ferrite lithium-substituted powders by the thermomagnetometry method. J. Therm. Anal. Calorim. **112**, 739–745 (2013)
22. Franco, V., Dodrill, B.: Magnetic Measurement Techniques for Materials Characterization. Springer, Switzerland (2021)
23. Hale, R.C., Seeley, M.E., King, A.E., Yu, L.H.: Analytical chemistry of plastic debris: sampling, methods, and instrumentation. In: Bank, M.S. (ed.) Microplastic in the Environment: Pattern and Process. Environmental Contamination Remediation and Management. Springer, Cham (2022)

24. Dierkes, G., Lauschke, T., Földi, C.: Analytical methods for plastic (microplastic) determination in environmental samples. In: Stock, F., Reifferscheid, G., Brennholt, N., Kostianaia, E. (eds.) Plastics in the Aquatic Environment—Part I. The Handbook of Environmental Chemistry. Springer, Cham (2021)
25. Alves, A.K.: Technological Applications of Nanomaterials. Springer, Switzerland (2022)
26. Hossain, M.A.F., Al-Mamun, M., Rahman, M.R., Hoque, S.M.: Effect of in-situ growth and separate addition method in hydrothermal process on the structural and magnetic properties of $CoNiFe_2O_4$@functionalized CNTs nanocomposite. J. Appl. Phys. A **127**, 896 (2021)
27. Sun, K., Li, Y., Feng, S., Gao, Q., Wang, Z., Wei, X., Fun, R.: Optimizing the soft magnetic properties of mn-zn ferrite by a proper control of sintering process. J. Electron. Mater. **50**, 1467–1473 (2021)
28. Yao, L., Wang, Y., Li, Y., Jiang, Zh., Qiu, D.: Controlled preparation of Fe_3O_4/PLA composites and their properties. Chem. Pap. **75**, 6399–6406 (2021)
29. Kahkhaie, M.R., Hazeri, N., Maghsoodlou, M.T., Yazdani-Elah-Abadi, A.: Synthesis and characterization of a novel and reusable Fe_3O_4@THAM-CH_2CH_2-SCH_2CO_2H magnetic nanocatalyst for highly efficient preparation of xanthenes and 3-aminoisoxazoles in green conditions. Res. Chem. Intermed. **47**, 5007–5025 (2021)
30. Zhang, X., Qian, M.: Magnetic Shape Memory Alloys. Springer, Harbin Institute of Technology Press, Singapore (2022)
31. Lysenko, E.N., Surzhikov, A.P., Astafyev, A.L.: Thermomagnetometric analysis of lithium ferrites. J. Therm. Anal. Calorim. **136**, 441–445 (2019)
32. Surzhikov, A.P., Lysenko, E.N., Astafyev, A.L.: Estimation of thermomagnetometry method sensitivity for magnetic phase determination. IOP Conf. Ser. Mater. Sci. Eng. **23**, 012090 (2016)
33. Legendre, B., Sghaier, M.: Curie temperature of nickel. J. Therm. Anal. Calorim. **105**, 141–143 (2011)
34. Murlieva, Z.K., Palchaev, D.K., Borzov, E.D.: Linear relationship between resistivity and isobaric thermal strain of nickel above and below the Curie temperature. Tech. Phys. Lett. **28**, 773–775 (2002)
35. Coey, J.M.D., Parkin, S.S.P.: Handbook of Magnetism and Magnetic Materials. Springer, Switzerland (2021)

Accumulation of Volume Electric Charge in Ferrites Under Electron Irradiation

Anatoliy Surzhikov, Elena Lysenko, and Oldřich Starý

Abstract The study investigates the effect of the flow of charged particles on sintering of lithium-titanium ferrite ceramics under a combined effect of high temperatures and powerful radiation fluxes. For radiation-thermal sintering, the pulsed electron accelerator ILU-6 was used. It was established that the charge accumulation in non-sintered and as-sintered samples of ferrite ceramics under radiation-thermal effects is insignificant. The electric field strength in the samples does not exceed 10^3 V cm^{-1} at room temperature, it decreases when temperature grows, and at T = 600 °C it is less than 1 V cm^{-1}. The electric field of the volume charge virtually does not affect the passage of injected electrons in samples and physicochemical processes of their sintering cannot provide the effect of radiation intensification of the compaction of powder materials during radiation-thermal sintering. Moreover, the influence of the volume charge electric field accumulated in ferrite upon electron beam irradiation on the injected electrons parameters was not found. No violation of the continuity of sintered ferrite due to spontaneous electrical discharges during radiation electrization was established.

1 Introduction

Sintering, the most frequently used technique for producing ferrite ceramics, is complex, time- and energy-consuming; it is characterized by high material and human costs, severe environmental conditions and insufficient product quality. Attempts to

A. Surzhikov · E. Lysenko (✉)
Tomsk Polytechnic University, 30 Lenin av., Tomsk 634050, Russia
e-mail: lysenkoen@tpu.ru

A. Surzhikov
e-mail: surzhikov@tpu.ru

O. Starý
Czech Technical University, Prague, Czech Republic
e-mail: oldstar@zentrum.cz

E. Lysenko et al. (eds.), *Recent Developments in the Field of Non-Destructive Testing, Safety and Materials Science*, Studies in Systems, Decision and Control 433,
https://doi.org/10.1007/978-3-030-99060-2_19

improve the existing techniques were and are reduced to mechanical modernization [1–10] and do not affect its physical basis.

It is apparent that not the modernization approach but fundamentally new, non-traditional ones should be developed to solve the problem and radically change the existing situation.

One of these approaches was the use of the effect of intense fluxes of ionizing radiation. The fundamental phenomenon of a multiple increase in synthesis of multi-component powder materials [11–14] and sintering [15–21] in the radiation-thermal (RT) mode was discovered.

The processes of lithium-titanium ferrite sintering have been most fully studied under the combined effect of high temperatures and intense electron fluxes [22, 23]. The compaction patterns for ferrite compacts were established, and a multiple increase in the compaction rate of lithium-titanium ferrite compacts under these sintering conditions was shown [24, 25].

On the other hand, it is known that irradiation with charged particles causes accumulation of the electric charge and formation of internal electric fields in the bulk of dielectric materials and high-resistance semiconductors [26].

To improve the radiation-thermal technology for sintering powder materials, the role of the electric field of the volume charge in stimulation of ferrite compaction under high-temperature electron irradiation should be specified.

In addition, a number of "technological" issues to be considered include the effect of the electric field of the volume charge on parameters of the injected electron flux, and the possibility of disrupting the continuity of sintered materials due to spontaneous electric discharges during radiation electrization.

This study is focused on the above issues.

2 Research Method

The objects under study was lithium-titanium ferrite that was synthesized from a mixture of oxides and lithium carbonate (in wt.%): Li_2CO_3—11.2, MnO—2.7, TiO_2—18.65, ZnO—7.6, and Fe_2O_3—59.81.

After drying and weighing the powders, they were mixed and milled in a vibrating mill for 1 h. Ferrite was synthesized in a thermal furnace at a temperature of 900 °C for 6 h. After that, the powder was crushed and milled in a vibrating mill for 45 min. Bi_2O_3 dissolved in concentrated nitric acid in an amount of 0.22 wt.% was added to the milled powder, after which mixing in a ball mill was carried out for 4 h. The synthesized lithium-titanium ferrite was of the chemical formula $Li_{0.649}Fe_{1.598}Ti_{0.5}Zn_{0.2}Mn_{0.051}O_4$.

To prepare compacts, a 10% solution of polyvinyl alcohol in an amount of 12 wt.% was added to the synthesized powder. The ferrite samples were compacted in the shape of tablets with 18 mm in diameter and 2–3 mm in thickness by cold single-side pressing in a hydraulic press.

For RT sintering, the pulsed electron accelerator ILU-6 was used, which was developed in Institute of Nuclear Physics of the SB RAS (Novosibirsk, Russia). The electron energy was equal to 2 meV, the beam current pulse was 0.5–0.9 A, the pulse duration was 500 μs and the pulse repetition frequency was 5–50 Hz.

3 Results and Considerations

3.1 Phenomenological Analysis of Charge Accumulation in Dielectric Materials Under Electron Irradiation

The accumulation of volume charge in materials is described by the system of continuity equations for the convection current and Poisson's equation [27]:

$$\frac{dq(x,t)}{dt} = G(x,t) - \frac{d}{dt}\left[\gamma_{r},(x,T,t)E(x,t)\right] \tag{1}$$

$$\varepsilon\varepsilon_0(T)\frac{dE(x,t)}{dx} = q(x,t) \tag{2}$$

where: x is the coordinate on the axis perpendicular to the surface (Fig. 1a); q is the volume charge density; G is the volume density of the charge injected into the elementary volume of the sample by irradiation (generation function); γ_r is the specific volume electrical conductivity of the material under irradiation at temperature T (radiation conductivity); ε_0 is the electrical constant; ε is the relative dielectric constant of the substance.

The generation function with a sufficiently good approximation is determined by the transmission function of injected electrons $J(x,t)$ if the electric field of the dielectric does not significantly affect the parameters of the electron flux [28]:

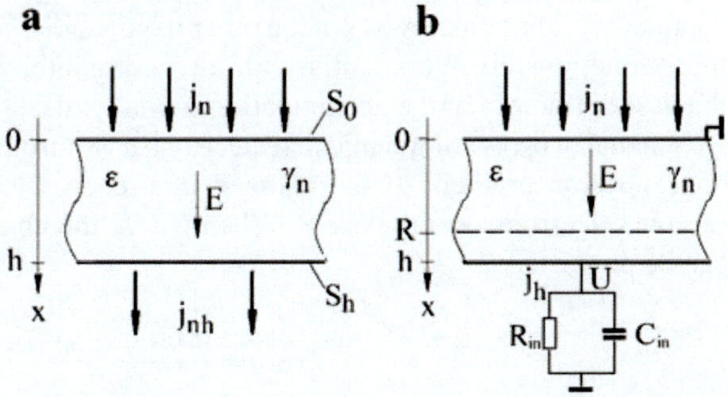

Fig. 1 Scheme of the sample irradiation

$$G(x,t) = j_{n0}(t)\frac{dJ(x,t)}{dx} \tag{3}$$

where: $j_{n0} = j_n\ (1 - K_0)$; j_n is the current density of ionization injected electrons, K_0 is the electron reflection coefficient.

According to [29], at electron energy exceeding 1 meV, $K_0 \ll 1$ and $j_{n0} \approx j_n$.

The solution to the system of Eqs. (1) and (2) depends on the boundary conditions. The samples irradiated at normal atmospheric pressure are short-circuited due to the conductivity of the ionized gas medium [30]

$$\int_o^h E(x,t)d = 0 \tag{4}$$

and the boundary conditions imposed on the solution of Poisson's equation have the form [29]:

$$E(o,t) = \frac{1}{\varepsilon\varepsilon_0}\int_o^h (h-x)q(x,t)dx;\quad E(h,t) = \frac{1}{\varepsilon\varepsilon_0}\int_o^h xq(x,t)dx \tag{5}$$

The boundary conditions imposed on the solution of the continuity equation are determined by the state of the metal–sample contact. Under irradiation, the contact can be considered neutral, and the accumulation of charge in the near-electrode regions of the sample due to the contact can be neglected [29].

Under the indicated boundary conditions, the system of Eqs. (1) and (2) is analytically resolved only for the case when the values of j_n and γ_r are constant in time [27, 31]. However, even under these conditions, the equations obtained cannot be used for numerical calculations of the volume charge density and the electric field strength E, since the electron energy spectrum obtained using the types of accelerators employed in the study is known approximately. Therefore, the function G and the change in the coordinate power of the absorbed radiation energy (Φ) (determining the radiation conductivity) are unknown. On the other hand, under penetrating irradiation ($R > h$), these functions change slightly with the coordinate, which significantly simplifies the description of charge accumulation in materials.

When $R > h$, part of the injected electrons pass through the sample. The maximum concentration gradient of thermalized (lost energy to thermal and stopped in the sample) electrons occurs near R. When $R > h$, the change in G with the coordinate can be neglected.

$$G(t) = \frac{j_{nh}(t) - j_n(t)}{h} \tag{6}$$

where j_{nh} is the current density of electrons passing through the sample.

Neglecting the change in Φ and, accordingly, γ_r with the coordinate, from Eqs. (1), and (2) for a given temperature, we obtain:

$$\frac{dq(t)}{dt} = G(t) - \gamma r(t)\frac{dE(x,t)}{dx}; \quad E(x,t) = \frac{(x - 0.5h)}{\varepsilon\varepsilon_o}q(t) \tag{7}$$

Solution of the equations with zero initial conditions yields:

$$q(t) = \exp\left[-\int_o^t dt'/\tau_r(t')\right]\int_o^t G(t')\exp\left[\int_o^t dt''/\tau_r(t'')\right]dt', \tag{8}$$

where $\tau_r(t) = \varepsilon\varepsilon_o/\gamma_r(t)$.

Under continuous irradiation of the sample, when the values of j_n and, hence, G and Φ are constant in time, and the change at increased absorbed dose $D(t)$ can be neglected, Eq. (8) yields:

$$q(t) = G\tau_r(1 - e^{-t/\tau_r}). \tag{9}$$

In a quasi-stationary state (at $t \geq t_k$), when the value of q virtually does not change during irradiation, we have:

$$q(t) = G\tau_r(t_k). \tag{10}$$

The maximum electric field strength of the sample in the near-electrode regions is determined by the equation:

$$E = 0.5hq/\varepsilon\varepsilon_o = 0.5hG/\gamma_r \tag{11}$$

In practice, it is necessary to know the behavior of γ_r under the given irradiation conditions to calculate q and E using Eqs. (8)–(11). If irradiation is continuous, when γ_r changes insignificantly at increased irradiation dose, q and E are calculated using relatively simple Eqs. (9)–(11). In our case, in a pulsed irradiation mode, it is necessary to know both the time (dose) dependence of γ_r and the change in the material conductivity during the interpulse period.

Under pulsed irradiation of samples with the amplitude value of the current density j^U_{nm}, the value of q is apparently less than that under continuous irradiation with the current density $j_n = j^U_{nm}$, due to partial relaxation of the accumulated charge during the interpulse period. On the other hand, if charge relaxation during the interpulse period is insignificant, the values of $q(t_k)$ under continuous and pulsed irradiation will differ insignificantly, but in the latter case, the accumulation time of the maximum charge will be longer. If the pulses are rectangular, the charge accumulation time under pulsed irradiation is ψ times longer than that under continuous irradiation, where $\psi = (\tau_U - \tau_n)/\tau_n$ is the pressure ratio.

3.2 Method for Determining the Value of the Accumulated Volume Charge in Materials Under Electron Irradiation

The irradiation scheme presented in Fig. 1a shows the conditions of radiation sintering of ferrite ceramics.

However, this scheme cannot be used for experimental determination of the charge during irradiation. Charge measurements after irradiation do not show the efficiency of the sample electrization during irradiation, since the Maxwellian charge relaxation time is less than a second even at conductivity values of $10^{-10}\ \Omega\ \mathrm{m}^{-1}$. Indeed, during acoustic probing of sintered and non-sintered samples of ferrite ceramics after electron irradiation by MIRA-2D pulsed accelerator, the volume charge was not detected after (5–7) seconds after sample irradiation [32]. The sensitivity threshold of the acoustic sounding method based on the inverse piezoelectric effect is less than $10^2\ \mathrm{V\ cm}^{-1}$.

If $R < h$ (Fig. 1b) and the material conductivity in the $R < x < h$ region is much less than that in the irradiated part of the sample, the charge accumulation can be determined experimentally from the potential U of the measuring electrode on the S_n surface [33]. The density j_n of the current flowing in the region $R < x < h$ and the electric field strength $E(h, t)$ in this region are related by the equation:

$$j_n(t) = \varepsilon\varepsilon_o \frac{dE(h,t)}{dt} = \frac{\varepsilon\varepsilon_o}{h-R}\frac{d}{dt}[\varphi(R,t) - U(t)], \tag{12}$$

where $\varphi(R, t)$ is the potential of the electric field in the dielectric at the interface between the irradiated and non-irradiated parts of the sample.

The U behavior depends on the input resistances R_{in} and the capacitance C_{in} of the recording equipment:

$$J_n(t) = \frac{U(t)}{S_{el}R_{in}} + \frac{C_{in}}{S_{el}}\frac{dU(t)}{dt}, \tag{13}$$

where S_{el} is the area of the measuring electrode.

Equations (12) and (13) yield:

$$\frac{d\varphi(R,t)}{dt} = \left(1 - \frac{C_{in}}{C_h}\right)\frac{dU(R,t)}{dt} \tag{14}$$

where $C_h = \varepsilon\varepsilon_o/(h - R)S_{el}$.

At $C_{in} \gg C_h$ in the current integration mode and $t_U \ll C_{in}R_{in}$ (the measurement time should be less than the relaxation time of the input circuit), the equation takes the form:

$$\frac{d\varphi(R,t)}{dt} = \frac{C_{in}}{C_h}\frac{dU(t)}{dt} = (h-R)\frac{dE(h,t)}{dt} = \frac{(h-R)}{\varepsilon\varepsilon_o}\frac{d\sigma_h(t)}{dt} \tag{15}$$

where

$$\sigma_{\text{h}}(t) = \int_{o}^{t} j_{\text{n}}(t')dt' = C_{\text{in}}\frac{U(t)}{S_{\text{el}}} \tag{16}$$

is the charge density on the capacitor plates.

Thus, the potential at the interface between the irradiated and non-irradiated parts of the sample and, accordingly, the field strength in the non-irradiated part, are determined by the potential of the measuring electrode.

Denote the volume charge per unit area as

$$Q = \int_{o}^{t} q(x)dx \tag{17}$$

and the volume charge centroid as

$$\overline{x} = \int_{o}^{h} xq(x)x/Q \tag{18}$$

The Q value can be determined using the equation:

$$Q(t) = \frac{\varepsilon\varepsilon_{\text{o}}}{x}E_{\text{h}}(t) = \frac{h}{x}\sigma_{\text{h}}(t) = \frac{C_{\text{in}}hU}{S_{\text{el}}\overline{x}} \tag{19}$$

For the considered scheme, the charge accumulates in the samples more efficiently than in case of $R > h$. Indeed, at $R < h$, all injected electrons are thermalized in the sample, while at $R > h$, a number of electrons pass through the sample. At $R < h$, the charge relaxation occurs mainly due to the conductivity of the irradiated part of the sample, while at $R > h$, the charge “leakage” occurs through both surfaces.

The charge measured according to the scheme in Fig. 1b can significantly exceed that for the scheme presented in Fig. 1a and act as an upper limit of the values above which the charge in the samples cannot exceed at $R > h$.

3.3 Radiation Conductivity of Ferrite Ceramics

The radiation conductivity of the ferrite ceramic samples was measured in a vacuum measuring chamber. The electron flux with particle beam energy of 2 meV from the ILU-6 accelerator is directed in the chamber through a 50 μm thick titanium foil. The current of electrons injected into the chamber is recorded using a Faraday cylinder.

The electron flux falls on the measured sample through a 8 mm diameter hole in the collimator. The collimator is attached to the sample holder, which is a massive copper parallelepiped with four holes for heaters and a reservoir for coolant. The sample holder is attached to the chamber base using two thin-walled stainless steel tubes. The sample with platinum electrodes deposited on its surface is located on a 200 μm thick boron pyronitride plate, which isolates the measuring electrode from the sample holder. The specific resistance of boron pyronitride is about 10^9 Ω cm at T = 500 °C and 10^7 Ω cm at 1000 °C. The sample is clasped to the holder using a metal ring. The voltage is applied to the ring and, accordingly, to the potential electrode, from an external source through the electrical connector at the base. The electrical signal from the measuring electrode was applied to the input of the current–voltage converter. The measured signal was recorded using a storage oscilloscope.

The measured sample is a signal source with a high internal resistance (close to an ideal current source). This allows using a fast charge meter scheme. The input amplifier has low input currents (less than 10^{-9} A) at high speed (1 μs). The output stage is a transistor power amplifier, which provides the required signal power to run the converter with a 50 Ω load. In order to eliminate interference, filters are installed in the supply circuits of the amplifier. In addition, power is supplied from an autonomous +9 V source installed in the measuring chamber.

Measurements showed that at room temperature the shape of the $\gamma_r(t)$ curve virtually reproduces the shape of the current pulse (Fig. 2).

Non-sintered samples exhibit a delayed conductivity component (Fig. 2, curve 3). After irradiation, during the pause τ_n, the conductivity of non-sintered samples was about two orders of magnitude lower than that before irradiation. For sintered samples, the longer sintering time τ_c for the samples at 1000 °C decreases the conductivity during the interpulse period. At $\tau_c = 1$ h, the conductivity after the pause is about an order of magnitude lower than that for sintered samples. This difference in conductivity is likely due to the presence of a gas phase in non-sintered samples.

In the range of amplitude values of the electron current density, which is ($3 \times 10^{-4} - 1 \times 10^{-2}$) A cm^{-2}, the γ_{rm} and j_{nm}^{U} values correspond to $\gamma_{rm} = A(j_{nm}^{U})^{\Delta}$, where $\Delta = (0.97–1.0)$. The A value for non-sintered samples is 3.2×10^{-6} m V^{-1}. For sintered samples, A increases as τ_c increases and $\tau_c = 1$ h, and it equals $A =$

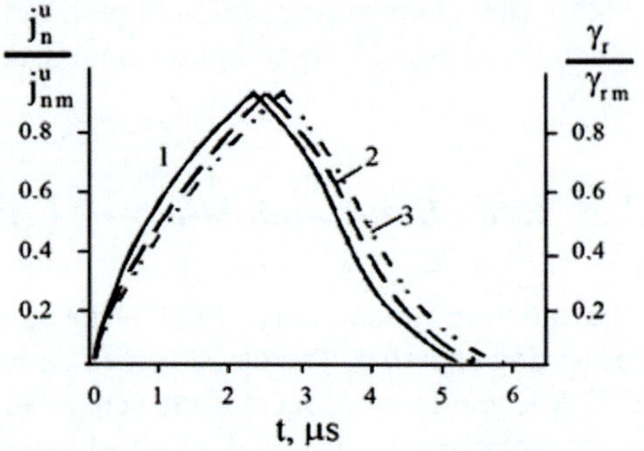

Fig.2 Relative change in $\gamma_r(t)$ for ferrite ceramic samples at T = (20–30) °C: 1—current pulse shape, 2—sample sintered at 1000 °C for 30 min, 3—non-sintered sample

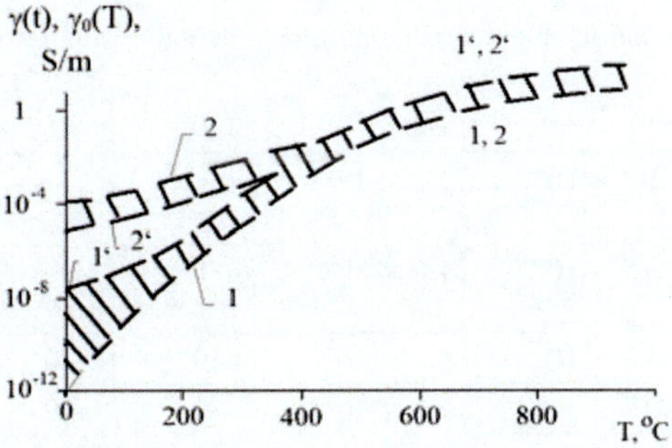

Fig. 3 Temperature dependence boundaries for the intrinsic conductivity (γ_0) and radiation conductivity (γ_r) of ferrites: 1,1′ – γ_0 for non-sintered samples and those pre-sintered at 1000 °C for (2–60) min; 2,2′ – γ_r for samples subject to electron irradiation at beam current $J_n = 5 \times 10^{-3}$ A cm^{-2}

4×10^{-6} m V^{-1}. The power of the absorbed radiation energy in the samples is proportional to their density and j_n. For sintered samples with a density of 4 g cm^{-3} at $j_n = 5 \times 10^{-3}$ A cm^{-2}, Φ corresponds to 1.5×10^7 Gy s^{-1}.

With increasing T from 20 to 400 °C γ_r increases by more than 1.5 order of magnitude (Fig. 3). At $j_n = 5 \times 10^{-3}$ A cm^{-2} and T = 400 °C, the sample intrinsic conductivity is significantly higher due to the irradiation effect. That is, at $j_n = 5 \times 10^{-3}$ A cm^{-2} and T = 400 °C, irradiation has an insignificant impact on the conductivity of ferrite ceramics.

3.4 *Results of Calculations and Experimental Determination of Electrification Characteristics*

During sample irradiation, $\gamma_r \sim j_n$ and, accordingly, $\gamma_r \sim G$, therefore the change in γ_r with the irradiation time can be neglected at a constant electron flux intensity. At constant j_n, the kinetics of charge accumulation can be described by Eq. (9). The maximum q and E values are determined using Eqs. (10) and (11).

Determine q and E for continuous electron irradiation with a constant current density j_n equal to the amplitude value during pulsed irradiation. It should be noted that if γ_r is proportional to G, q and E in the quasistationary state do not depend on j_n. Note that q is inversely proportional to ε, and E does not depend on ε (Eq. 11). Measurements showed that ε varies slightly in the frequency range of (1–100) kHz F at 20–250 °C and decreases as the sintering time of the samples at 1000 °C increases from (130–140) for non-sintered samples to (20–25) for those sintered for 1 h at 1000 °C.

Table 1 summarizes the calculation results for q and E in a quasi-stationary state at different temperatures.

Table 1 Charge accumulation characteristics in ferrite ceramic samples under uniform electron irradiation

T, [°C]		20	100	300	600	800	1000
$q(t)$,	Max	1.2×10^{-8}	1.3×10^{-8}	1.3×10^{-10}	2.3×10^{-12}	4.6×10^{-13}	3.5×10^{-13}
[C cm^{-3}]	Min	2.5×10^{-9}	1.6×10^{-9}	4.0×10^{-10}	6.0×10^{-13}	9.0×10^{-14}	1.0×10^{-13}
$E(t_k)$,	Max	6.3×10^3	4.1×10^3	410	0.70	0.14	0.11
[V cm^{-1}]	Min	7.7×10^2	5.1×10^2	130	0.19	0.28	0.03

The upper two rows in the table provide the q values for $G = 0.5j_n/h$ and $\varepsilon = 20$, and the minimum q values (curve 2, Fig. 3); the bottom two rows show the maximum q values (curve 1, Fig. 3). In the calculations, the thickness of the non-sintered samples was 3 mm, and that of the sintered samples amounted to 2.0 mm, which corresponds to the conditions of the studies carried out on the radiation sintering of ceramics. For the indicated thicknesses, $j_{nh} > 0.5j_n$.

The above results yield the conclusion that even under continuous irradiation of the samples at room temperature, the electric field strength of the volume charge is insignificant. Under pulsed irradiation of the samples, the E values can be significantly lower than those obtained due to the charge relaxation during the pause. As noted, $\gamma_r \sim j_0$ and, hence, the maximum q and E values do not depend on the electron current density and the accelerator type. It should be noted that even if the conductivity values after irradiation are equal to the initial ones (without irradiation), the relaxation time of the volume charge at $\tau_n = 5 \times 10^{-3}$ s at T 200 °C is less than the interpulse period. In this case, the maximum charge corresponds to that accumulated in one pulse. At sample temperatures exceeding 400 °C, the q and E values are negligible.

It should be noted that at $E = 10^4$ V cm^{-1}, the electron energy in the electric field of the volume charge changes by only 1 keV when passing a distance of 1 mm, which is negligible compared to the energy of injected electrons. Thus, the electric field of the volume charge cannot significantly affect the parameters of the flux of injected electrons.

For non-sintered samples (2.9–3.0) mm thick with a density of $\rho = 2.8$ g cm^{-3} and those sintered at T = 1000 °C for 1 h (h = 2 mm, $\rho = 4.12$ g cm^{-3}), the charge was measured according to the scheme presented in Fig. 1b. The samples were irradiated with electrons with electron beam energy of 2 meV using an ILU-6 accelerator with an amplitude value of the current density of 5×10^{-3} A cm^{-2}. To reduce the range of electrons and implement the $R < h$ condition, aluminum absorbers were placed in front of the samples. The thickness of the absorber located in front of the non-sintered samples was 2 mm. The amplitude of the current density of electrons passing through the absorber attained 3×10^{-3} A cm^{-2}. The thickness of the absorber in front of the sintered samples was 2.2 mm. In this case, the electron current density behind

Table 2 Electric field strength in samples, E, [V cm^{-1}]

Sample	T, [°C]				
	20	100	200	300	400
Non-sintered	2.9×10^3	1.3×10^3	8.2×10^2	2×10^2	1×10^2
Sintered	8.9×10^2	4.1×10^2	2.9×10^2	1×10^2	–

the absorber was 1.2×10^{-3} A cm^{-2}. The electron flux behind the sample was not recorded.

The E_h values obtained from the results of measuring the maximum U values (Eq. 19) are provided in Table 2.

Likewise, in the case of uniform irradiation, the maximum electric field strength in non-sintered samples is about 10^3 V cm^{-1}. When the range of electrons exceeds 0.5 h, the electric field strength near the surface facing the radiation source and E differ less than fivefold.

4 Conclusion

The effect of the flow of charged particles on sintering of lithium-titanium ferrite ceramics under a combined effect of high temperatures and powerful radiation fluxes was studied.

It was found that the charge accumulation in non-sintered and sintered samples of ferrite ceramics under radiation-thermal effects is negligible. The electric field strength in the samples does not exceed 10^3 V cm^{-1} at room temperature, decreases with increasing temperature, and is less than 1 V cm^{-1} at T = 600 °C. In this case, the electric field of the volume charge has insignificant impact on the passage of injected electrons in the ferrite samples, and physicochemical processes of their sintering cannot provide the effect of radiation intensification of the compaction of powder materials during radiation-thermal sintering.

Acknowledgements The research is funded by The Ministry of Science and Higher Education of the Russian Federation as part of the Science Program, Project FSWW-2020-0008. The authors are grateful to M.V. Korobeynikov (Budker Institute of Nuclear Physics, Novosibirsk) for radiation-thermal experiments using ILU accelerator.

References

1. Inoue, A., Kong, F.: Soft magnetic materials, reference module in materials science and materials engineering, Elsevier. (2020). https://doi.org/10.1016/B978-0-12-803581-8.11725-4
2. Dmitriev, M.V., Letyuk, L.M., Shipko, M.N.: Study of oxygen diffusion in the surface layers of Mn-Zn ferrites. Tech. Phys. **27**, 338–339 (1982)

3. Letyuk, L.M.: Recrystallization of ferrites and its effect on the processes of microstructure formation in ferrospinels. Sov. Powder Metall. Met. Ceram. **19**(5), 359–364 (1980)
4. Zinovik, M.A., Zinovik, E.V.: Ferrites with rectangular and square hysteresis loops. Powder Metall. Met. Ceram. **44**, 66–74 (2005)
5. Xie, F., Chen, Y., Bai, M., Wang, P.: Co-substituted LiZnTiBi ferrite with equivalent permeability and permittivity for high-frequency miniaturized antenna application. Ceram. Int. **45**, 17915–17919 (2019)
6. Xu, F., Shi, X., Liao, Y., Li, J., Hu, J.: Investigation of grain growth and magnetic properties of low-sintered LiZnTi ferrite-ceramics. Ceram. Int. **46**, 14669–14673 (2020)
7. Surzhikov, A.P., Lysenko, E.N., Vasendina, E.A., Sokolovskii, A.N., Vlasov, V.A., Pritulov, A.M.: Thermogravimetric investigation of the effect of annealing conditions on the soft ferrite phase homogeneity. J. Therm. Anal. Calorim. **104**(2), 613–617 (2011)
8. Guo, J., Zhang, H., He, Z., Li, S., Li, Z.: Electrical properties and temperature sensitivity of Mo-modified $MnFe_2O_4$ ceramics for application of NTC thermistors. J. Mater. Sci. Mater. Electron. **29**, 2491–2499 (2018)
9. Zahir, R., Chowdhury, F.-U.-Z., Uddin, M.M., Hakim, M.A.: Structural, magnetic and electrical characterization of Cd substituted Mg ferrites synthesized by double sintering technique. J. Magn. Magn. Mater. **410**, 55–62 (2016)
10. Manjura, H.S., Abdul, H.M., Mamun, Al.: Study of the bulk magnetic and electrical properties of $MgFe_2O_4$ synthesized by chemical method. Mater. Sci. Appl. **2**, 1564–1569 (2011)
11. Surzhikov, A.P., Galtseva, O.V., Vasendina, E.A., Vlasov, V.A., Nikolaev, E.V.: Processing line for industrial radiation-thermal synthesis of doped lithium ferrite powders. IOP Conf. Ser. Mater. Sci. Eng. **110**(1), 012002 (2016)
12. Surzhikov, A.P., Frangulyan, T.S., Ghyngazov, S.A., Lysenko, E.N., Galtseva, O.V.: Physics of magnetic phenomena: investigation of electroconductivity of lithium pentaferrite. Russ. Phys. J. **49**(5), 506–510 (2006)
13. Yurov, V.M., Baltabekov, A.S., Laurinas, V.C., Guchenko, S.A.: Dimensional effects and surface energy of ferroelectric crystals. Eurasian Phys. Tech. J. **16**(1), 18–23 (2019)
14. Zhuravlev, V.A., Naiden, E.P., Minin, R.V., Itin, V.I., Suslyaev, V.I., Korovin, E.Yu.: Radiation-thermal synthesis of W-type hexaferrites. IOP Conf. Ser. Mater. Sci. Eng. **81**, 012003 (2015)
15. Kostishyn, V., Isaev, I., Scherbakov, S., Nalogin, A., Belokon, E., Bryazgin, A.: Obtaining anisotropic hexaferrites for the base layers of microstrip SHF devices by the radiation-thermal sintering. East. Eur. J. Enterp. Technol. **5**, 32–39 (2016)
16. Kostishin, V.G., Andreev, V.G., Korovushkin, V.V., Chitanov, D.N., Yudanov, N.A., Morchenko, A.T., Nikolaev, A.N.: Preparation of 2000NN ferrite ceramics by a complete and a short radiation-enhanced thermal sintering process. Inorg. Mater. **50**, 1317 (2014)
17. Malyshev, A.V., Lysenko, E.N., Vlasov, V.A., Nikolaeva, S.A.: Electromagnetic properties of $Li_{0.4}Fe_{2.4}Zn_{0.2}O_4$ ferrite sintered by continuous electron beam heating. Ceram. Int. **42**, 16180–16183 (2016)
18. Neronov, V.A., Voronin, A.P., Tatarintseva, M.I., Melekhova, T.E., Auslender, V.L.: Sintering under a high-power electron beam. J. Less Common Met. **117**, 391–394 (1986)
19. Naiden, E.P., Zhuravlev, V.A., Minin, R.V., Suslyaev, V.I., Itin, V.I., Korovin, E.Yu.: Structural and magnetic properties of SHS-produced multiphase W-type hexaferrites: influence of radiation-thermal treatment. Int. J. Self Propag. High Temp. Synth. **24**, 148–151 (2015)
20. Ancharova, U.V., Mikhailenko, M.A., Tolochko, B.P., Lyakhov, N.Z., Korobeinikov, M.V., Bryazgin, A.A., Shtarklev, E.A.: Synthesis and staging of the phase formation for strontium ferrites in thermal and radiation thermal reactions. IOP Conf. Ser. Mater. Sci. Eng. **81**, 012122 (2015)
21. Salimov, R.A., Cherepkov, V.G., Kuksanov, N.K., Kuznetzov, S.A.: The use of electron accelerators for radiation disinfestation of grain. Radiat. Phys. Chem. **57**, 625–627 (2000)
22. Surzhikov, A.P., Frangulyan, T.S., Ghyngazov, S.A., Koval, N.N.: Structural-phase transformations in near-surface layers of alumina-zirconium ceramics induced by low-energy high-current electron beams. Nucl. Instrum. Methods Phys. Res. Sect. B **267**(7), 1072–1076 (2009)

23. Stary, O., Malyshev, A.V., Lysenko, E.N., Petrova, A.: Formation of magnetic properties of ferrites during radiation-thermal sintering. Eurasian Phys. Tech. J. **17**(2), 6–10 (2020)
24. Surzhikov, A.P., Frangulyan, T.S., Ghyngazov, S.A., Vasil'ev, I.P., Chernyavskii, A.V.: Sintering of zirconia ceramics by intense high-energy electron beam. Ceram. Int. **42**(12), 13888–13892 (2016)
25. Nikolaev, E.V., Astafyev, A.L., Nikolaeva, S.A., Lysenko, E.N., Zeinidenov, A.K.: Investigation of electrical properties homogeneity of Li-Ti-Zn ferrite ceramics. Eurasian Phys. Tech. J. **17**(1), 5–12 (2020)
26. Gromov, V.V.: Electric charge in irradiated dielectrics, and their properties. Russ. Chem. Rev. **62**(11), 991–1003 (1993)
27. Boev, S.G.: Charge accumulation by insulators irradiated with electrons. Sov. Phys. J. **24**(2), 143–150 (1981)
28. Dyrkov, V.A., Evdokimov, O.B., Yagushkin, N.I.: Space-charge dynamics in dielectrics on irradiation with fast electrons. Radiat. Phys. Chem. **23**(3), 331–340 (1984)
29. Bronshtein, I.M., Fraiman, B.S.: Secondary Electron Emission. Nauka, Moscow, Russia (1964)
30. Boev, S.G., Sigaev, G.I.: Generation of external electric fields by insulators as a result of electron irradiation at normal atmospheric pressure. Sov. Phys. J. **23**(11), 968–972 (1980)
31. Tyutnev, A.P., Saenko, V.S., Vannikov, A.V., Oskin, V.E.: Radiation induced dielectric effect in polymers. Phys. Status Solidi (a) **86**(1), 363–374 (1984)
32. Rozno, A.G., Gromov, V.V.: Acoustic probing: a new method of measuring the electrical charge and field in the volume of dielectrics. Bull. Acad. Sci. U.S.S.R. Phys. Ser. **50**(3), 28–36 (1985)
33. Boev, S.G.: Kinetics of charge formation in electron-irradiated dielectrics. Sov. Surf. Eng. Appl. Electrochem. **2**, 69–72 (1987)

Control of the Object Temperature and Beam Parameters Under Combined Exposure to High Temperatures and High-Energy Charged Particle Fluxes

Anatoliy Surzhikov, Elena Lysenko, and Oldřich Starý

Abstract High temperatures, high heating rates and electric charge caused by a beam of accelerated particles in a bulk material during its processing pose additional requirements to the accuracy of temperature control under these specific conditions. The study experimentally investigated the temperature distribution in the bulk CaO pellets in heating by high-energy electron beam. The ILU-6 pulsed electron accelerator was used with the following beam parameters: electron beam energy of 1.5–2.0 meV, pulse beam current of 0.5–0.9 A, irradiation pulse duration of 500 μs, the irradiation pulse repetition rate of 5–50 Hz. The reliability of temperature measurements was estimated from the melting temperatures of individual reference compounds. It was found that the temperature distribution over the volume of the irradiated material at the heating stage is characterized by nonuniformity. Use of radiation-transparent thermal shields in the isothermal heating mode virtually completely eliminates temperature gradients. The performance of the ILU-6 electron accelerator most often used for practical implementation of radiation thermal processes was evaluated, and characteristics of the accelerator radiation field were determined.

1 Introduction

A number of methods have been developed to improve sintering processes in order to increase the efficiency of ceramic technologies for obtaining materials and products within economically reasonable duration of thermal treatment. These include a

A. Surzhikov · E. Lysenko (✉)
Tomsk Polytechnic University, 30 Lenin av., Tomsk 634050, Russia
e-mail: lysenkoen@tpu.ru

A. Surzhikov
e-mail: surzhikov@tpu.ru

O. Starý
Czech Technical University, Prague, Czech Republic
e-mail: oldstar@zentrum.cz

E. Lysenko et al. (eds.), *Recent Developments in the Field of Non-Destructive Testing, Safety and Materials Science*, Studies in Systems, Decision and Control 433,
https://doi.org/10.1007/978-3-030-99060-2_20

two-stage addition of components, doping of ferrite with ferrite powder of similar composition, the presence of a liquid phase, forced sintering, mechanical activation of the initial powders, and superposition of ultrasound on sintering [1–16].

In recent years, the effects of ionizing radiation fluxes have been widely used in production and modification of materials. A fundamental phenomenon of a multiple increase in the efficiency of synthesis of multi-component powder materials [17–19] and sintering [20–24] under radiation thermal treatment was discovered.

Sintering of lithium ferrites have been most fully studied under specific conditions of combined exposure to high temperatures and intense electron fluxes [25, 26]. The compaction patterns of ferrite compacts were found, and a multiple increase in the compaction rate of lithium ferrite compacts under these sintering conditions was shown [27–29].

The main feature of the radiation thermal treatment, including radiation thermal sintering, is a simultaneous effect of both high temperatures and high-energy electron fluxes on the treated material. In this case, the target is heated to 1500 °C and above at an extremely high speed (within units of seconds) using pulsed electron accelerators, which excludes additional heat sources from the process [30–34].

High temperatures, high heating rates and introduction of an electric charge into the bulk target by accelerated particle beam pose additional, previously insignificant, requirements to the accuracy of temperature control and beam dosimetry under these specific conditions [35–37]. The issue concerns the development of practical object thermometry and dosimetry under combined exposure to high temperatures and high-energy charged particles fluxes.

In this work, the methods for measuring the temperature and the temperature distribution in bulk CaO pellets in heating by high-energy electron beam were investigated. The performance of the ILU-6 electron accelerator was evaluated, and characteristics of the accelerator radiation field were determined.

2 Research Method

The study used pressed CaO pellets with various mass thicknesses.

The ILU-6 pulsed electron accelerator, which was developed in Institute of Nuclear Physics of the SB RAS (Novosibirsk, Russia), was used with the following beam parameters: electron beam energy of 1.5–2.0 meV, pulse beam current of 0.5–0.9 A, irradiation pulse duration of 500 μs, the irradiation pulse repetition rate of 5–50 Hz [38–40].

The platinum/platinum–rhodium (Pt-PtRh) thermocouple was used for temperature control. One can notice that such thermocouples are most preferable in this case due to their high temperature chemical inertness to most substances and in an aggressive ionized environment, plasticity (to provide various methods of supply to the test object), ease of obtaining a sufficiently clean measuring junction, and a wide range of operating temperatures.

The reliability of temperature measurements was estimated from the melting temperatures of individual reference compound based on Na_2CO_3.

3 Results and Considerations

3.1 Temperature Distribution in the Irradiated Bulk Material

When accelerated electrons pass through the matter, most of the beam radiation energy is converted into heat energy, which increases the irradiated material temperature. The heating rate and the object temperature are determined by the power of the accelerated electron beam. The temperature field in the irradiated material is significantly inhomogeneous due to both uneven distribution of the absorbed dose along the material depth and heat transfer processes.

It should be noted that the temperature drop in the target exposed to low-energy electron beams (which most often occurs in practice) was not given sufficient attention since, first, the target temperature did not exceed tens of degrees; second, low heating rates caused temperature equalization in the bulk material. However, using high-energy electron accelerators capable of heating the target to 2500–3000 °C at a rate of up to 1200 °C/min can provide significant temperature drops in the irradiated bulk material (particularly on the heated side).

Analytical calculations of the temperature field in irradiated objects pose significant difficulties, particularly in complex multi-component systems, for which accurate values of thermophysical characteristics and thermochemical effects are not always available. Therefore, experimental data of this type are preferable. In this case, the focus should be placed on the stage of material heating (non-isothermal mode).

A package ("sandwich") of CaO pellets with a mass thickness of 0.1 g/cm^2 each with plain junctions of thermocouples inserted into the gaps between the pellets was heated on a corundum substrate by electron beam. The obtained heating curves are shown in Fig. 1. The temperature distribution along the package depth at different times for three heating modes is presented in Fig. 2.

As can be seen, this geometry shows a significant temperature difference along the irradiated material depth, and an increased absorbed dose rate (P) causes an increase in the temperature gradient. Increased exposure time, and therefore increased temperature, causes a depth gradual shift in the temperature maximum.

On the one hand, this shift is due to increased heat transfer from the outer surface of the package at high temperatures, and, on the other hand, it is due to decreased heat transfer from the sample to the substrate when the latter is heated during irradiation.

Figure 3 shows temperature distribution in the sample composed of two CaO pellets with a variable thickness ratio and heated by beam at $P = 60$ kGy/s. The readings of the thermocouple inserted between the pellets and the corundum substrate were used as the reference.

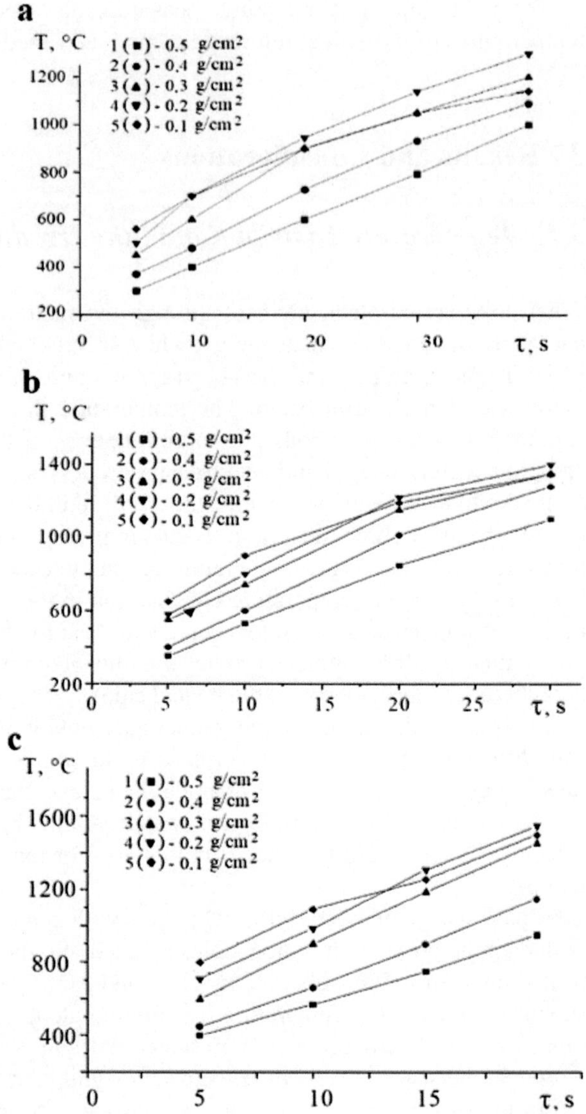

Fig. 1 Heating curves for the CaO pellet package at different depths (*r*) and at different absorbed dose rates (*P*): 40 kGy/s (**a**), 60 kGy/s (**b**), 100 kGy/s (**c**)

As can be seen from the Fig. 3, the temperature difference along the sample depth significantly decreased in comparison with the previous version.

The described experiments indicate that the gaps between the pellets play the role of "thermal resistance" due to lower thermal conductivity of air in comparison with that of the pressed sample. A decreased number of "thermal resistance" in the last experiment (from 4 to 1) leads to temperature equalization in the bulk sample.

To study the role of heat transfer from the outer surfaces of the material in temperature field formation, a package of pellets was heated by irradiation in a quartz foam

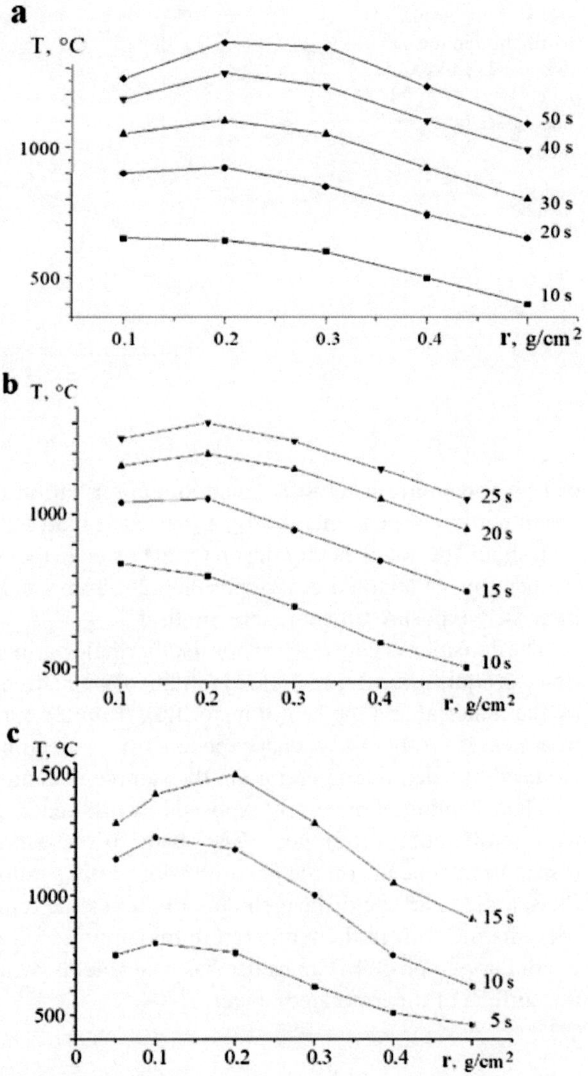

Fig. 2 Temperature distribution along the depth of the CaO pellet package at different times and at different absorbed dose rates (*P*): 40 kGy/s (**a**), 60 kGy/s (**b**), 100 kGy/s (**c**)

cell with a density of 0.1 g/cm^3. The cell walls 3 mm thick adhered tightly to the sample surface. Due to its low density, the cell virtually cannot be an external heat source, but it creates a sufficient heat shield for heat transfer from the material surface.

It was found that patterns of the temperature profile formation did not basically change. However, the temperature gradients significantly decreased, and the temperature profile smoothed. The time, when the temperature at the maximum distribution was attained, reduced by 23%. During heating in the cell, the temperature drop from the package surface to the temperature maximum decreased by 20%. This value

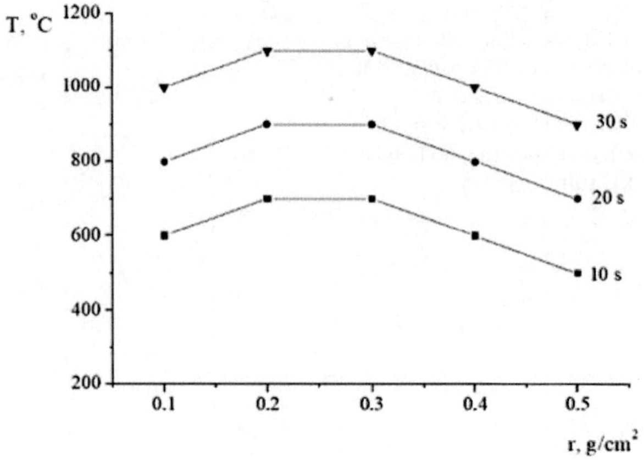

Fig. 3 Temperature distribution in the sample composed of two CaO pellets with a variable thicknesses ratio (*r*)

can be apparently used to estimate the contribution of the surface heat transfer to accumulation of heat energy of the irradiated material.

It should be noted that for the material heated in the cell (heat shield), the maximum temperature is attained at a depth of 0.25 g/cm^2, and it virtually does not shift at increased exposure time and temperature.

The considered data refer to non-isothermal heating. At sufficiently long exposure time, an equilibrium temperature is attained, when the energy fluxes initially absorbed by the material and the heat transfer flux from the sample are equal, i.e. isothermal treatment is set. In this case, at exposure time exceeding (4–8) min when heat shields are used, the depth temperature in the sample is completely equalized (Table 1).

Thus, heating of materials exposed to a high-energy electron beam causes significant temperature differences. The shape of the curve of the in-depth temperature distribution depends on the absorbed dose rate, heating rate, and heat transfer from the sample. The use of heat shields improves the temperature field uniformity and prevents the shift in the temperature maximum. Refractory foams or powder cover of similar composition can be used as heat shields, which improve heat transfer from the surface of the irradiated object.

Table 1 Temperature distribution (in °C) in the CaO pellet "sandwich" in the cell. Isothermal exposure is 8 min

P_{equ} [kGy/s]	r [g/cm^2]		
	0.17	0.33	0.5
7	820	830	835
12	1070	1075	1075
13	1140	1140	1140
18	1400	1415	1405

The electron energy is 1.5 meV. P_{equ} is the absorbed dose rate, which provides a given equilibrium temperature; r is thermocouple insertion depth

The maximum temperature falls on a thickness of $0.3R_e$, where R_e is the extrapolated electron range. This means that the thermocouple measuring junction should be inserted at a depth of $0.3R_e$ from the irradiated surface to measure the maximum temperature of an object in the non-isothermal mode.

In the isothermal mode of radiation thermal treatment with heat shields used, an almost complete equalization of temperature along the depth of the material is attained.

3.2 Dosimetry of High Energy Electron Beam Under Radiation Thermal Conditions

The radiation fields of electron accelerators can be characterized by the distribution of current density (J) over the surface of the irradiated region and the distribution of absorbed dose (D) and the absorbed dose rate along the depth (r) of the irradiated material.

It should be noted that the absorbed dose and dose rate can hardly be directly determined under radiation thermal conditions, since it requires small-size sensors with a large capacity (thousands of kGy) and reliable operation in a wide range of temperatures (up to 1500 °C) and absorbed dose rates (up to 200–500 kGy/s). The lack of chemical dosimeters that meet these requirements necessitates calculation of the P and D values based on measurements of the electron beam parameter J.

In the study, the beam current density was measured using the Faraday cup method. The pulsed electron beam parameters (amplitude, shape, duration, irradiation pulse repetition rate) were determined using an oscilloscope, and the electron energy and the total beam current were measured using the accelerator control panel.

The Faraday cup method widely used in diagnostics of high-current electron beams implies measuring the current or beam charge in the electrically conductive detector. To determine J in any region of the accelerator radiation field, a detector is placed in this region with a collimating plate sometimes installed in front of it, which isolates part of the beam of a specified area. Average current density is

$$J = \frac{J_g}{S} \tag{1}$$

where J_g is the beam current passing through the detector, S is the effective detector area (from which electrons are collected) or the cross section of the collimating hole.

The error of the Faraday cup method is mainly determined by the factors such as electron backscattering from the detector surface, ionization (plasma formation) over the detector, electron absorption (or emission) at the ends of the collimating hole. These factors should be considered to choose the detector design.

The sensor was made of graphite with high thermal stability and low backscattering coefficient (less than 3% at normal incidence of electrons with energy of

1.5 meV. Backscattering correction was performed with no regard to the spectral-angular distribution of the electron flux. The error due to the neglect of this factor did not exceed +1%.

To assess the error associated with air ionization, the recordings of one detector in the evacuated and unpressurized versions were compared. To create a vacuum, the collimating hole was hermetically sealed with a thin (30 μm) aluminum foil, and air was evacuated from the detector to a residual pressure of 10^{-3} Torr. The experiment showed that the role of air ionization is negligible up to current densities of 60–80 μA/cm^2 in case of the grounded sensor body that serves as a shield. At current densities exceeding 100 μA/cm^2, the error caused by ionization increases sharply (up to +30%) and becomes the main one, which is due to increased air temperature near the beam heated collector. The evacuated version cannot be implemented under these conditions due to instability of metal foils or polymer films. The measurement accuracy can be improved by cooling the Faraday cup with a compressed air jet that reduces the error to +15% (at beam current densities less than 150 μA/cm^2), or by quick shift of the detector through the irradiated region.

It is rather difficult to accurately determine the error related to absorption and emission of electrons at the ends of the collimating hole. For the hole with a circular cross-section, it, in the first approximation, is proportional to the collimating plate thickness and inversely proportional to the hole radius. The minimum allowable thickness of the collimator corresponds to the electron range in the material it is made from (the range of electrons with energy of 1.5 meV in the graph is 3.1 mm, and in steel it is 0.9 mm). A significant increase in the hole radius is undesirable, since it distorts data on current density local values, particularly in inhomogeneous regions of the beam. In the detectors employed in the study, the hole area did not typically exceed 1–2 cm. In general, the maximum error caused by the considered factor was 3–5% (for average incidence angles $\alpha < 30°$).

An insignificant increase in the error by the Faraday cup method during measurements in the peripheral regions of the beam caused by the shift in the angular electron distribution towards large angles increases backscattering and absorption losses in the end parts of the collimator. In these conditions, collimation-free versions of the method are more appropriate. High measurement accuracy is provided by the Faraday cup design without a collimation plate with a small gap between the cylindrical detector and the surrounding annular shield, which results in a total error of less than 2% at relatively low beam intensities. This version was used at $J \leq 20$ μA/cm^2.

The scheme for recording signals from the Faraday cylinder block depends on the accelerator type. A pulsed accelerator ILU-6 can be used to measure the magnitude of charge supplied to the detector per irradiation pulse using an *RC* integrator circuit. An oscilloscope is used to record the voltage from the capacitor of known capacity *C*, which is charged by current pulse from the Faraday cup. The average beam current density is calculated using the formula:

$$J = \frac{CUf}{S} \tag{2}$$

where C is the capacitor capacitance, μF; U is the voltage across the capacitor plates, V; f is the pulse repetition rate, Hz; S is the collimating hole area, cm^2; J is the current density, $\mu A/cm^2$.

Several types of current sensors—Faraday cylinders—have been developed for radiation thermal experiments. The simplest design—a single-section sensor—is advisable for operational control in case of known beam current density distribution.

This single-section current sensor can also be used to determine the beam current density distribution. A single-section Faraday cup used to obtain the beam current density pattern requires a well-calibrated three-axis moving device with the sensor mounted on it; otherwise, it is uninformative. Inconvenience of the moving device is due to the predominant use of multi-section sensors for this purpose.

To determine the in-plane distribution of the beam current density using the ILU-6 accelerator with a small-size exhaust device, 9- and 12-section current sensors were employed.

The signal is measured alternately from each detector. In this case, other detectors are grounded to eliminate charge transfer due to air ionization. The detector unit in the experiments performed using accelerators with a concentrated release covered almost the entire beam, which provided a comprehensive pattern of the current density distribution in the sensor location plane. The region with the minimum inhomogeneity of the beam current density was chosen as a working one.

In most experiments, the total absolute measurement error of the Faraday cup method did not exceed 10%.

To determine the absorbed dose rate P at a given point of the irradiated sample based on the current density, the distribution of the absorbed energy along the material depth should be obtained ("dose curve").

In our experiments, the dose distribution was measured using film dosimeters placed between the irradiated material layers of known thickness. The resulting "dose curves" for calcium oxide are presented in Fig. 4.

The average absorbed dose rate along the depth was calculated based on the values of the current density and radiation intensity using the following equation:

$$\overline{P} = \frac{I}{R_e} = \frac{JE}{R_e} \tag{3}$$

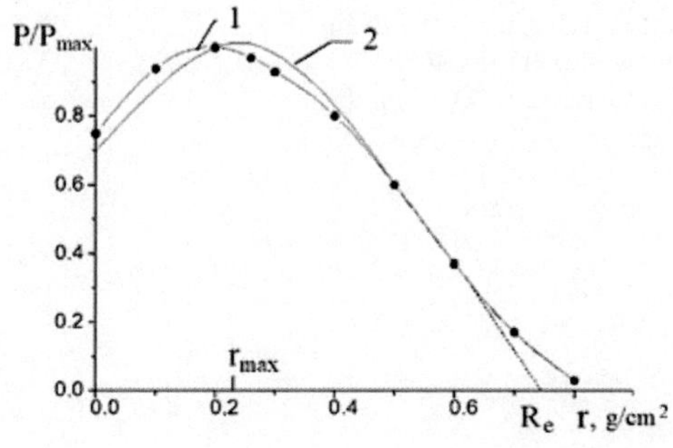

Fig. 4 Distribution of the absorbed energy of 1.5 meV electrons: 1—along the CaO depth, 2—along the Al depth

where $\overline{P}$ is the average absorbed dose rate, kGy/s; I is radiation intensity, W/cm^2; R_e is extrapolated electron range, g/cm^2; J is the beam current density, μA/cm^2; E is the electron energy, MeV.

Since $E/R_e \approx 2.2$ in the energy range of 1.2–1.8 meV, the relationship between $\overline{P}$ and J can be simplified: $\overline{P} \approx 2.2\ J$.

3.3 The Capabilities of the ILU-6 Accelerator in Relation to the Study and Implementation of Radiation Thermal Processes.

The developed methods of current-dose and temperature measurements enabled investigation of the characteristics of the ILU-6 accelerator radiation fields and the limits of the accelerator applicability to radiation thermal analysis.

This high-energy accelerator is capable of releasing a high intensity beam into the atmosphere and, which is most important, of adjusting the radiation parameters.

As a rule, a foil release system of small standard size 400 × 60 mm^2 is employed. In this version, due to a high level of organization of heat removal, the maximum beam density at the foil level exceeds 200 μA/cm^2. At a distance from the foil, the radiation intensity decreases due to electron scattering in air, and the beam size increases.

The beam current density distribution along and across the foil is shown in Fig. 5. The majority of radiation thermal experiments require a uniform distribution of the beam current density over the irradiated material surface. In kinetic experiments, the value 2–3% is taken as the permissible limit of the current density non-uniformity over the sample area, which corresponds to temperature fluctuations at the level of 20–25 °C at about 1000 °C.

The irradiated samples are typically placed in the center of the irradiated region, where the current density non-uniformity is minimal. The release foil of the accelerator is blown by a powerful jet of compressed air; therefore an additional shielding foil is placed between the accelerator horn and the irradiation plane to prevent the

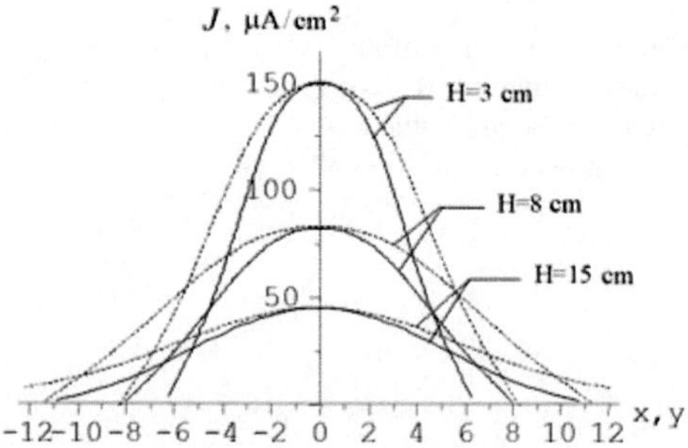

Fig. 5 Distribution of current density in the beam generated by ILU-6 accelerator: (– –) along the scan, (—) across the scan

blowing effect on the irradiated material. The additional foil increases electron scattering and smoothest the current density distribution, but it decreases its maximum value. In this geometry, in the center of the irradiated region at a distance of 12 cm from the outlet window, current densities of 120–150 μA/cm^2 are achieved, which corresponds to the absorbed dose rate in the irradiated material 250–330 kGy/s.

Typical heating curves for calcium oxide samples obtained at different absorbed dose rates are presented in Fig. 6.

The heating rate $\Delta T/\Delta\tau$ in the initial, close to adiabatic, regions of the curves is maximum, and it is determined by the approximate ratio $\Delta T/\Delta\tau \approx P/C$, where P is the absorbed dose rate, kGy/s; C is the heat capacity of the material, J/g·K.

At increased temperature, the heat transfer from the irradiated material increases, the heating rate gradually decreases to zero, and the temperature reaches a plateau and attains an equilibrium value. The dependence of the maximum (equilibrium) temperature on the absorbed dose rate is shown in Fig. 7.

Figures 6 and 7 show that the ILU-6 accelerator can provide heating rates up to 250 °C/s (in adiabatic regions) and maximum temperatures up to 2000–2300 °C. The radiation parameters and, consequently, the thermal treatment mode of the IILU-6 accelerator are adjusted in different ways: by changing the repetition rate of irradiation pulses, beam current in the pulse, energy, and the scanning width. The transition

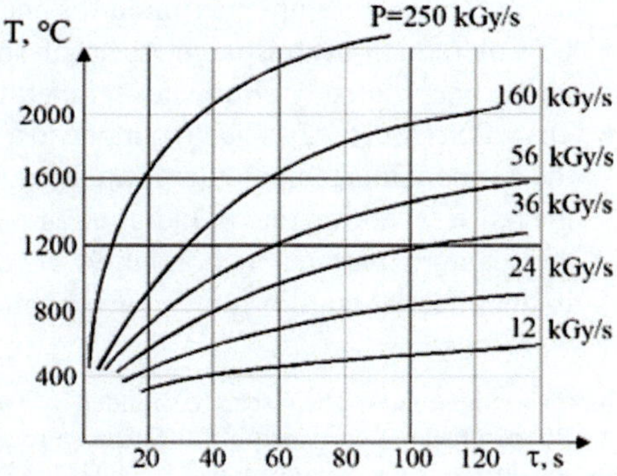

Fig. 6 Heating curves for calcium oxide samples exposed to electron beam at different absorbed dose rates

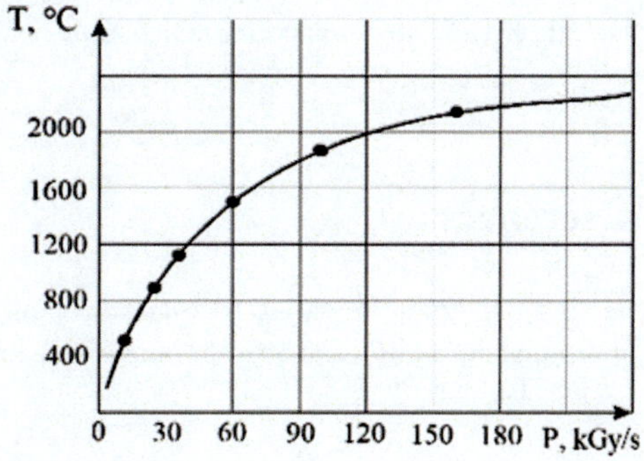

Fig. 7 Dependence of the "equilibrium" temperature of calcium oxide on the absorbed dose rate

from one mode to another one is carried out within less than a second. Adjustability of the radiation parameters ensures any mode of temperature variation. In particular, to switch from the fast heating mode to a given temperature level *T*, it is sufficient, at the moment this temperature is attained, to reduce the beam current (by changing the pulse repetition rate of radiation or pulse current) or its intensity (by changing the scanning parameters or energy) to values that provide the dose rate absorbed in the material equilibrium for temperature *T*.

4 Conclusion

The temperature distribution in the bulk CaO pellets in heating by high-energy electron beam were experimentally investigated. The performance capabilities of the ILU-6 electron accelerator have been experimentally studied to investigate and practically implement radiation thermal processes in high-energy electron beams, and the radiation field characteristics of the ILU-6 accelerator with the foil release of the electron beam into the atmosphere have been determined. Based on the results obtained, the following conclusions were drawn.

- At the heating stage, temperature distribution in the irradiated bulk material exhibits a temperature maximum for one third of the extrapolated electron range.
- Use of radiation-transparent thermal shields in the isothermal sintering mode virtually completely eliminates temperature gradients.
- To switch from the fast heating mode to a given temperature level, it is sufficient, at the moment this temperature is attained, to reduce the beam current (by changing the pulse repetition rate of radiation or pulse current) or its intensity (by changing the scanning parameters or energy) to values that provide the dose rate absorbed in the material equilibrium for given temperature.

Acknowledgements The research is funded by The Ministry of Science and Higher Education of the Russian Federation as part of the Science Program, Project FSWW-2020-0014. The work was carried out using the equipment of the Center for Sharing Use "Nanomaterials and Nanotechnologies" of Tomsk Polytechnic University supported by the RF Ministry of Education and Science, project 075-15-2021-710. The authors are grateful to M.V. Korobeynikov (Budker Institute of Nuclear Physics, Novosibirsk) for radiation-thermal experiments using ILU accelerator.

References

1. Wang, R., Zhou, T., Zhong, Z.: Low-temperature processing of LiZn-based ferrite ceramics by co-doping of V_2O_5 and Sb_2O_3: composition, microstructure and magnetic properties. J. Mater. Sci. Technol. **99**, 1–8 (2022)

2. Surzhikov, A.P., Frangulyan, T.S., Ghyngazov, S.A.: A thermoanalysis of phase transformations and linear shrinkage kinetics of ceramics made from ultrafine plasmochemical ZrO_2(Y)-Al_2O_3 powders. J. Therm. Anal. Calorim. **115**(2), 1439–1445 (2014)
3. Guo, J., Zhang, H., He, Z., Li, S., Li, Z.: Electrical properties and temperature sensitivity of Mo-modified $MnFe_2O_4$ ceramics for application of NTC thermistors. J. Mater. Sci. Mater. Electron. **29**, 2491–2499 (2018)
4. Mazen, S.A., Abu-Elsaad, N.I.: Structural, magnetic and electrical properties of the lithium ferrite obtained by ball milling and heat treatment. Appl. Nanosci. **5**, 105–114 (2015)
5. Amin, A.M.M., Soliman, Y.M.M., El-Dek, S.I. Ahmed, Y.M.Z., Zaki, A.H.: Valorization of industrial iron and zinc sludges for the synthesis of $ZnFe_2O_4$ ceramics. J. Magn. Magn. Mater. **544**, 168681 (2022)
6. Kumar, S., Sarkar, K., Mondal, R., Dey, S., Majumder, S.: Presence of mixed magnetic phase in mechanically milled $Co_{0.5}Zn_{0.5}Fe_2O_4$: a study on structural, magnetic and hyperfine properties. J. Magn. Magn. Mater. **487**, 165303 (2019)
7. Xu, F., Shi, X., Liao, Y., Li, J., Hu, J.: Investigation of grain growth and magnetic properties of low-sintered LiZnTi ferrite-ceramics. Ceram. Int. **46**, 14669–14673 (2020)
8. Nili-Ahmadababdi, M., Sarraf-Mamoory, R., Yourdkhani, A., Diaconu, A., Rotaru, A.: Magnetic and electrical properties of $Mg_{1-x}Co_xFe_2O_4$ ($x=0-0.15$) ceramics prepared by the solid-state method. J. Eur. Ceram. Soc. **42**(2), 442–447 (2022)
9. Kavanlooee, M., Hashemi, B., Maleki-Ghaleh, H., Kavanlooee, J.: Effect of annealing on phase evolution, microstructure, and magnetic properties of nanocrystalline ball-milled LiZnTi ferrite. J. Electron. Mater. **41**, 3082–3086 (2012)
10. Berbenni, V., Bruni, G., Milanese, C., Girella, A., Marini, A.: Synthesis and characterization of $LaFeO_3$ powders prepared by a mixed mechanical/thermal processing route. J. Therm. Anal. Calorim. **133**, 413–419 (2018)
11. Zahir, R., Chowdhury, F.-U.-Z., Uddin, M.M., Hakim, M.A.: Structural, magnetic and electrical characterization of Cd substituted Mg ferrites synthesized by double sintering technique. J. Magn. Magn. Mater. **410**, 55–62 (2016)
12. Xie, F., Chen, Y., Bai, M., Wang, P.: Co-substituted LiZnTiBi ferrite with equivalent permeability and permittivity for high-frequency miniaturized antenna application. Ceram. Int. **45**, 17915–17919 (2019)
13. Hu, J., Yan, M., Luo, W., Wu, J.M.: Effects of microstructure on the temperature dependence of relative initial permeability of NiCuZn ferrites. Physica B **400**, 119–123 (2007)
14. Widatallah, H.M., Ren, X.L., Al-Omari, I.A.: The influence of TiO_2 polymorph, mechanical milling and subsequent sintering on the formation of Ti-substituted spinel-related $Li_{0.5}Fe_{2.5}O_4$. J. Mater. Sci. **41**, 6333–6338 (2006)
15. Widatallah, H.M., Johnson, C., Berry, F.J.: The influence of ball milling and subsequent calcination on the formation of $LiFeO_2$. J. Mater. Sci. **37**, 4621–4625 (2002)
16. Teo, M.L.S., Kong, L.B., Li, Z.W., Lin, G.Q., Gan, Y.B.: Development of magneto-dielectric materials based on Li-ferrite ceramics I. Densification behavior and microstructure development. J. Alloys Compd **459**, 557–566 (2008)
17. Ancharova, U.V., Mikhailenko, M.A., Tolochko, B.P., Lyakhov, N.Z., Korobeinikov, M.V., Bryazgin, A.A., Shtarklev, E.A.: Synthesis and staging of the phase formation for strontium ferrites in thermal and radiation thermal reactions. IOP Conf. Ser. Mater. Sci. Eng. **81**, 012122 (2015)
18. Yurov, V.M., Baltabekov, A.S., Laurinas, V.C., Guchenko, S.A.: Dimensional effects and surface energy of ferroelectric crystals. Eurasian Phys. Tech. J. **16**(1), 18–23 (2019)
19. Zhuravlev, V.A., Naiden, E.P., Minin, R.V., Itin, V.I., Suslyaev, V.I., Korovin, E.Yu.: Radiation-thermal synthesis of W-type hexaferrites. IOP Conf. Ser. Mater. Sci. Eng. **81**, 012003 (2015)
20. Kostishin, V.G., Andreev, V.G., Korovushkin, V.V., Chitanov, D.N., Yudanov, N.A., Morchenko, A.T., Nikolaev, A.N.: Preparation of 2000NN ferrite ceramics by a complete and a short radiation-enhanced thermal sintering process. Inorg. Mater. **50**, 1317 (2014)
21. Kostishyn, V., Isaev, I., Scherbakov, S., Nalogin, A., Belokon, E., Bryazgin, A.: Obtaining anisotropic hexaferrites for the base layers of microstrip SHF devices by the radiation-thermal sintering. East. Eur. J. Enterp. Technol. **5**, 32–39 (2016)

22. Naiden, E.P., Zhuravlev, V.A., Minin, R.V., Suslyaev, V.I., Itin, V.I., Korovin, E.Y.: Structural and magnetic properties of SHS-produced multiphase W-type hexaferrites: influence of radiation-thermal treatment. Int. J. Self Propag. High Temp. Synth. **24**, 148–151 (2015)
23. Surzhikov, A.P., Frangulyan, T.S., Ghyngazov, S.A., Koval, N.N.: Structural-phase transformations in near-surface layers of alumina-zirconium ceramics induced by low-energy high-current electron beams. Nucl. Instrum. Methods Phys. Res. Sect. B **267**(7), 1072–1076 (2009)
24. Surzhikov, A.P., Frangulyan, T.S., Ghyngazov, S.A., Vasil'ev, I.P., Chernyavskii, A.V.: Sintering of zirconia ceramics by intense high-energy electron beam. Ceram. Int. **42**(12), 13888–13892 (2016)
25. Malyshev, A.V., Lysenko, E.N., Vlasov, V.A., Nikolaeva, S.A.: Electromagnetic properties of $Li_{0.4}Fe_{2.4}Zn_{0.2}O_4$ ferrite sintered by continuous electron beam heating. Ceram. Int. **42**, 16180–16183 (2016)
26. Surzhikov, A.P., Lysenko, E.N., Vasendina, E.A., Sokolovskii, A.N., Vlasov, V.A., Pritulov, A.M.: Thermogravimetric investigation of the effect of annealing conditions on the soft ferrite phase homogeneity. J. Therm. Anal. Calorim. **104**(2), 613–617 (2011)
27. Surzhikov, A.P., Frangulyan, T.S., Ghyngazov, S.A., Lysenko, E.N., Galtseva, O.V.: Physics of magnetic phenomena: investigation of electroconductivity of lithium pentaferrite. Russ. Phys. J. **49**(5), 506–510 (2006)
28. Nikolaev, E.V., Astafyev, A.L., Nikolaeva, S.A., Lysenko, E.N., Zeinidenov, A.K.: Investigation of electrical properties homogeneity of Li-Ti-Zn ferrite ceramics. Eurasian Phys. Tech. J. **17**(1), 5–12 (2020)
29. Stary, O., Malyshev, A.V., Lysenko, E.N., Petrova, A.: Formation of magnetic properties of ferrites during radiation-thermal sintering. Eurasian Phys. Tech. J. **17**(2), 6–10 (2020)
30. Salimov, R.A., Cherepkov, V.G., Kuksanov, N.K., Kuznetzov, S.A.: The use of electron accelerators for radiation disinfestation of grain. Radiat. Phys. Chem. **57**, 625–627 (2000)
31. Ivanov, A.S., Maznev, V.P., Ovchinnikov, V.P., Svinin, M.P., Tolstun, N.G.: Present state of works on development of electron accelerators for energy consuming processes at Efremov research institute. https://citeseerx.ist.psu.edu/viewdoc/download?doi=10.1.1.730.9688&rep=rep1&type=pdf
32. Salimov, R.A., Cherepkov, V.G., Golubenko, J.I., Krainov, G.S., Korabelnikov, B.M., Kuznetsov, S.A., Nemytov, P.I.: High power electron accelerators of ELV-series: status, development, applications. Radiat. Phys. Chem. **57**, 661–665 (2000)
33. Matsuyama, S., Ishii, K., Fujisawa, M., Kawamura, Y., Tsuboi, S., Yamanaka, K., Kikuchi, Y.: Upgrading of the 4.5 MV dynamitron accelerator at Tohoku university for microbeam and nanobeam applications. Nucl. Instrum. Methods Phys. Res. Sect. B Beam Interact. Mater. Atoms **267**(12–13), 2060–2064 (2009)
34. Cleland, M.R., Parks, L.A.: Medium and high-energy electron beam radiation processing equipment for commercial applications. Nucl. Instrum. Methods Phys. Res. Sect. B **208**, 74–89 (2003)
35. Mehnert, R.: Review of industrial applications of electron accelerators. Nucl. Instrum. Methods Phys. Res. Sect. B **113**, 81–87 (1996)
36. Abramyan, E.A.: Industrial Electron Accelerators and Applications. Hemisphere Publishing Corporation, New York (1988)
37. Zhuravlev, V.A., Meshcheriakov, V.A., Suslyaev V.I.: Computer simulation of processes of radiation-thermal heating. IOP Conf. Ser. Mater. Sci. Eng. **81**, 012054 (2015)
38. Auslender, V.L.: ILU-type electron accelerator for industrial technologies. Nucl. Instrum. Methods Phys. Res. Sect. B **89**, 46–48 (1994)
39. Bryazgin, A.A.: Electron accelerator for energy up to 5.0 MeV and beam power up to 50 kW with X-ray converter. Probl. Atomic Sci. Technol. **1**, 21–23 (2004)
40. Auslender, V.L., Bryazgin, A.A., Faktorovich, B.L., Gorbunov, V.A., Kokin, E.N., Korobeinikov, M.V., Voronin, L.A.: Accelerators for E-beam and X-ray processing. Radiat. Phys. Chem. **63**, 613–615 (2002)

Practical Thermometry of Materials Irradiated by Powerful Beams of Accelerated Electrons

Anatoliy Surzhikov, Elena Lysenko, and Oldřich Starý

Abstract The thermometry of materials irradiated with powerful beams of accelerated electrons was investigated using various methods for measuring temperature with a thermocouple sensor. For materials annealing, the ILU-6 pulsed electron accelerator was used with the following beam parameters: electron beam energy of 1.5–2.0 meV, pulse beam current of 0.5–0.9 A, irradiation pulse duration of 500 μs, the irradiation pulse repetition rate of 5–50 Hz. The platinum/platinum–rhodium thermocouple was used for the thermometry. The reliability of temperature measurements was estimated from the melting temperatures of individual reference compounds. The calcium oxide was used as the object of research. The effect of the geometry of the thermocouple measuring junction introduction into the test object on accuracy of temperature measurement was studied. The application limits were set to each of the geometries, and the methods for measuring the thermocouple EMF under radiation thermal treatment were tested. The schemes were proposed that ensure the measurement of the true temperature value during electron beam treatment.

1 Introduction

The published results of experimental studies indicate that the use of high-energy electron beams is promising in the production of materials for various functional purposes [1–19]. This was due to the development of relatively small but powerful (up to 400 kW) electron accelerators providing electron energies up to 5 meV. Among them, ELV accelerator generating continuous electron beam and ILU accelerator

A. Surzhikov · E. Lysenko (✉)
Tomsk Polytechnic University, 30 Lenin av., Tomsk, Russia 634050
e-mail: lysenkoen@tpu.ru

A. Surzhikov
e-mail: surzhikov@tpu.ru

O. Starý
Czech Technical University, Prague, Czech Republic
e-mail: oldstar@zentrum.cz

E. Lysenko et al. (eds.), *Recent Developments in the Field of Non-Destructive Testing, Safety and Materials Science*, Studies in Systems, Decision and Control 433,
https://doi.org/10.1007/978-3-030-99060-2_21

generating pulse electron beam, which were developed in Institute of Nuclear Physics of the SB RAS (Novosibirsk, Russia) [20, 21]. Other similar modifications are the industrial electron accelerators by Efremov Scientific Research Institute of Electrophysical Apparatus (St. Petersburg, Russia) [22], and Dynamitron accelerators by IBA (Belgium) [23, 24], etc. [25–28]. The geometrical parameters of these accelerators are suitable, and the irradiation mode can be adjusted in a wide range, which enables heating of test objects to a given temperature without other additional heat sources employed. It should be noted that when using electron energies up to 5 meV, the electron accelerator does not create induced radioactivity [29].

It is known that when processing a material with accelerated electrons, most of the radiation energy is converted into thermal energy, which leads to an increase in the temperature of the irradiated object [30–32]. The heating rate and material temperature are determined by the parameters of the electron beam and heat transfer conditions [33]. Due to the fact that the absorption of radiation energy by the material is uneven in depth, then the temperature distribution over the thickness of the sample will be uneven. In this case, the region of the maximum temperature in the sample found to be determined by the electron energy in the beam. The higher the electron energy in the beam, the greater the electron range and, thus, the deeper the maximum heat release from the irradiated sample surface [34].

When carrying out experiments on the processing of materials with electron beams, it is extremely important to control the maximum temperature of the heated material in order to avoid overheating of the samples. Thus, in this work, the methods for correctly determining the temperature of samples heated by an electron beam were proposed.

2 Research Method

The ILU-6 pulsed electron accelerator, which was developed in Institute of Nuclear Physics of the SB RAS (Novosibirsk, Russia), was used with the following beam parameters: electron beam energy of 1.5–2.0 meV, pulse beam current of 0.5–0.9 A, irradiation pulse duration of 500 μs, the irradiation pulse repetition rate of 5–50 Hz [35–38].

The fundamental question that determines all further methodology is the choice of the thermal sensor type. Based on the analysis of general practical thermometry, we have chosen a thermocouple method, which allows measurement of local temperatures in the bulk material and is considered important for such studies. Pt–PtRh thermocouples are most preferable in this case due to their high temperature chemical inertness to most substances and in an aggressive ionized environment, plasticity (to provide various methods of supply to the test object), ease of obtaining a sufficiently clean measuring junction, and a wide range of operating temperatures [39, 40].

The choice of the thermocouple as a temperature sensor, in turn, poses a number of problems, which should be solved to employ practical thermometry in the field of high-energy electron beams:

(i) temperature distribution in the bulk material;
(ii) measurement of the object temperature using a thermocouple;
(iii) measurement of the thermocouple EMF.

The reliability of temperature measurements was estimated from the melting temperatures of Na_2CO_3, sodium monoferrite $NaFeO_2$, and calcium fluoride. The material under study was bulk CaO.

3 Results and Considerations

3.1 *Measuring Temperature of Irradiated Objects with a Thermocouple Sensor*

The feature of the thermocouple method under irradiation is that the thermocouple measuring junction inserted into an object is heated directly by electron beam. Therefore, significantly different specific thermal conductivity of the junction and that of the material, which typically occurs, can cause deviation of the recorded temperature from that of the material in the region of the measuring junction. For example, the specific heat capacity of platinum and its alloys is 0.5 J/g K, and the heat capacity, for example, of oxide materials is close to 1.0 J/g K. At such significant differences, the measuring junction is expected to overheat. However, in real conditions, overheating does not always occur due to factors, which reduce or eliminate it, namely, due to heat transfer from the material to the junction and heat dissipation along the thermocouple wires. Under certain conditions, the latter factor can lead to an opposite effect, when the junction temperature will be lower than the material temperature.

In case of significant difference between the junction temperature and the material one, the measurement error ΔT can be quantified in a number of cases by short-term

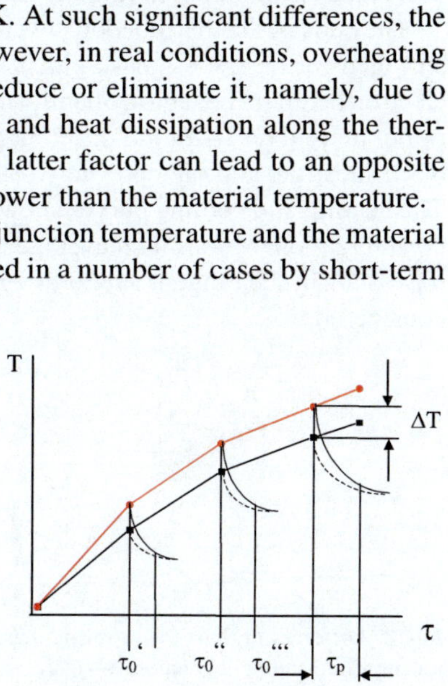

Fig. 1 Method for evaluating the temperature measurement error and plotting the true heating curve (dotted line) in case of overheating of the thermocouple measuring junction

shutdown of the electron beam during object heating. Figure 1 shows the case for $\Delta T > 0$.

After beam shutdown at the moment τ_0, the thermocouple is cooled to the temperature of the object during the time τ_p and, upon further cooling, the temperature reaches its true value. The duration of the temperature equalization depends on the value of the thermocouple measuring junction and the conditions of the contact between the junctions; it ranges from 0.1 s to 1.0 s. To determine the object temperature at the moment the accelerator is turned off, the high-temperature section of the cooling curve is approximated at the moment τ_0. Calibration of different values of the final temperature can yield a true heating curve (Fig. 1 (dotted line)).

However, this technique is laborious; it cannot be used to promptly adjust the temperature regime, and requires high repeatability of experimental results. In addition, the calculated ΔT is sufficiently rigorous only for the simplest "geometries" of the irradiated object, for example, for an isolated cylinder. In real conditions, when heat transfer processes are superimposed by the effect of the substrate, heat removal through the thermocouple wires and other factors, a sufficiently strict solution of the problem is problematic.

Therefore, temperature measurement techniques that do not require a quantitative calculation of corrections, where $\Delta T = 0$, are more appropriate. When choosing such techniques, consider different options for inserting thermocouples into the sample; the reliability of temperature measurements was estimated from the melting temperatures of individual reference compounds Na_2CO_3, sodium monoferrite $NaFeO_2$, and calcium fluoride.

The first of the considered options for inserting a thermocouple implies pressing of its measuring junction into the sample center (Fig. 2a).

The ends of the thermocouple were led through a corundum tube to the thermostat and then to the EMF measurement circuit. The samples were installed below the electron beam in the region of uniform distribution of the current density (fluctuations of not more than 1–2%) in different geometries: in a hanging position, on a fireclay brick substrate and on a powder cover of composition similar to that of the sample. The heating and cooling curves show the areas corresponding to sample melting and crystallization. Modes with average rates from 15 to 150 °C/s, provided by absorbed dose rates in the range from 20 to 150 kGy/s and different cooling conditions, were considered.

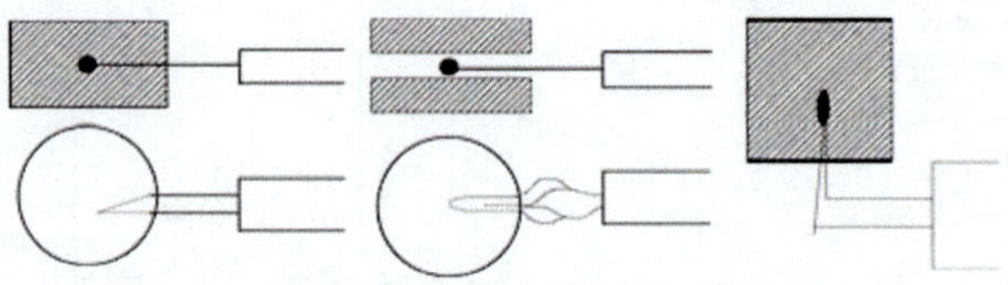

Fig. 2 Different options for inserting thermocouples into the sample: **a**—pressing the measuring junction into the pellet center, **b**—plain junction between two pellets, **c**—vertical insertion of the junction

In all the experiments, the melting temperature coincided with the true melting point of the compounds used with an accuracy of 20 °C. In a number of experiments, when reaching melting temperature, heating was stopped, and the sample was subjected to microscopic examination, which showed no melted zones in the region of the junction–sample contact.

The performed calibration experiments showed the reliability of the temperature measurement results for a pressed-in thermocouple in all three geometries, i.e. under conditions that ensure tight contact between the sample material and the thermocouple junction. However, well-known laboriousness of the thermocouple press-in technique should be emphasized.

The second, less laborious option, involves the use of thermocouples with a plain measuring junction. A platinum thermocouple junction with a diameter of 100–250 μm exhibits high ductility and, therefore, it can be rolled into foil with a thickness of 30–70 μm without violating its integrity and distorting temperature measurement readings. The rolled junction is inserted between parts of the sample (Fig. 2b). The gap between the pellets does not exceed 100 μm. Calibration experiments similar to those described above in relation to the pressed-in thermocouple showed that the level of temperature measurement reliability in both cases is similar, but the procedure for inserting thermocouples with plain junction is much simpler. They can be used if no pellet warping occurs during heating, which leads to deterioration of the contact between the junction and the sample and distortion of the measurement results.

The third, the simplest and most convenient insertion option is vertical insertion (Fig. 2c), when the thermocouple measuring junction is inserted into the hole along the sample axis to the depth corresponding to the position of the temperature maximum. Calibration experiments under conditions of relatively high absorbed dose rates of 50 kGy/s showed that heating of the reference samples to melting temperature (according to thermocouple readings), does not cause melting of the sample region adjacent to the vertically inserted junction, whereas, outside this region, melting can be distinctly observed. Junction cooling is caused by heat removal through the thermocouple wires under a significant gradient.

To estimate the size of the gap ΔT between the readings of the vertically inserted thermocouple and the temperature values of the material at the same depth (but not in the region adjacent to the junction), a thermocouple with plain junction inserted horizontally to the same depth with a vertical junction was used as the reference (Fig. 3).

Experiments on irradiation of the reference compounds showed that the junction indicates melting, whereas, the vertically inserted junction does not indicate it. A particularly significant difference in readings of the thermocouples can be observed at high dose rate (Fig. 4). An increase in ΔT at increased P is attributed to the fact that an increased heating rate causes an increase in the temperature difference between the sample and the substrate, and, accordingly, heat removal along the wires of the vertically inserted thermocouple.

As follows from Fig. 4, at increased irradiation duration, when the heat removal of the thermocouple decreases due to substrate heating, ΔT decreases. When the beam is shutdown, the readings of both thermocouples are aligned, ΔT decreases to zero,

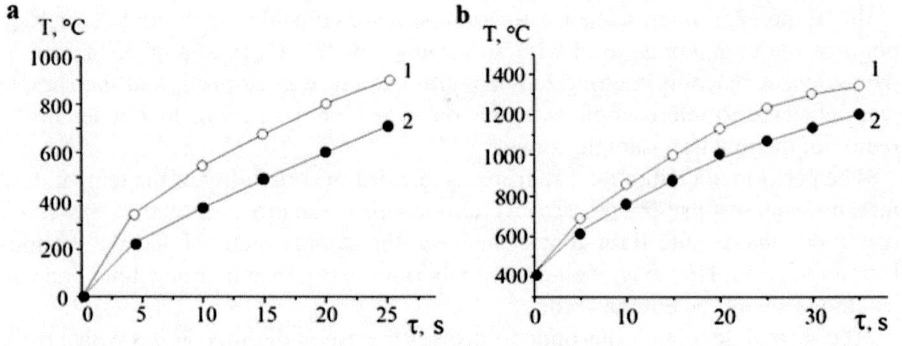

Fig. 3 Curves of CaO sample heating by electron beam, where $P = 45$ kGy/s: 1—thermocouple with plain junction, 2—vertically inserted junction, **a** cold substrate, **b** substrate preheated to 400 °C

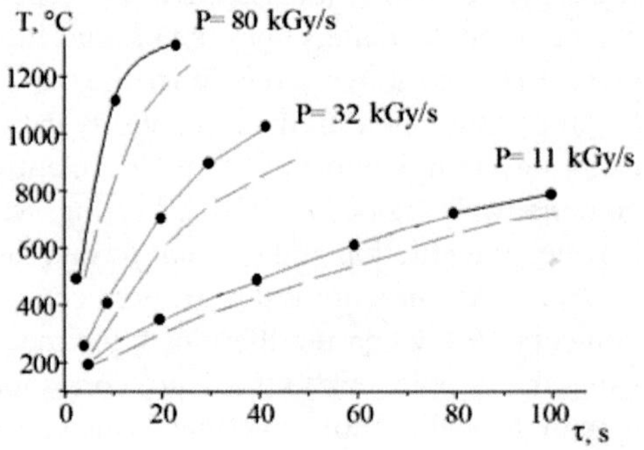

Fig. 4 Effect of the absorbed dose rate and irradiation duration on the temperature measurement error for Na_2CO_3: (—)—a thermocouple with a vertically inserted measuring junction, (—)—a thermocouple with a plain measuring junction

and upon further cooling it changes its sign due to slow cooling of the substrate as compared to that of the sample. It should be noted that pre-heating of the substrate to 400 °C decreases the limiting value of ΔT (Fig. 3).

Under isothermal conditions, the readings of the vertical thermocouple and horizontal thermocouple with plain junction are aligned (Table 1).

The data presented show that the vertically inserted thermocouple provides reliable results only at low heating rates (low irradiation intensities) or long isothermal

Table 1 Isothermal readings of the vertically inserted thermocouple and the horizontally inserted one, which were placed in CaO samples to a depth of 0.25 g/cm^2

P (kGy/s)	*T* (°C)	
	Horizontal junction	Vertical junction
6	700	680
7.5	860	880
15	1120	1110
25	1460	1440

The electron beam energy was 1.5 meV, the exposure time was 10 min exposures. At high absorbed dose rates, a systematic error ΔT occurs due to heat transfer through the wires to the "cold" substrate. The absolute ΔT value depends on the absorbed dose rate, substrate temperature, conditions of the contact between the measuring junction and the sample, the thermocouple wire diameter, the electron beam energy, etc.

3.2 Measuring the Thermocouple EMF

Various devices can be used to measure the thermocouple EMF: analog and digital voltmeters, electronic recording potentiometers, etc. However, in radiation thermal experiments, a number of specific factors impose special requirements on measuring instruments and on the whole EMF measurement circuit.

In the radiation field, the thermocouple is affected by the electromagnetic field of the accelerator and especially by the space charge, both supplied directly to the thermocouple and accumulated in the samples during irradiation; in cable lines (up to several tens of meters) and in measuring instruments, different types of interference can be generated. As a result, parasitic EMF in radiation thermal treatment can attain up to 3 mV (for platinum thermocouples). In addition, during irradiation, the thermocouple potential reaches hundreds of volts relative to the ground, which poses danger to both the service personnel and the measuring equipment. These features of thermometry under electron beam are well illustrated in the following experiment (Fig. 5).

The measuring junction of the Pt–PtRh thermocouple (1) calibrated by Na_2CO_3 melting ($T_{melt} = 853 \pm 1$ °C) was placed in crucible (2) with Na_2CO_3, and its free ends (3) were connected to temperature measuring device (4) with an input resistance of 90 Ω. The device input was ground isolated; the temperature of the thermocouple ends was monitored using a mercury thermometer. Grounded voltmeter (5) with an input resistance of 10^7 Ω was connected to one of the input terminals to measure the thermocouple potential relative to the ground, which significantly exceeds the input resistance of the thermal measuring device. The crucible with soda and thermocouple

Fig. 5 Scheme for measuring the temperature of molten salt under powerful electron beam irradiation

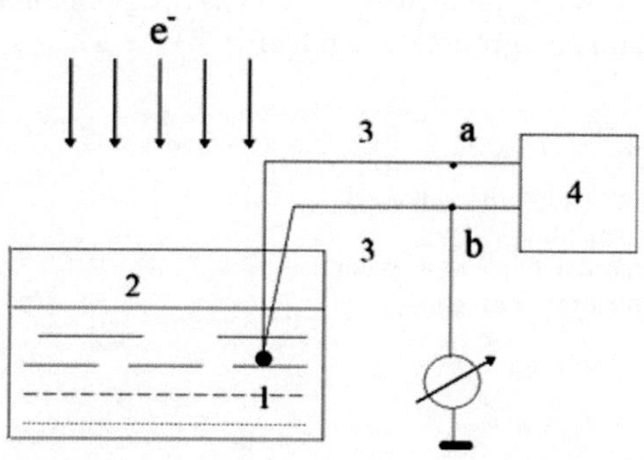

was placed at a distance of 16 cm from the entrance window of the accelerator. The electron beam parameters were as follows: electron energy in the beam of 2 meV, beam current in the pulse of 600 mA, pulse duration of 500 μs, and pulse repetition rate of 15 Hz. The rate of soda heating was 400 °C/min.

In general, thermocouple grounding is sufficient to equalize the electrical potentials of the thermocouple and the ground, and the grounding location does not have any effect on the process. Thus, the voltmeter did not detect any signal in alternate grounding of the input terminals *a* and *b* (Fig. 5) of the thermal measuring device.

However, thermocouple grounding is not sufficient for eliminating parasitic (radiation-induced) EMF. In this case, the grounding location and the pattern of its implementation are of relevance. Grounding of one of the thermocouple ends distorts the thermo-EMF value by the radiation-induced EMF value. The distortion pattern (reduction or gain) depends on the choice of the grounded end, and the distortion value is similar in both cases. Indeed, when the platinum–rhodium end of the thermocouple was grounded, the thermal measuring device showed 8.27 mV (896 °C), which is 43 °C higher than the true temperature of molten salt. When the platinum end of the thermocouple was grounded, the thermometer showed 7.31 mV (810 °C), which is 43 °C less than the true temperature of molten salt.

A number of modifications of the thermocouple method have been developed for comprehensive solution of the problem, namely, measurement of the true temperature of an object in powerful beams of accelerated electrons and equalization of electrical potentials of the thermocouple and the ground.

The first modification suggests grounding of the thermocouple measuring junction. In this case, a simple mechanical contact of the grounded measuring junction with the grounding conductor is not sufficient (tightening, twisting, etc.), since radiation thermal effects significantly oxidize the surfaces of contacting materials. Therefore, an additional grounding conductor should be introduced into the thermocouple junction during its manufacturing to improve the grounding quality. In addition, to preserve the composition of the measuring junction, an additional grounding conductor is manufactured from the material similar to that of one of the thermocouple free ends.

Figure 6 shows the proposed temperature measurement scheme that comprises the thermocouple with measuring junction (1) located in test area (2), which consists of three conductors: two measuring conductors (3) connected to recording unit (4), and one grounded conductor (5).

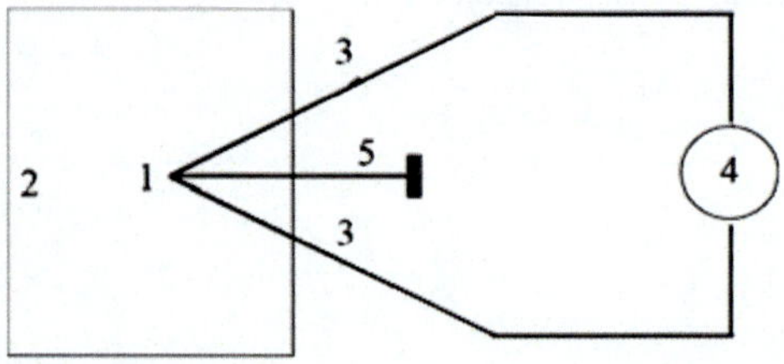

Fig. 6 Scheme for measuring the object temperature using a thermocouple with grounded measuring junction

The device operates as follows. An increased temperature in test area (2) causes heating of measuring junction (1). As a result, a thermo-EMF arises between thermocouple free ends (3) recorded by unit (4), and the EMF value indicates the temperature in control area (2). By means of grounded conductor (5), radiation-induced EMF is eliminated and potentials of the thermocouple and the ground are equalized.

Consider the advantages of the proposed method for recording temperature in the field of ionizing radiation on the example of the Pt–PtRh thermocouple with an additional platinum conductor introduced into the measuring junction during welding.

The thermocouple measuring junction was placed in the crucible with soda Na_2CO_3, and the additional platinum conductor was grounded. The thermocouple free ends were connected to the input of the electrical measuring device. In addition, to measure the thermocouple potential relative to the ground, the grounded microvoltmeter with an input resistance of 10^7 Ω was connected to one of the terminals. The electron accelerator was turned on and soda was heated by irradiation. The heating rate for soda and the electron beam parameters are similar to those indicated earlier.

When soda was melted by the beam, the thermal measuring device showed 7.80 mV, which corresponds to the value of the soda melting point (853 °C). At the same time, the microvoltmeter did not record any signal.

Thus, the proposed measurement scheme ensures measurement of the true temperature value for soda during irradiation and eliminates the difference in electrical potentials between the thermocouple and the ground.

The second modification of the thermocouple method implies grounding of the thermocouple free ends via fixed resistors. In this case, a fixed resistor is connected to the thermocouple free ends to form a resistance bridge. The middle point of the bridge opposite to the measuring junction is grounded, and the resistor resistance is chosen with regard to the bridge equilibrium condition.

Figure 7 presents a device for measuring temperature. The thermocouple measuring junction (1) is in test area (2). Thermocouple free ends (3, 4) are connected to unit (5) for recording the electrical signal and grounded via fixed resistors (6, 7). The device operates as follows.

An increased temperature in test area (2) causes heating of thermocouple measuring junction (1). As a result, thermo-EMF arises between thermocouple free ends (3, 4), which is controlled by recording unit (5), and the obtained value is used to determine the temperature in the test area. Grounding of thermocouple free ends (3,

Fig. 7 Scheme for measuring the object temperature using a thermocouple with free grounded ends

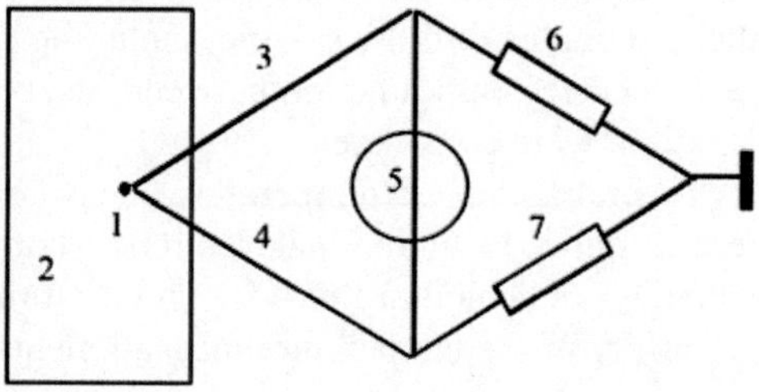

4) via resistors (6, 7) equalizes the radiation-induced potential of the thermocouple to the ground potential value, and complete mutual compensation of the voltage drops of the radiation-induced currents across these resistors is performed by choosing values of resistors (6, 7).

Consider the validity of the above arguments on the example of the calibrated thermocouple Pt–PtRh. The thermocouple measuring junction is placed in the crucible with Na_2CO_3. The electrical resistance of the thermocouple free ends is measured from the measuring junction to the terminals of the electrical measuring microvoltmeter (input resistance of 10^7 Ω). The resistance of the platinum wire is $R_p = 1.18$ Ω, and the resistance of the platinum–rhodium wire attains $R_{pr} = 0.27$ Ω. At the point of the microvoltmeter connection to the terminal, the thermocouple free end is grounded via the fixed resistor with resistance $R_1 = 10^3$ Ω. Providing the equilibrium of the bridge $R_{pr} \cdot R_1 = R_p \cdot R_2$, the resistance R_2 is chosen for grounding the free platinum–rhodium end of the thermocouple at the point of its connection to the microvoltmeter input terminal. Hence, $R_2 = 1500$ Ω.

A grounded voltmeter is connected to one of the microvoltmeter terminals to measure the radiation-induced electric potential of the thermocouple. The crucible with soda and the thermocouple junction are placed in front of the operating accelerator. The beam parameters remain similar. When soda was melted under the beam, the microvoltmeter showed 7.80 mV, which is similar to the temperature of molten soda (853 °C). In this case, the voltmeter did not record any signal. The proposed measurement scheme measures the true temperature of the exposed target and eliminates a potential difference of 137 V between the thermocouple and the ground.

Both of the scheme modifications of the thermocouple EMF measurement are suitable for radiation thermal treatment of large batches of products, when technological modes are tried and tested. For research or search work which requires frequent or abrupt change in the irradiation modes or thermocouple replacement, it is recommended to ground fixed resistors via a variable resistor to quickly correct the equilibrium conditions of the bridge (according to the second modification).

4 Conclusion

The thermometry of materials irradiated with powerful beams of accelerated electrons was investigated using various methods for measuring temperature with the platinum/platinum–rhodium thermocouple sensor. The different options for inserting thermocouples into the sample, including pressing the measuring junction into the pellet center, plain junction between two pellets, as well as vertical insertion of the junction, were considered.

The reliability of temperature measurements was estimated from the melting temperatures of individual reference compounds Na_2CO_3, sodium monoferrite $NaFeO_2$, and calcium fluoride. The performed calibration experiments showed the reliability of the temperature measurement results for a pressed-in thermocouple in all three geometries.

Temperature measurement methodology tested on CaO samples. The data presented show that the vertically inserted thermocouple provides reliable results only at low heating rates (low irradiation intensities) or long isothermal exposures.

At high absorbed dose rates, a systematic temperature error occurs, which depends on the absorbed dose rate, substrate temperature, contact conditions of the measuring junction and the sample, the thermocouple wire diameter, the electron beam energy, etc.

Methods for measuring the thermocouple EMF under radiation thermal treatment have been developed. Grounding of the thermocouple measuring junction via a conductor of one of the components is most appropriate for solving the problems caused by frequent changes in the irradiation mode.

Acknowledgements The research is funded by The Ministry of Science and Higher Education of the Russian Federation as part of the Science Program, Project FSWW-2020-0014. The work was carried out using the equipment of the Center for Sharing Use "Nanomaterials and Nanotechnologies" of Tomsk Polytechnic University supported by the RF Ministry of Education and Science, project 075-15-2021-710. The authors are grateful to M.V. Korobeynikov (Budker Institute of Nuclear Physics, Novosibirsk) for radiation-thermal experiments using ILU accelerator.

References

1. Aisala, H., Nygren, H., Seppänen-Laakso, T., Heiniö, R.-L., Kießling, M., Aganovic, K., Ritala, A.: Comparison of low energy and high energy electron beam treatments on sensory and chemical properties of seeds. Food Res. Int. **148**, 110575 (2021)
2. Lee, S.B., Yoo, S.J., Jung, J., Han, H.N.: Surface roughening and filling of nanoholes in Au surfaces under electron-beam irradiation. Acta Materialia **220**, 117280 (2021)
3. Wang, M., Zhang, G., Qiu, G., Cai, D., Wu, Z.: Degradation of herbicide (glyphosate) using sunlight-sensitive MnO_2/C catalyst immediately fabricated by high energy electron beam. Chem. Eng. J. **306**, 693–703 (2016)
4. Ning, K., Lu, J., Xie, P., Hu, J., Huang, J., Sheng, K.: Study on surface modification of silicone rubber for composite insulator by electron beam irradiation. Nucl. Instrum. Methods Phys. Res. Sect. B **499**, 7–16 (2021)
5. Malyshev, A.V., Lysenko, E.N., Vlasov, V.A., Nikolaeva, S.A.: Electromagnetic properties of $Li_{0.4}Fe_{2.4}Zn_{0.2}O_4$ ferrite sintered by continuous electron beam heating. Ceram. Int. **42**, 16180–16183 (2016)
6. Costantini, J.-M., Ogawa, T., Bhuian, A.K.M.S.I., Yasuda, K.: Cathodoluminescence induced in oxides by high-energy electrons: effects of beam flux, electron energy, and temperature. J. Luminescence **208**, 108–118 (2019)
7. Surzhikov, A.P., Malyshev, A.V., Lysenko, E.N., Vlasov, V.A., Sokolovskiy, A.N.: Structural, electromagnetic, and dielectric properties of lithium-zinc ferrite ceramics sintered by pulsed electron beam heating. Ceram. Int. **43**, 9778–9782 (2017)
8. Chettri, P., Deka, U., Rao, A., Nagaraja, K.K., Okram, G.S., Petwal, V.C., Dwivedi, J.: Effect of high energy electron beam irradiation on the structural properties, electrical resistivity and thermopower of $La_{0.5}Sr_{0.5}MnO_3$ manganites. Phys. B Condensed Matter **585**, 412119 (2020)
9. Stephen, S.K., Ganesh, S., Varghese, T.: Modifications of structural and optical properties of nanophase $BaWO_4$ phosphors: dose dependent effect of high energy electron beam irradiation. Radiat. Phys. Chem. **180**, 109317 (2021)

10. Chaudhary, N., Singh, A., Aswal, D.K., Bharti, M., Sharma, A., Tillu, A.R., Debnath, A.K.: High energy electron beam induced improved thermoelectric properties of PEDOT:PSS films. Polymer **202**, 122645 (2020)
11. Ancharova, U.V., Mikhailenko, M.A., Tolochko, B.P., Lyakhov, N.Z., Korobeinikov, M.V., Bryazgin, A.A., Shtarklev, E.A.: Synthesis and staging of the phase formation for strontium ferrites in thermal and radiation thermal reactions. IOP Conf. Ser. Mater. Sci. Eng. **81**, 012122 (2015)
12. Yang, S., Cai, J., Lv, P., Zhang, C., Huang, W., Guan, Q.: Microstructures and properties of zirconium-702 irradiated by high current pulsed electron beam. Nucl. Instrum. Methods Phys. Res. Sect. B **358**, 151–159 (2015)
13. Yurov, V.M., Baltabekov, A.S., Laurinas, V.C., Guchenko, S.A.: Dimensional effects and surface energy of ferroelectric crystals. Eurasian Phys. Tech. J. **16**(1), 18–23 (2019)
14. Zhuravlev, V.A., Naiden, E.P., Minin, R.V., Itin, V.I., Suslyaev, V.I., Korovin, E.Yu.: Radiation-thermal synthesis of W-type hexaferrites. IOP Conf. Ser. Mater. Sci. Eng. **81**, 012003 (2015)
15. Kostishin, V.G., Andreev, V.G., Korovushkin, V.V., Chitanov, D.N., Yudanov, N.A., Morchenko, A.T., Nikolaev, A.N.: Preparation of 2000NN ferrite ceramics by a complete and a short radiation-enhanced thermal sintering process. Inorg. Mater. **50**, 1317 (2014)
16. Kostishyn, V., Isaev, I., Scherbakov, S., Nalogin, A., Belokon, E., Bryazgin, A.: Obtaining anisotropic hexaferrites for the base layers of microstrip SHF devices by the radiation-thermal sintering. Eastern-Eur. J. Enterp. Technol. **5**, 32–39 (2016)
17. Naiden, E.P., Zhuravlev, V.A., Minin, R.V., Suslyaev, V.I., Itin, V.I., Korovin, E.Yu.: Structural and magnetic properties of SHS-produced multiphase W-type hexaferrites: influence of radiation-thermal treatment. Int. J. Self-Propagat. High-Temp. Synth. **24**, 148–151 (2015)
18. Surzhikov, A.P., Frangulyan, T.S., Ghyngazov, S.A., Koval, N.N.: Structural-phase transformations in near-surface layers of alumina-zirconium ceramics induced by low-energy high-current electron beams. Nucl. Instrum. Methods Phys. Res. Sect. B **267**(7), 1072–1076 (2009)
19. Solodkyi, I., Bogomol, I., Loboda, P.: High-speed electron beam sintering of WC-8Co under controlled temperature conditions. Int. J. Refract. Metals Hard Mater. **102**, 105730 (2022)
20. Salimov, R.A., Cherepkov, V.G., Kuksanov, N.K., Kuznetzov, S.A.: The use of electron accelerators for radiation disinfestation of grain. Radiat. Phys. Chem. **57**, 625–627 (2000)
21. Salimov, R.A., Cherepkov, V.G., Golubenko, J.I., Krainov, G.S., Korabelnikov, B.M., Kuznetsov, S.A., Nemytov, P.I.: High power electron accelerators of ELV-series: status, development, applications. Radiat. Phys. Chem. **57**, 661–665 (2000)
22. Ivanov, A.S., Maznev, V.P., Ovchinnikov, V.P., Svinin, M.P., Tolstun N.G.: Present state of works on development of electron accelerators for energy consuming processes at Efremov research institute. https://citeseerx.ist.psu.edu/viewdoc/download?doi=10.1.1.730.9688&rep=rep1&type=pdf
23. Matsuyama, S., Ishii, K., Fujisawa, M., Kawamura, Y., Tsuboi, S., Yamanaka, K., Kikuchi, Y.: Upgrading of the 4.5 MV dynamitron accelerator at tohoku university for microbeam and nanobeam applications. Nuclear Inst. Methods Phys. Res. Sect. B Beam Interact. Mater. Atoms **267**(12–13), 2060–2064 (2009)
24. Cleland, M.R., Parks, L.A.: Medium and high-energy electron beam radiation processing equipment for commercial applications. Nucl. Instrum. Methods Phys. Res. Sect. B **208**, 74–89 (2003)
25. Mehnert, R.: Review of industrial applications of electron accelerators. Nucl. Instrum. Methods Phys. Res. Sect. B **113**, 81–87 (1996)
26. Abramyan, E.A.: Industrial Electron Accelerators and Applications. Hemisphere Publishing Corporation, New York (1988)
27. Benny, P.G., Khader, S.A., Sarma, K.S.S.: Evaluation of various operational and dosimetric parameters of an industrial electron beam accelerator of 2 MeV energy. Nucl. Instrum. Methods Phys. Res. Sect. A **751**, 88–94 (2014)
28. Yang, J., Zhang, Z., Yang, H., Zhang, J., Liu, J., Yin, Ju, J.: Compact intense electron-beam accelerators based on high energy density liquid pulse forming lines. Matter Radiat. Extremes **3**(6), 278–292 (2018)

29. Prez, L., Heukelom, S., Jansen, B., Jansen, W., Kamer, J., Klink, W., Wittkämper, F.: An on-site dosimetry audit for high-energy electron beams. Phys. Imag. Radiat. Oncol. **5**, 44–51 (2018)
30. Malyshev, A.V., Lysenko, E.N., Vlasov, V.A.: Microstructure, electromagnetic and dielectric properties of zinc substituted lithium ferrites prepared by radiation-thermal heating. Ceram. Int. **41**, 13671–13675 (2015)
31. Surzhikov, A.P., Frangulyan, T.S., Ghyngazov, S.A., Vasil'ev, I.P., Chernyavskii, A.V.: Sintering of zirconia ceramics by intense high-energy electron beam. Ceram. Int. **42**(12), 13888–13892 (2016)
32. Surzhikov, A.P., Lysenko, E.N., Vasendina, E.A., Sokolovskii, A.N., Vlasov, V.A., Pritulov, A.M.: Thermogravimetric investigation of the effect of annealing conditions on the soft ferrite phase homogeneity. J. Therm. Anal. Calorim. **104**(2), 613–617 (2011)
33. Oane, M., Toader, D., Iacob, N., Ticos, C.M.: Thermal phenomena induced in a small tungsten sample during irradiation with a few MeV electron beam: experiment versus simulations. Nucl. Instrum. Methods Phys. Res. Sect. B **337**, 17–20 (2014)
34. Zhuravlev, V.A., Meshcheriakov, V.A., Suslyaev V.I.: Computer simulation of processes of radiation-thermal heating. IOP Conf. Ser. Mater. Sci. Eng. **81**, 012054 (2015)
35. Auslender, V.L.: ILU-type electron accelerator for industrial technologies. Nucl. Instrum. Methods Phys. Res. Sect. B **89**, 46–48 (1994)
36. Bryazgin, A.A.: Electron accelerator for energy up to 5.0 MeV and beam power up to 50 kW with X-ray converter. Prob. Atomic Sci. Technol. **1**, 21–23 (2004)
37. Auslender, V.L., Bryazgin, A.A., Faktorovich, B.L., Gorbunov, V.A., Kokin, E.N., Korobeinikov, M.V., Voronin, L.A.: Accelerators for E-beam and X-ray processing. Radiat. Phys. Chem. **63**, 613–615 (2002)
38. Auslender, V.L., Bryazgin, A.A., Bukin, A.D., Voronin, L.A., Lukin, A.N., Sidorov, A.V.: Online measurement of dose and dose distribution at bremsstrahlung facilities. Radiat. Phys. Chem. **71**(1–2), 381–383 (2004)
39. Lysenko, E.N., Malyshev, A.V., Vlasov, V.A., Nikolaev, E.V., Surzhikov, A.P.: Microstructure and thermal analysis of lithium ferrite pre-milled in a high-energy ball mill. J. Therm. Anal. Calorim. **134**(1), 127–1331 (2018)
40. Surzhikov, A.P., Frangulyan, T.S., Ghyngazov, S.A.: A thermoanalysis of phase transformations and linear shrinkage kinetics of ceramics made from ultrafine plasmochemical $ZrO_2(Y)$-Al_2O_3 powders. J. Therm. Anal. Calorim. **115**(2), 1439–1445 (2014)

Acoustic-Electrical Testing of Changes in the Stress–Strain State on the Example of Rock Samples

Anatoly Bespalko, Denis Dann, Maxim Petrov, Evgeny Pomishin, Gregory Utsyn, and Pavel Fedotov

Abstract A complex method of acoustic-electrical testing of crack formation under stress–strain state changes is discussed on the example of dielectric rock samples. The paper deals with the results of numerical and experimental studies of changes in the parameters of electromagnetic responses and their spectra under pulsed deterministic acoustic excitation of rock samples of different composition and texture. The results of mathematical calculations of the stress concentration on cracks located along and across the sample axis, across which deterministic acoustic pulses were input, are presented. The results of experimental studies of the electromagnetic emission of the calcite and magnetite containing sample under uniaxial compression to fracture are shown. The patterns of changes in the amplitudes of electromagnetic signals during acoustic sounding under stepwise uniaxial compressive loading and shear to failure are shown. It is assumed that the methods presented in the article can be used to detect and localize different types of defects in solid dielectric materials and products.

A. Bespalko · D. Dann (✉) · M. Petrov · E. Pomishin · P. Fedotov
Research School of High-Energy Physics, Tomsk Polytechnic University, 30 Lenin Avenue, 634050 Tomsk, Russia
e-mail: dddann@tpu.ru

A. Bespalko
e-mail: besko48@tpu.ru

M. Petrov
e-mail: mvp17@tpu.ru

P. Fedotov
e-mail: pif@tpu.ru

G. Utsyn
Department of Mechanics and Graphics, Tomsk State University of Control Systems and Radioelectronics, 40 Lenin Avenue, 634050 Tomsk, Russia
e-mail: grigorii.e.utsyn@tusur.ru

E. Lysenko et al. (eds.), *Recent Developments in the Field of Non-Destructive Testing, Safety and Materials Science*, Studies in Systems, Decision and Control 433,
https://doi.org/10.1007/978-3-030-99060-2_22

1 Introduction

The development of destruction of solid dielectric structures is accompanied by occurrence of alternating electromagnetic fields. This phenomenon is being addressed to employ the parameters and characteristics of electromagnetic emission for testing imperfection and to study development of destruction of solid dielectric materials. Alternating electromagnetic fields arise from mechanical–electrical transformations (MET) in solid-state structures when their stress–strain state changes under mechanical load with subsequent acoustic emission inside the test material. External acoustic excitation of the solid-state structure can also generate electromagnetic signals. Electromagnetic radiation due to various types of destruction mechanisms has been proven to emerge in all dielectric materials [1–12]. During MET processes, electromagnetic signals (EMS) are generated due to the changed charge state of materials or their interfaces. The presence, occurrence, and quantitative or qualitative change of electric charges are a prerequisite for EMS generation.

Electrification of the interfaces between dissimilar materials and media can result from: the difference in the work function of electrons from the surfaces of contacting materials; movement of dislocations and accumulation of charged point defects, which localize at media interfaces; friction of layers; delamination or growth of cracks [13–19]. In [20], the issues of charging layers of rocks of different texture and composition are considered in detail. In case of disintegration of contacts between blocks, texture layers, minerals and crystallites, electrification of dielectric materials intensifies. In [21], Vishnevskaya and Zashchinsky considered charged cracks in the form of a system of dipoles with a disordered orientation of the dipole moments. At some stage of mechanical impact, these dipoles acquire a preferential orientation. The authors calculated the magnitude of the electric field strength, which is a function of both elementary sources and the average value of their electric moment. The minimum concentration of elementary sources should be proportional to the crack formation energy and inversely proportional to the electric dipole moment of the elementary source. This is a necessary condition for a self-consistent field.

In [3], Perelman and Khatiashvili experimentally and theoretically proved that acoustic wave induced generation of EMS, which is a consequence of the oscillatory movements of double electric layers and fluctuation-charged edges of microcracks, and release and vibration of charged dislocations. EMS parameters depend on the amplitude-frequency spectrum of the acoustic field excited by linear deformations of the sources.

Mechanisms of EMS generation and source types were summarized to distinguish processes that induce separation of electric charges and form electric dipoles. These processes include: uneven distribution of electric charges on crack edges when ionic or other types of bonds are broken; in inhomogeneous field of mechanical stresses, charged defects start to migrate in the area of crack formation; at the interface of mineralized water released during heating with the rock; friction-induced electrification during movement of structural elements. On the other hand, the change in the dipole moment is determined by the breakdown between the charged sides of the crack, charge relaxation if the current flows through the dielectric body outside the crack; electron emission; vibrations of charged surfaces.

Thus, EMS generation is due to the change in the charge state of the test material during development of cracks of various scales or double electric layers at the interface of media, minerals, and aqueous solutions. During crack development, acoustic emission methods can be employed, which use acoustic pulses emitted by growing cracks of different lengths and record electromagnetic signals when the charged edges of these cracks vibrate. These are mechanical–electrical methods. Other methods for testing crack development under strength loading of dielectric materials or rocks is complex non-destructive testing techniques that use proved testing algorithms and newly created ones. For example, acoustic-electrical transformations in solid-state structures [3, 6, 11] can be successfully used to create such complex testing techniques. Acoustic-electrical testing uses contact acoustic sensing of the test object and contactless recording of the electromagnetic response to this impact, followed by the amplitude-frequency analysis of the electromagnetic signal. External acoustic deterministic video pulses or radio pulses can be used as a vibration source [3, 6, 9, 11, 20, 22]. Under this impact, charges or double electric layers at the interfaces of media, inclusions or blocks, on crack edges or on other structural defects emit electromagnetic signals.

In [11], mathematical and physical justification of testing dielectric heterogeneous materials by parameters of the electromagnetic signal is reported. It is indicated that disturbance of mechanical vibrations by a single normalized impact should cause displacement current. Experimental studies [2–12] indicate that the passage of acoustic waves causes EMS generation, associated by the authors with vibrations of double electrical layers. In this case, EMS amplitude-frequency parameters depend on characteristics of acoustic pulses and a charge state of the defects in the form of inclusions.

Thus, under external deterministic acoustic impact, crack development can be successfully tested at increased mechanical compressive or shear load, in particular, at uniaxial compression and shear of the layers of the test dielectric material or rock.

2 Materials and Methods

For experimental studies of mechanical–electrical and acoustic-electrical transformations, samples with a diameter of $(42 \pm 1) \times 10^{-3}$ m and a length of $(80 \pm 5) \times 10^{-3}$ m were made from rock cores of the Tashtagol iron ore deposit. The test magnetite ore samples contained mainly magnetite and calcite. The magnetite content in the samples ranged from 60 to 77%. The percentage of magnetite was determined from the equation [23]:

$$V_m = V_0[1 - (\rho_0 - \rho_m)/(\rho_v - \rho_m)], \tag{1}$$

Fig. 1 Appearance of the rock samples from the Tashtagol iron ore deposit used to measure the parameters of electromagnetic signals and characteristics of electromagnetic emission under acoustic excitation and destruction by uniaxial compression or shear. Number 1 denotes magnetite ore samples

where V_0 is the sample volume; V_m is the calculated volume of magnetite in the sample; ρ_0 is the average specific gravity of the sample; ρ_m is specific gravity of magnetite; ρ_v is specific gravity of the rock (calcite). For 75% magnetite ore, ρ_v was 4150 kg/m^3. For specific samples of magnetite ore, the longitudinal sound velocity c_l obtained experimentally equalled 6870 m/s. As a result, the acoustic impedance of the ore samples $z_0 = \rho_v c_l$ was within 28.51 × 10^6 kg/s × m^2. For calcite, ρ_c is 2710 kg/m^3 and c_c is 6300 m/s [24]. The acoustic impedance z_c is less than that of the ore and is equal to 17.07 × 10^6 kg/s m^2. For magnetite, ρ_m is 6150 kg/m^3 and c_m is 7400 m/s [24]. Then, for magnetite, acoustic impedance z_m will be 45.51 × 10^6 kg/s × m^2, which exceeds z_0 of the ore. Figure 1 shows a photograph of the rock samples used to study mechanical–electrical and acoustic-electrical transformations during their destruction by compressive or shear stresses. Magnetite ore samples are marked with number 1. The end faces of the samples are perpendicular to the axis of the cylindrical surface, and they were ground to parallel each other.

Experimental studies of EMS parameters and characteristics of electromagnetic emission (EME) under compressive and shear loads were performed using a base testing bench. The block diagram of the base testing bench is shown in Fig. 2a, b. In the experiment, the sample was placed between support (1) and movable plate (2) of the IP500.1 automatic press, which developed force P up to 500 kN. Under uniaxial compression or shear stresses, this force is sufficient for failure of any known heterogeneous dielectric material and rock with a loading area within 20 × 10^{-4} m^2. The load and loading rate were set from PC2 computer (10) through the control system CS to the servo valve actuator using a specialized program. The changes set in the load could be linear, stepwise or cyclical. The recorded changes in sample deformation caused by applied forces and pattern of changes were displayed on PC2 computer. Uniaxial compression was performed with a movable crosshead of the press at a constant speed of 0.3 Pa/s.

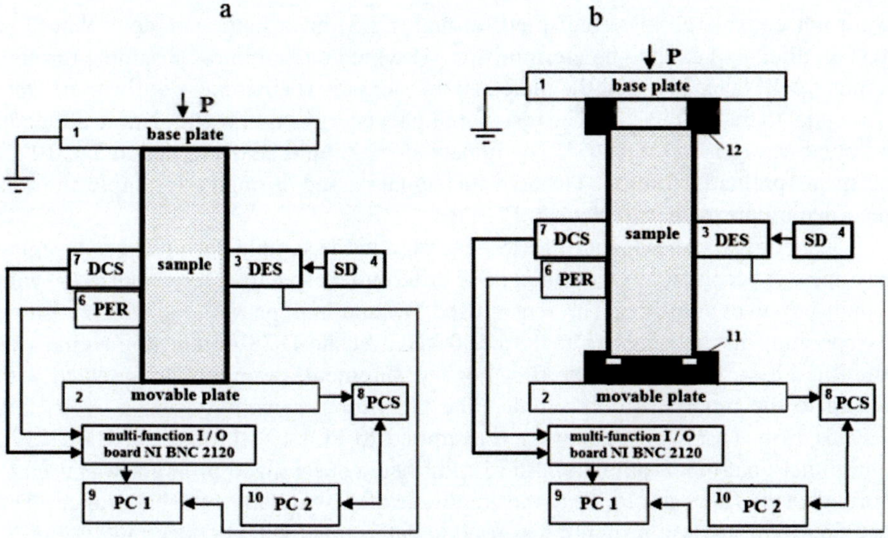

Fig. 2 Block diagram of the base testing bench for uniaxial compression **a** and shear induced by uniaxial compression **b** of test samples: (1) base plate of the IP500.1 press; (2) movable press of the IP500.1 press; (3) dynamic excitation system to control the ball-impact energy; (4) spring device SD for accelerating the ball to input a deterministic acoustic pulse in the sample; (5) NI BNC 2120 multifunctional input–output board; (6) piezoelectric receiver PER of acoustic pulses; (7) electromagnetic differential capacitive sensor DCS; (8) control system of the press with a servo valve PCS; (9) personal computer PC1 for visualization and amplitude-frequency analysis of acoustic and electromagnetic signals; (10) personal computer PC2 for job entry to the press servo drive and output of data on the applied stresses and sample deformation; (11), (12) sample holders with protrusion diameter 11 smaller than that of the hole in 12

Acoustic pulse (AP) excitation of the samples by ball impact was performed through a dynamic excitation system DES (3) [12, 23] using spring device SD (4) for ball acceleration. The AP was close to a bell-shaped one, and the base duration was 50×10^{-6} s. DES system consisted of a metal tube with two optical pairs embedded at a distance of 5×10^{-2} m, each of which included LED and a photodiode. A point impact with a ball weighing 8.6×10^{-4} kg was applied in the middle of the free cylindrical part of the sample. The ball flying through the optical pairs indicated marks on the monitor of PC1 (9). These marks were used to calculate the impact velocity V_1 and rebound velocity V_2 of the ball.

The obtained values of the velocity and ball mass (m), as well as the approximation of elastic collision of the ball with the grounded metal plate were used to determine the acoustic impact energy transmitted to the sample as

$$E_{exc} = \frac{m}{2}(V_1^2 - V_2^2), \tag{2}$$

where E_{exc} is the energy injected into the sample by the impact of the ball.is the energy input in the sample upon ball impact. Energy losses of the AP in the plate were not considered. The ball flied through DES, hit a hardened steel plate 2 $\times$ 10^{-3} m thick and excited a deterministic AP, which passed into the sample through a mineral oil layer between the plate and the sample. The hardness of the plate steel was equal to that of the ball. The residual impact energy input in the sample after ball rebound was (25–30) $\times$ 10^{-3} J. The impact energy could be decreased to 5 $\times$ 10^{-3} J by spring preload variation. The acoustic signal passed through the sample and was recorded by piezoelectric receiver PER (6).

The electrical component of EMS generated in the sample during acoustic signal passage was received by electromagnetic differential capacitive sensor DCS (7) with a built-in power amplifier. The sensor used low and high pass filters, which ensured its operation in the range from 1 to 100 kHz. At the DCS output, the signal was amplified by a factor of 10 or 100. For measurement, the gain was selected with regard to the input EMS amplitude. The DCS input sensitivity was 5 $\times$ 10^{-4} V. Signals from DCS and PER were transmitted to PC1 (9) through NI BNC 2120 multifunctional input–output board (5). Further, a specialized program was used to normalize the EMS amplitude to the amplitude of AP excited by the ball impact. The fast Fourier transform program was used to determine the EMS spectra [23, 25].

Shear stresses under uniaxial compression were created using special holders. The diameter of the protrusion of lower holder (11) was 4 $\times$ 10^{-3} m less than that of the hole in upper holder (12). In addition, holders 11 and 12 ensured sample centring, and, hence, the coincidence of the protrusion and hole axes.

3 Theoretical and Experimental Research

This paper considers the results of numerical and experimental study of changes in the parameters of electromagnetic signals generated under deterministic acoustic excitation of rock samples. For calculations, we analysed wave propagation in elastic inhomogeneous medium of limited dimensions with given physical and mechanical properties under pulse exposure of part of its surface. The parameters of the stress–strain state (displacement, deformation, stress) were calculated. The simplest classical rigid body model was used [26–28].

The problems of wave mechanics are solved by hypotheses conventional for classical elastic medium, such as the approximation of homogeneity and continuity. Any inhomogeneities caused by defects or inclusions are taken into account explicitly, i.e. they are included in the problem statement.

For a given cylindrical magnetite ore sample, this model can be applied with some assumptions. For example, for problems related to wave propagation, it was assumed that the samples used were free of porosity, empty cavities, or other clearly distinguishable parts in the form of inclusions. The study of the sample texture showed that calcite and magnetite minerals are in continuous contact and smoothly mix with each other.

The boundary conditions correspond to the laboratory experiment. Zero displacement was specified at the flat boundaries of the sample, since in the laboratory experiment the sample was clamped between the punches. The side faces are free, therefore, the stresses at the boundary equal zero. For a three-dimensional case, the free boundaries show no components of the stress vector:

$$\sigma_{\tau 1} = \sigma_{\tau 2} = \sigma_n = 0, \tag{3}$$

where σ_n is normal stresses and $\sigma_{\tau 1}, \sigma_{\tau 2}$ are tangential stresses.

In the middle of the vertical dimension of the sample along the normal, load is applied in the form:

$$\begin{aligned} \sigma_n &= T(t) \times F(x, z), \\ \sigma_{\tau 1} &= \sigma_{\tau 2} = 0, \end{aligned} \tag{4}$$

where $T(t)$ is a function of t that determines the law of load variation with time, $F(x, z)$ is a function that determines the law of load variation over the surface (in a two-dimensional case along the contour line), σ_n is normal stresses and $\sigma_{\tau 1}, \sigma_{\tau 2}$ are tangential stresses, the last components of the load vector are neglected. For calculations, the dependence $T(t)$ was used in the form of a half exponentially decaying sinusoid:

$$\begin{aligned} T(t) &= e^{-\beta t} \sin\left(\frac{2\pi t}{\tau}\right) \times \eta(t), \\ \eta(t) &= \begin{cases} 1, 0 \le t \le \tau \\ 0, t > \tau \end{cases}, \end{aligned} \tag{5}$$

where τ is load exposure time that depends on the duration of the deterministic AP.

At the face end, the load was specified as a gradually increasing and then gradually decreasing function dependent on temporal and spatial coordinates. After pulse exposure, stresses at the upper boundary were set to zero. In calculations, the acoustic excitation pulse was close to the experimental one in shape, amplitude, and duration. In the three-dimensional case, when the load is applied to the surface, the AP in the form of a bell-shaped function is represented as:

$$F(x, z) = A \times \exp\left(-\beta \frac{(x - x_0)^2 + (z - z_0)^2}{2d_0^2}\right), \tag{6}$$

where x_0, z_0 are the coordinates of the center of the AP source, and d_0 is the value that determines the distance from the pulse centre at which the normal stress decreases by a factor of e^{β}.

In the general case, the system of equations that describes the behaviour of a deformable solid body in the spatial case includes the equations of motion

$$\rho \dot{U}_i = \rho G_i + \sigma_{ij,j}, \tag{7}$$

where ρ is density, U_i is displacement, σ_{ij} is stresses, G_i is mass force vector components, $i, j = 1, 2, 3$. The dot above the symbol indicates the time derivative, and the comma after the index indicates the derivative of the corresponding coordinate. Summation is performed over the repeated indices, for example:

$$\sigma_{ij,j} = \frac{\partial \sigma_{i1}}{\partial x_1} + \frac{\partial \sigma_{i2}}{\partial x_2} + \frac{\partial \sigma_{i3}}{\partial x_3}$$

the continuity equation

$$\frac{\partial \rho}{\partial t} + \rho\left(\frac{dU_1}{dx_1} + \frac{dU_2}{dx_2} + \frac{dU_3}{dx_3}\right) = 0, \tag{8}$$

where x_i is the coordinate axis.

The relation between the strain tensor components and displacements from the Cauchy relation is represented as

$$\varepsilon_{ij} = 0.5\left(U_{i,j} + U_{j,i}\right) \tag{9}$$

and the constitutive relations specify the relations between stress and strain tensor components as:

$$\sigma_{ij} = f\left(\varepsilon_{ji}\right) \tag{10}$$

In the elastic case, relations (10) are taken in the form of Hooke's law. In the applied problems of wave mechanics, the processes are described by hyperbolic equations when formulated correctly. Hyperbolic equations and systems make up a significant part of the mathematical models used to solve applied problems. A distinctive feature of hyperbolic equations and systems is that they describe wave processes with the finite rate of disturbance (wave) propagation in the considered region.

For each equation of the form

$$\frac{\partial U}{\partial t} = \frac{\partial F}{\partial x} \tag{11}$$

the difference scheme used corresponds to the simplest version of McCormack non-central schemes [29, 30].

A volumetric design scheme was used for equation solution. McCormack's method is a finite-difference predictor–corrector method, which refers to shock-capturing schemes. In accordance with the scheme, the position of the wave front is not identified, and the computational domain is analysed for each moment of time, which is needed for subsequent use of the simulation results. McCormack's scheme assumes the use of a rectangular computational grid, which has a number of advantages: simplified mathematical calculations, less computing time, simplified specification of boundary conditions, and simplified processing of results without data loss. Results obtained by square-cell grids are best since the hourglass effect—twisting of the grid cells—is minimized. The numerical algorithm of the method is similar to the Runge–Kutta method [30] used to solve ordinary differential equations. To obtain solutions at the next time step $U(t + \Delta t)$ by the Runge–Kutta method using the known $U(t)$, one or several iterations are required.

Before using the method similar to the Runge–Kutta method of the second order of accuracy, spatial derivatives in the equations were replaced by the corresponding ratios of finite differences. Non-central difference operators were used, for example, alternately left or right differences instead of central ones. This approach forms the basis for the effective non-central second-order scheme proposed by McCormack.

Non-central schemes are superior to most conventional central schemes since the program logic is simplified, non-uniform terms are easily included, and generalization to multidimensional problems is performed directly. The advantage of non-central schemes is the absence of half-integer indices, which simplifies realization of the boundary conditions. The scheme has a second order of approximation in both spatial and temporal variables.

Thus, the boundary conditions can be specified by displacements. The size of elements in the finite element model was $(10^{-3} \times 10^{-3})$ m^2. The calculation involves 237,500 points at a sampling rate of 10^{-6} s. The calculations were performed for an elastic sample of magnetite ore used in the experiments of stepwise compression and subsequent excitation at steps by deterministic AP. Cracks specified were of different sizes with a length along the sample axis or along the AP propagation at the point of the ball impact. Numerical simulation was carried out using a special graphics package. The results of numerical simulation are visualized in the form of stress intensity propagation regions [31, 32]. The constructed mathematical model was used to analyse elastic wave propagation in the sample under pulsed mechanical action. This enables detailed distributions of parameters that describe the material behaviour in contrast to experimental studies, when the measured quantities have predominantly integral meaning. Elastic disturbances were simulated numerically in a cylindrical region, where the lateral surface was exposed to pulse excitation. It is reasonable to consider the problem in flat formulation, namely, with the axial section of the cylinder with dimensions of $(42 \times 80) \times 10^{-6}$ m. Figure 3 shows examples of numerical modelling for this region at the AP propagation time of 25×10^{-6} s.

In numerical modelling, changes in stress intensity over time from zero to 35×10^{-6} s were studied. For example, Fig. 3a shows changes in stress intensity in the bulk of the sample on a crack 10×10^{-3} m long located along the sample axis at a distance of 20×10^{-3} m from the excited cylindrical surface within 25×10^{-6} s from the moment of the AP induced by the ball collision with the cylindrical surface. As can be seen in the Fig. 3a, stresses are concentrated on the crack, and then they drop sharply along the line of AP propagation. As reported in the introduction, the amplitude-frequency spectrum of EMS is directly related to the parameters of the time-varying AP. As a result, the EMS spectrum also changes.

Figure 3b shows changes in stress intensity in the bulk of the sample on the crack 10×10^{-3} m long located across the sample axis at a distance of 40×10^{-3} m from the

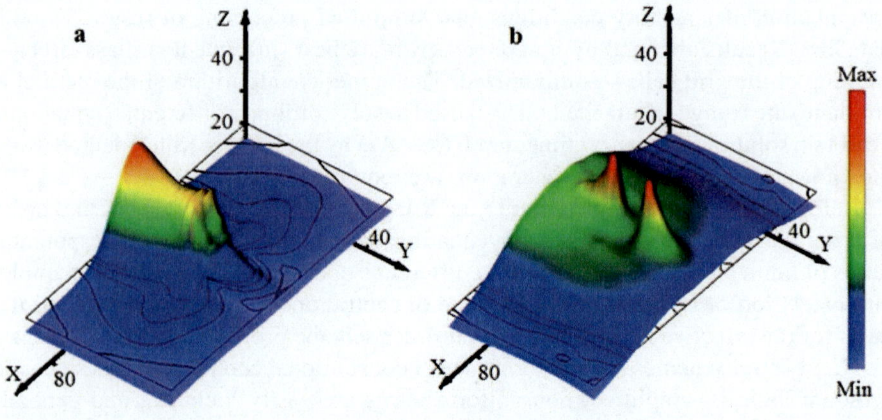

Fig. 3 Calculated values of stress intensity in the sample bulk within 25×10^{-6} s from the moment of the acoustic pulse input by the ball impact on the cylindrical surface of the sample: **a** crack 10×10^{-3} m long along the sample axis at a distance of 20×10^{-3} m from the excited cylindrical surface; **b** crack 10×10^{-3} m long across the sample at a distance of 40×10^{-3} m from the face end of the sample under similar external impact

sample end face under similar external action from the excited cylindrical surface. The Fig. 3b also presents the calculation of stresses within 25×10^{-6} s from the moment of the AP input after the ball impact on the cylindrical surface. As can be seen in the Fig. 3b, stresses are concentrated on the crack, but their shape differs from the previous one (Fig. 3a). Therefore, the EMS amplitude-frequency spectrum will be different. The form of the amplitude-frequency spectrum under deterministic acoustic excitation can help determine the location of existed or emerging cracks during sample fracture propagation caused by different types of strength loading.

In the experiments, electromagnetic emission (EME) of rock samples under compressive loading was studied first. Data on measuring EME during fracture propagation in the magnetite ore sample before the growth of the tensile crack are presented in Fig. 4. The magnetite ore sample is elastic rock. Figure 4 shows the stages of loading and propagation of the sample fracture: (1) compaction stage; (2) stage of the fracture focus initiation; (3) stage of sustainable strength; (4) pre-fracture stage; (5) growth of the tensile crack. The second and third stages of fracture propagation are of interest. The test ore sample consists of magnetite and 30% calcite. Figure 2 shows the time of sample compaction and the onset of crack propagation in the region of 0.4 of the ultimate load P_{limit}.

Two EME peaks are observed in the region of fracture focus propagation. The first peak corresponds to destruction of calcite inclusions, since their strength is lower than that of magnetite. In addition, the intensity of electromagnetic emission from calcite is higher, since its specific electrical resistance is higher than that of magnetite. For calcite $\rho_c = 10^7$ to 10^{14} $\Omega \times$ m, and for magnetite and hematite $\rho_m = 10^{-5}$ to 10^3 $\Omega \times$ m [23]. This significantly enhances accumulation of charge

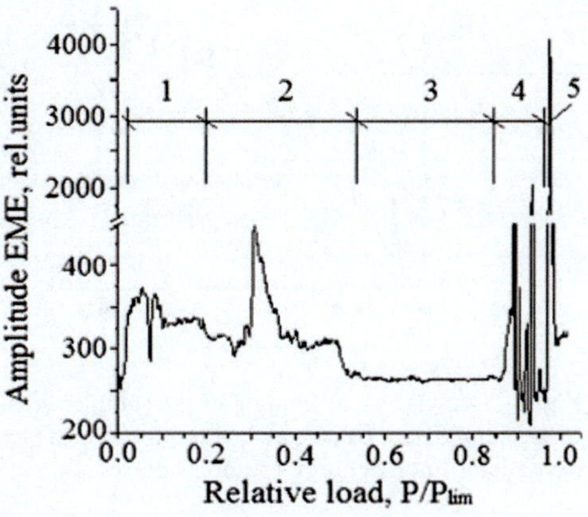

Fig. 4 Changes in the amplitudes averaged over one second at a frequency of 100 kHz of electromagnetic signals averaged over one second at a frequency of 100 kHz during EME at different relative load: (1) compaction stage; (2) stage of the fracture focus initiation; (3) stage of sustainable strength; (4) pre-fracture stage; (5) growth of the tensile crack

on defects and crack sides in calcite compared to that in magnetite and, as a result, increases EME intensity.

In [9, 11], the intensity and amplitude-frequency characteristics of EME depend on duration and amplitude of acoustic pulses formed during crack growth, as well as on electrical properties of the sample material. At the third stage, cracks develop and accumulate in the bulk of the sample without reducing its strength. Subsequently, the pre-fracture stage and the stage of growth of the tensile crack or sample fracture can be monitored by sharp changes in EME intensity under increased compressive stresses.

It is of interest to determine the possibility of using acoustic-electrical transformations to test formation of destructive zones under increased compressive or shear stresses. Figures 5 and 6 present the results of external acoustic sounding of magnetite ore samples under stepwise compression and shear loading to fracture and measurements of the electromagnetic response to this impact.

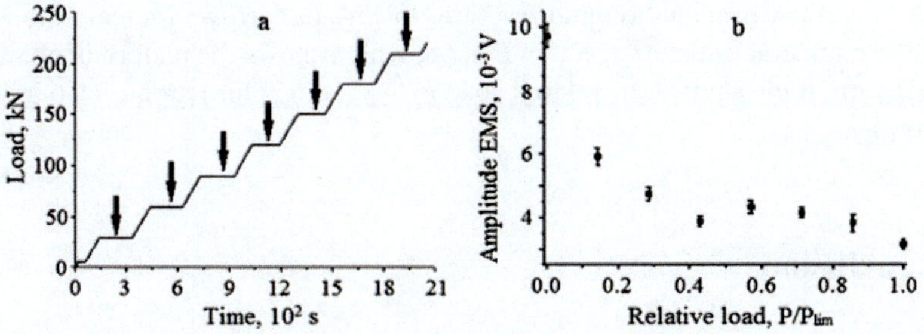

Fig. 5 Stepwise uniaxial compressive loading of the sample (arrows indicate the deterministic acoustic excitation of the sample) (**a**); changes in the amplitude of electromagnetic signals during deterministic acoustic excitation of magnetite ore samples under uniaxial compression, where P_{limit} is destructive load (**b**)

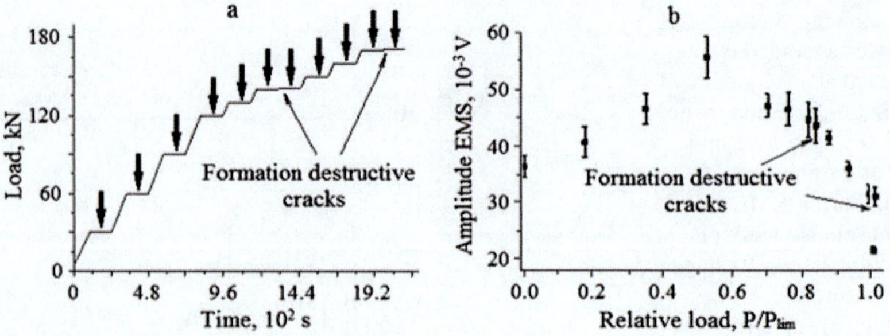

Fig. 6 Stepwise shear loading of the sample (arrows indicate the deterministic acoustic excitation of the sample) (**a**); **b** changes in the amplitude of electromagnetic signals upon deterministic acoustic excitation of magnetite ore samples during uniaxial shear

In the Figs. 5 and 6, arrows indicate the moments of deterministic acoustic excitation by the ball impact, and a continuous line with shelves, which show the constancy of values, indicates stepwise loading under compressive and shear stresses. The test bench configuration presented in Fig. 2a was used to create uniaxial compressive stresses. Figure 5b shows that the onset of an increase in the EMS amplitude coincides with the onset of destructive crack formation and corresponds to 0.42 of ultimate load P_{limit} at which the sample fractured. Subsequently, the destruction increased and started to decrease at about 0.7 of ultimate load P_{limit}.

Under stepwise loading (Fig. 6a) during creation of shear stresses and deterministic acoustic excitation, another type of changes in the EMS amplitudes could be observed. The test bench configuration presented in Fig. 2b was used to implement shear stresses. Figure 6b shows the experimentally observed change in the EMS amplitude upon deterministic acoustic excitation of magnetite ore samples during uniaxial shear. This type of loading also caused an increase in the amplitude of the electromagnetic response to external acoustic excitation during propagation of the destructive zone in the region of (0.3–0.53) P_{limit} and a decrease in the EMS amplitude during crack propagation and fracture. In Fig. 6a, arrows indicate the load at which the main cracks emerge, and in Fig. 6b, which shows the pattern of changes in the EMS amplitude against the relative load P/P_{limit}, crack initiation is also indicated by arrows.

4 Conclusions

The observed types of loading through uniaxial compression and shear are accompanied by the propagation of destructive zones and fracture caused by the development of tensile cracks. Electromagnetic responses to sounding deterministic acoustic pulses at constant values of applied stresses under uniaxial compression or shear differ significantly in amplitude and form. However, in both cases, zones of the onset of fracture initiation and propagation can be clearly observed as the EMS amplitude changes. Figures 4b and 5b show that propagation of compressive fracture and shears features certain patterns. This is due to the fact that under compression the fracture propagates in the bulk of the entire sample. Under shear stresses, created as shown in Fig. 2b, fracture mostly develops at the edges of the sample. Numerical modelling shows that changes in stress intensity in the sample bulk on cracks located across and along the sample axis differ significantly from each other. This results in different electromagnetic responses to deterministic acoustic excitation of the samples on these cracks.

Thus, the form of the amplitude-frequency spectrum under deterministic acoustic excitation helps determine the location of existed or emerging cracks during propagation of the sample fracture under various types of strength loading. Methods of mechanical–electrical and acoustic-electrical transformations under force compressive or shear loading can be efficiently used for testing and monitoring of the processes of crack initiation and fracture propagation in solid dielectric materials and products. In the future, the considered methods can be used to detect different types of defects in dielectric materials.

Acknowledgements The study was supported by the Russian Science Foundation, project No. 20-79-10156 (TPU – 19.0066.RNF.2020).

References

1. Vorobyov, A.A., Zavodovskaya, E.K., Sal'nikov, V.N.: Changes in electrical conductivity and radio emission of rocks and minerals during physical and chemical processes in them. Dokl. Akad. Nauk SSSR **220**(1), 82–85 (1975)
2. Misra, A., Gosh, S.: Electromagnetic radiation characteristics during fatigue crack propagation and failure. Appl. Phys. **23**, 387–390 (1980)
3. Khatiashvili, N.G.: Generation of electromagnetic radiation when acoustic waves pass through crystal dielectrics and some rocks. Dokl. Akad. Nauk SSSR **263**(4), 839–842 (1982)
4. Surzhikov, A.P., Lysenko, E.N., Vasendina, E.A., Sokolovskii, A.N., Vlasov, V.A., Pritulov, A.M.: Thermogravimetric investigation of the effect of annealing conditions on the soft ferrite phase homogeneity. J. Therm. Anal. Calorim. **104**(2), 613–617 (2011)
5. Ogawa, T., Oike, K, Miura, T.: Electromagnetic radiations from rocks. J. Geophys. Res. **90**, 6245–6249 (1985)
6. Gordeev, V.F., Malyshkov, Yu.P., Chakhlov, V., Baumbach, H., Kapphan, G., Slovik, V., Birkholts, V.: Electromagnetic emission control of concrete strength. Defectoscopiya **7**, 76–80 (1992)
7. Surzhikov, A.P., Frangulyan, T.S., Ghyngazov, S.A., Vasil'ev, I.P., Chernyavskii, A.V.: Sintering of zirconia ceramics by intense high-energy electron beam. Ceram. Int. **42**(12), 13888–13892 (2016)
8. O'Keefe, S.G., Thiel, D.V.: A mechanism for the production of electromagnetic radiation during fracture of brittle materials. Phys. Earth Planet. Inter. **89**(11), 127–135 (1995)
9. Bespal'ko, A.A., Gold, R.M., Yavorovich, L.V.: Influence of calcite electrification on parameters of electromagnetic signals under pulsed acoustic influence. Phys. Mesomech. **7**(5), 95–99 (2004)

10. Lacidogna, G., Carpinteri, A., Manuello, A., Durin, G., Schiavi, A., Niccolini, G., Agosto, A.: Acoustic and electromagnetic emissions as precursor phenomena in failure processes. Strain **47**, 44–152 (2010)
11. Bespal'ko, A.A., Isaev, Y.N., Yavorovich, L.V.: Transformation of acoustic pulses into electromagnetic response in stratified and damaged structures. J. Min. Sci. **52**(2), 279–285 (2016)
12. Fursa, T.V., Utsyn, G.E., Dann, D.D., Petrov, M.V.: Development prospects for nondestructive testing of heterogeneous nonmetallic materials by the parameters of electrical response to a shock action. Rus. J. Nondestr. Test. **53**(2), 104–110 (2017)
13. Oreshkin, P.T.: Physics of Semiconductors and Dielectrics. Higher School, Moscow (1977)
14. Ashcroft, N.W., Mermin, N.D.: Solid State Physics. Saunders College Publishing, New York (1976)
15. Cherepanov, G.P.: Mechanics of Destruction of Composite Materials. Nauka, Moscow (1983)
16. Seeger, K.: Semiconductor Physics. Springer, Wien (1973)
17. Eshelby, J.D.: Solid State Physics, vol. 3. Academic Press, New York (1956)
18. Mashkov, Yu.K., Kropotin, O.V.: Tribophysics and Structural Modification of Tribosystem Materials. Omstu Publishing House, Omsk (2009)
19. Rabinovitch, A., Shay, A., Liraz, R., Frid, V., Bahat, D.: Electromagnetic radiation emitted during friction process. Int. J. Fract. **131**(2), 21–27 (2005)
20. Bespal'ko, A., Surzhikov, A., Fedotov, P., Pomishin, E., Stary, O.: Polarization and electromagnetic emissions of natural crystalline structures upon acoustic excitation. Mater. Sci. Forum **970**, 153–166 (2019)
21. Vishnevskaya, N.L., Zashchinsky, L.A.: Calculation of the strength of a self-consistent electric field that occurs in a dielectric under mechanical stress. Izvestiya VUZov. Phys. **5**, 71–74 (1977)
22. Khorsov, P., Laas, R., Surzhikov, A.P.: The application of reverberation in method of mechano-electrical transformations for estimation of stress-strain state in solid dielectrical matter. Mater. Sci. Forum **970**, 47–54 (2019)
23. Yavorovich, L.V., Bespal'ko, A.A., Fedotov, P.I.: Experimental research of the influence of the strength of ore samples on the parameters of an electromagnetic signal during acoustic excitation in the process of uniaxial compression. IOP Conf. Ser.: Mater. Sci. Eng. **289**, 012025 (2018)
24. Dortman, N.B. (ed.): Petrophysics. GIS Equipment and Equipment: Reference Book, vol. 1. Nedra, Moscow (1992)
25. Fursa, T.V., Utsyn, G.E., Petrov, M., Dann, D.D., Sokolovskiy, A.N.: Detecting degradation in reinforced concrete subjected to uniaxial compression, using the parameters of electric response to mechanical. Res. Nondestruct. Eval. **30**(2), 317–333 (2019)
26. Ziman, J.M.: Principles of the Theory of Solids. Cambridge University Press, London (1972)
27. Davydov, A.S.: Theory of a Solid Body. Nauka, Moscow (1976)
28. Mason, W.P. (ed.): Physical Acoustics, vol. 1. Academic Press, New York (1964)
29. Hoffman, J.D.: Numerical Methods for Engineers and Scientists, 2nd edn. Marcel Dekker Inc., New York (2001)
30. Hairer, E., Wanner, G.: Solving Ordinary Differential Equations II: Stiff and Differential-Algebraic Problems, 2nd edn. Springer, Berlin (1996)
31. Gorshkov, A.G., Starovoitov, E.I., Tarlakovsky, D.V.: Theory of Elasticity and Plasticity. FIZMATLIT, Moscow (2002)
32. Molotnikov, V., Molotnikova, A.: Theory of Elasticity and Plasticity. A Textbook of Solid Body Mechanics. Springer International Publishing, Switzerland (2021)